Progress in Nonlinear Differential Equations and Their Applications
Volume 6

Kung-ching Chang

Infinite Dimensional Morse Theory and Multiple Solution Problems

Birkhäuser
Boston · Basel · Berlin

Kung-ching Chang
Department of Mathematics
Peking University
Beijing, 100871
People's Republic of China

Library of Congress Cataloging-in-Publication Data

Chang, Kung-ching
Infinite dimensional Morse theory and multiple solution problems / by Kung-ching Chang
p. cm. -- (Progress in nonlinear differential equations and their applications ; v. 6)
Includes bibliographical references and index.
ISBN 0-8176-3451-7 (acid free)
1. Morse Theory I. Title. II. Series.
QA331.C445 1991 91-12511
515--dc20 CIP

Printed on acid-free paper.

ISBN 0-8176-3451-7
ISBN 3-7643-3451-7

Typeset in TeX by Ark Publications, Inc., Newton Centre, MA
Printed and bound by Quinn-Woodbine, Woodbine, NJ.
Printed in the U.S.A.

9 8 7 6 5 4 3 2 1

Table of Contents

Preface vii

Introduction ix

Chapter I: Infinite Dimensional Morse Theory

1. A Review of Algebraic Topology 1
2. A Review of the Banach-Finsler Manifold 14
3. Pseudo Gradient Vector Field and the Deformation Theorems . 19
4. Critical Groups and Morse Type Numbers 32
5. Gromoll-Meyer Theory 43
6. Extensions of Morse Theory 54
   - 6.1. Morse Theory Under General Boundary Conditions . . 55
   - 6.2. Morse Theory on a Locally Convex Closed Set 60
7. Equivariant Morse Theory 65
   - 7.1. Preliminaries 66
   - 7.2. Equivariant Deformation 67
   - 7.3. The Splitting Theorem and the Handle Body Theorem for Critical Manifolds 69
   - 7.4. G-Cohomology and G-Critical Groups 74

Chapter II: Critical Point Theory.

1. Topological Link 83
2. Morse Indices of Minimax Critical Points 92
   - 2.1. Link 92
   - 2.2. Genus and Cogenus 96
3. Connections with Other Theories 99
   - 3.1. Degree theory 99
   - 3.2. Ljusternik-Schnirelman Theory 105
   - 3.3. Relative Category 109
4. Invariant Functionals 111
5. Some Abstract Critical Point Theorems 121
6. Perturbation Theory 131
   - 6.1. Perturbation on Critical Manifolds 131
   - 6.2. Uhlenbeck's Perturbation Method 136

Chapter III: Applications to Semilinear Elliptic Boundary Value Problems.

1. Preliminaries 140
2. Superlinear Problems 144
3. Asymptotically Linear Problems 153

3.1. Nonresonance and Resonance with the Landesman-Lazer Condition . . . 153
3.2. Strong Resonance . . . 156
3.3. A Bifurcation Problem . . . 161
3.4. Jumping Nonlinearities . . . 164
3.5. Other Examples . . . 169
4. Bounded Nonlinearities . . . 172
4.1. Functionals Bounded From Below . . . 172
4.2. Oscillating Nonlinearity . . . 173
4.3. Even Functionals . . . 176
4.4. Variational Inequalities . . . 177

Chapter IV: Multiple Periodic Solutions of Hamiltonian Systems
1. Asymptotically Linear Systems . . . 182
2. Reductions and Periodic Nonlinearities . . . 188
2.1. Saddle Point Reduction . . . 188
2.2. A Multiple Solution Theorem . . . 195
2.3. Periodic Nonlinearity . . . 198
3. Singular Potentials . . . 203
4. The Multiple Pendulum Equation . . . 209
5. Some Results on Arnold Conjectures . . . 215
5.1. Conjectures . . . 215
5.2. The Fixed Point Conjecture on (T^{2n}, ω_0) . . . 218
5.3. Lagrange Intersections for $(\mathbb{C}P^n, \mathbb{R}P^n)$ . . . 220

Chapter V: Applications to Harmonic Maps and Minimal Surfaces
1. Harmonic Maps and the Heat Flow . . . 229
2. The Morse Inequalities . . . 246
3. Morse Decomposition . . . 250
4. The Existence and Multiplicity for Harmonic Maps . . . 257
5. The Plateau Problem for Minimal Surfaces . . . 260

Appendix: Witten's Proof of the Morse Inequalities
1. A Review of Hodge Theory . . . 274
2. The Witten Complex . . . 282
3. Weak Morse Inequalities . . . 287
4. Morse Inequalities . . . 295
References . . . 298
Index of Notation . . . 310
Index . . . 311

Preface

The book is based on my lecture notes "Infinite dimensional Morse theory and its applications", 1985, Montréal, and one semester of graduate lectures delivered at the University of Wisconsin, Madison, 1987. Since the aim of this monograph is to give a unified account of the topics in critical point theory, a considerable amount of new materials has been added. Some of them have never been published previously.

The book is of interest both to researchers following the development of new results, and to people seeking an introduction into this theory. The main results are designed to be as self-contained as possible. And for the reader's convenience, some preliminary background information has been organized.

The following people deserve special thanks for their direct roles in helping to prepare this book.

Prof. L. Nirenberg, who first introduced me to this field ten years ago, when I visited the Courant Institute of Math Sciences.

Prof. A. Granas, who invited me to give a series of lectures at SMS, 1983, Montreal, and then the above notes, as the primary version of a part of the manuscript, which were published in the SMS collection.

Prof. P. Rabinowitz, who provided much needed encouragement during the academic semester, and invited me to teach a semester graduate course after which the lecture notes became the second version of parts of this book.

Professors A. Bahri and H. Brézis who suggested the publication of the book in the Birkhäuser series.

Professors E. Zehnder and A. Ambrosetti, who provided a favorable environment during the period in which this book was written.

Mrs. Ann Kostant, for aiding me in editing and typesetting the manuscript.

My teacher Prof. M. T. Cheng for his constant support and influence over the many years.

And, of course, I thank my wife and my children for their love, patience and understanding while I was writing this book.

Kung-Ching Chang
Mathematical Institute, Peking University, Beijing.

INTRODUCTION

This book deals with Morse theory as a way of studying multiple solutions of differential equations which arise in the calculus of variations. The theory consists of two aspects: the global one, in which existence, including the estimate of the number of solutions, is obtained by the relative homology groups of two certain level sets, and the local one, in which a sequence of groups, which we call critical groups, is attached to an isolated critical point (or orbit) to describe the local behavior of the functional. Morse relations link these two ideas.

In comparison with degree theory, which has proved very useful in nonlinear analysis in proving existence and in estimating the number of solutions to an operator equation, Morse theory has a great advantage if the equation is variational. Relative homology groups and critical groups are series of groups that provide both a finer structure and better estimate of the number of solutions than does the degree, which is only an integer. The relationship between the Leray-Schauder index and critical groups is established.

The minimax method is another important tool in critical point theory. In this volume it is treated in a unified manner from the Morse theoretic point of view. The mountain pass theorem, the saddle point theorem and multiple solution theorems, discussed in Ljusternik-Schnirelman theory, index theory and pseudo index theory, are studied by observing the relative homology groups for specific level sets. Critical groups for critical points are also estimated. The purpose of this treatment is to provide a unified framework which contains different theories so that various techniques are able to be combined in estimating the number of critical points.

Applications to semilinear elliptic boundary value problems, periodic solutions of Hamiltonian systems, and geometric variational problems are also emphasized. These problems are chosen for their own interest as well as for explaining how Morse theory is applied.

The book is organized into five chapters and an appendix. Chapter 1 is devoted to Morse theory. Sections 1 and 2 review the basic facts of algebraic topology and infinite dimensional manifolds, respectively. Two deformation theorems, which play a fundamental role in critical point theory, are proved in detail in Section 3. Morse relations and the Morse handle body theorem are studied in Section 4. Section 5 deals with Gromoll-Meyer theory and discusses the main properties of critical groups for isolated critical points, including homotopy invariance and a shifting lemma. The Marino-Prodi approximation theorem is also studied in this section. In the rest of the chapter, Morse theory is extended: in Section 6.1, to manifolds with boundaries together with certain boundary value conditions, and, in Section 6.2, to locally convex closed sets. The latter extension is motivated by variational inequalities. G-equivariant Morse theory is investigated in Section 7, where all the main results of Sections 4 and 5 are completely extended to invariant functions under a compact Lie group action.

Chapter 2 views critical point theory with respect to homology groups. Sections 1 through 4 are devoted to this study. The homological link, subordinate homology classes, and Cech-Alexander-Spanier cohomological rings are used to link up minimax principles with Morse theory. Morse index estimates in Minimax theorems are also presented. In Section 5, we give some abstract critical point theorems which will be applied in subsequent chapters. Two perturbation theories are studied in Section 6, one of which is concerned with the perturbation effect on a critical manifold, and the other with Uhlenbeck's perturbation theory.

Semilinear elliptic BVPs are considered to be models in the applications of critical point theory. The reader will find that there are many different and very interesting results presented in Chapter 3. Although some of them will be familiar, the proofs given here are new and are based on the above unified framework. Problems with superlinear, asymptotically linear and bounded nonlinear terms are studied by example in Sections 2–4. Variational inequalities are also discussed.

Chapter 4 deals with some topics on Hamiltonian systems. Since there are special books on this subject, we satisfy ourselves with introducing material that does not overlap. The following problems were selected: asymptotically linear systems, Hamiltonians with periodic nonlinearities, second order systems with singular potentials, the double pendulum equation, Arnold conjectures on symplectic fixed points and on Lagrangian intersections. Our treatment of these is limited to examples.

In the final chapter, we analyze two-dimensional harmonic maps and the Plateau problem for minimal surfaces as examples from geometric variational problems. Because of the lack of the Palais-Smale condition, Morse theory for harmonic maps is established by the heat flow. The Plateau problem is considered to be a function defined on a closed convex set in a Banach space. Extended Morse theory is applied to give a proof of the Morse-Tompkins-Shiffman theorem on unstable coboundary minimal surfaces.

In the appendix, Witten's proof of the Morse inequalities is presented in a self-contained way. Although the material is totally independent of the context of this book, we introduce Witten's idea because the proof is so beautiful and surprising; moreover, it is a good example of the interplay between analysis and topology.

This book is not intended to be complete, either as a systematic study of Morse theory or as the presentation of many applications. We do not deal with Conley theory [Con1], stratified Morse theory, and the beautiful applications in the study of closed geodesics. (For an overview of the literature, the reader is referred to the book by Klingenberg [Kli1]) as well as to the study of gauge theory [AtB1].

CHAPTER I

Infinite Dimensional Morse Theory

The basic results in Morse theory are the Morse inequalities and the Morse handle body theorem. They are established on the Banach Finsler manifolds or on the Hilbert Riemannian manifolds in Section 4. The tool in this study is the deformation theorem, which is introduced in Section 3. Some preliminaries on algebraic topology and on infinite dimensional manifolds are reviewed in Sections 1 and 2 respectively. Readers who are familiar with the background material may skip over these two sections. Gromoll-Meyer theory on isolated critical points plays an important role in the applications of Morse theory because the nondegeneracy assumption in the handle body theorem might not hold for concrete problems. Section 5 is devoted to introducing Gromoll-Meyer theory systematically and examines the splitting lemma, the homotopy invariance theorem, the shifting theorem, and the Marino Prodi approximation theorem. The rest of the chapter consists of the extensions of the basic results of Morse theory in different directions: in Section 6.1, to the extension to manifolds with boundaries as well as to the functions satisfying certain boundary value conditions, in Section 6.2, to the extension from manifolds to the locally convex closed subsets; and, in Section 7, to functions with symmetry under a compact Lie group action.

1. A Review of Algebraic Topology

The idea of algebraic topology is to assign algebraic data to topological spaces so that topological problems may be translated into algebraic ones. The singular homology group is an example of algebraic data. It is constructed of the maps of geometric simplexes into arbitrary topological spaces so that it is applicable to infinite dimensional problems.

Let X be a topological space, and let

$$\Delta_q = \left\{ \sum_{j=0}^{q} \lambda_j e_j \mid \lambda_j \geq 0,\ \sum \lambda_j = 1 \right\}$$

be the standard q-simplex, $q = 0, 1, \ldots$ where

$$\begin{aligned} e_0 &= (0, 0, \ldots 0, \ldots) \\ e_1 &= (1, 0, \ldots 0, \ldots) \\ \cdots & \quad \cdots \\ e_q &= (0, 0, \ldots \underset{q^{th}}{1}, \ldots) \\ \cdots & \quad \cdots \end{aligned}$$

are vectors in $\mathbb{R}^\infty$.

A singular q-simplex is defined as a continuous map $\varphi : \Delta_q \longrightarrow X$. Also, let $\sum_q$ denote the set of all singular q-simplexes.

Given an Abelian group G, we define the formal linear combinations: $\sigma = \sum g_i \sigma_i$, $g_i \in G$, $\sigma_i \in \sum_q$. These sums are called singular q-chains. The set of all singular q-chains is denoted by $C_q(X, G)$.

Suppose that X, X' are two topological spaces, and that

$$f : X \longrightarrow X'$$

is continuous, then

$$f : \sigma = \sum g_i \sigma_i \longrightarrow \sum g_i f(\sigma_i)$$

is a reduced homomorphism: $C_q(X, G) \longrightarrow C_q(X', G)$.

For each $\sigma \in \sum_q$, we define the boundary operator

$$\partial \sigma = \sum_{j=0}^{q} (-1)^j \sigma^{(j)}$$

where $\sigma^{(j)} = \varphi[\theta, e_1, \ldots, \hat{e}_j, \ldots, e_q]$, $[e_0, e_1, \ldots, \hat{e}_j, \ldots, e_q]$ denotes the $q-1$ simplex generated by the vectors $e_0, e_1, \ldots, e_q$ except $e_j, j = 0, 1, \ldots, q$. Then we extend the operator ∂ linearly onto $C_q(X, G)$, i.e.,

$$\partial \sum g_i \, \sigma_i = \sum g_i \, \partial \sigma_i.$$

It is not difficult to verify:

(1) $\partial : C_q(X, G) \longrightarrow C_{q-1}(X, G)$ is a homomorphism, $q = 1, 2, \ldots$.

(2) $\partial^2 c = \partial \partial c = 0 \; \forall \, c \in C_q(X, G)$.

A different boundary opertor $\partial^\#$ can be defined on 0-chains as follows:

$$\partial^\# \sum g_i \sigma_i = \sum g_i \qquad \forall \, \sigma_i \in C_0(X, G), \, \forall \, i.$$

The relation

$$\partial^\# \partial = 0$$

also holds.

Suppose that (X, Y) is a pair of topological spaces, with $Y \subset X$ (being a subspace of X). We call (X, Y) a topological pair.

For two topological pairs (X, Y) and (X', Y'), we say that a map $f : (X, Y) \longrightarrow (X', Y')$ is continuous if $f : X \longrightarrow X'$ is continuous with $f(Y) \subset Y'$.

Two maps $f, g : (X, Y) \longrightarrow (X', Y')$ are called homotopic if $\exists\ F : [0,1] \times X \longrightarrow X'$, which is continuous and satisfies

$$F(0, \cdot) = f,\ F(1, \cdot) = g,$$

and

$$F : [0,1] \times Y \longrightarrow Y'.$$

Let (X, Y) be a topological pair, since

$$\partial : C_q(X, G) \longrightarrow C_{q-1}(X, G)$$

implies

$$\partial : C_q(Y, G) \longrightarrow C_{q-1}(Y, G).$$

The boundary operator induces a homomorphism $\overline{\partial}$ which makes the diagram

$$\begin{array}{ccc} C_q(X,G) & \longrightarrow & C_q(X,G)/C_q(Y,G) \\ \partial\downarrow & & \overline{\partial}\downarrow \\ C_{q-1}(X,G) & \longrightarrow & C_{q-1}(X,G)/C_q(Y,G) \end{array}$$

commutative. Clearly $\overline{\partial}\,\overline{\partial} = 0$. We call

$$C_q(X, Y, G) = C_q(X, G)/C_q(Y, G)$$

the singular q-relative chain module. Then we define

$Z_q(X, Y, G) = \ker(\overline{\partial})$, the singular q-relative closed chain module,

$B_q(X, Y, G) = \mathrm{Im}(\overline{\partial})$, the singular q-relative boundary module, and

$H_q(X, Y, G) = Z_q(X, Y, G)/B_q(X, Y, G)$, the singular q-relative *homology* module. The rank of $H_q(X, Y, G)$ is called the singular q-Betti number.

In the case where $Y = \emptyset$, we write $H_q(X, Y, G) = H_q(X, G)$. For $q = 0$, $H_0^{\#}(X, G)$ is defined as the quotient of $\ker(\partial^{\#})$ by $\mathrm{Im}(\partial)$, and for $q > 0$, let $H_q^{\#}(X, G) = H_q(X, G)$. We call $H_q^{\#}(X, G)$ the q-reduced homology module. The 0-reduced relative homology module $H_0^{\#}(X, Y, G)$ is defined as $H_0(X, Y, G)$ if $Y \neq \emptyset$ and $H_0^{\#}(X)$ if $Y = \emptyset$. The basic properties of singular homology modules are summarized as follows. Their proofs can be found in the book of M. J. Greenberg [Gr 1].

1. Suppose that $f : (X,Y) \longrightarrow (X',Y')$ is continuous, then there is a reduced homomorphism

$$f_* : H_q(X,Y;G) \to H_q(X',Y';G) \;\forall\, q.$$

(a) If $f = \text{id}$, then $f_* = \text{id}$;
(b) If $g : (X',Y') \longrightarrow (X'',Y'')$ is another continuous map, then the reduced homomorphism g_* satisfies

$$(gf)_* = g_* f_*.$$

(c) $\overline{\partial} f_* = f_* \overline{\partial}$.

2. *Homotopy invariance*: If $f, g : (X,Y) \longrightarrow (X',Y')$ are homotopic, then $f_* = g_*$.

Two topological pairs (X,Y) and (X',Y') are called homotopically equivalent if there exist continuous maps

$$\phi : (X,Y) \longrightarrow (X',Y'),$$
$$\psi : (X',Y') \longrightarrow (X,Y),$$

satisfying

$$\psi \circ \phi = \text{id}_{(X,Y)}, \quad \phi \circ \psi \cong \text{id}_{(X',Y')}.$$

Thus, if (X,Y) and (X',Y') are homotopically equivalent, then

$$H_q(X,Y,G) \cong H_q(X',Y',G) \quad \forall\, q.$$

We say (X',Y') is a deformation retract of (X,Y) if $X' \subset X,\ Y' \subset Y$, and if $\exists\, \eta : [0,1] \times X \longrightarrow X$ satisfying

$$\eta(0,\cdot) = \text{id}_X,\ \eta(1,X) \subset X',\ \eta(1,Y) \subset Y',$$
$$\eta(t,Y) \subset Y \text{ and } \eta(t,\cdot)|_{X'} = id_{X'},\ \forall\, t \in [0,1].$$

Thus, if (X',Y') is a deformation retract of (X,Y), then

$$H_q(X',Y',G) \cong H_q(X,Y,G).$$

3. *Excision*: If $U \subset X$ satisfies $\overline{U} \subset \text{int}(Y)$, then

$$H_q(X \backslash U, Y \backslash U, G) \cong H_q(X,Y,G).$$

4. *Exactness*: If $Z \subset Y \subset X$ are three topological spaces, and we define the injections $i : (Y,Z) \longrightarrow (X,Z)$, and $j : (X,Z) \longrightarrow (X,Y)$, then we have the following exact sequence:

$$\cdots \to H_q(Y,Z,G) \xrightarrow{i_*} H_q(X,Z,G) \xrightarrow{j_*} H_q(X,Y,G)$$
$$\xrightarrow{\partial} H_{q-1}(Y,Z,G) \to \cdots .$$

In particular, since $H_q(X,G) = H_q(X,\emptyset,G)$, we have

$$\cdots \to H_q(Y,G) \xrightarrow{i_*} H_q(X,G) \xrightarrow{j_*} H_q(X,Y,G) \xrightarrow{\partial} H_{q-1}(Y,G) \to \cdots .$$

The same exact sequence also holds for reduced homology modules.

5. If X consists of a family of path-connected components $\{X_k\}$, then

$$H_q(X,Y;G) \cong \oplus \sum H_q(X_k, X_k \cap Y;G) \quad \forall\, q.$$

6. $H_q(X,X;G) \cong 0, \quad \forall\, q.$

7. $H_0(X,G)$ is a free group on as many generators as there are path components of X.

If $Y \neq \emptyset, Y \subset X$, and X is path-connected, then

$$H_0(X,Y;G) \cong 0.$$

8. *Künneth formula:* Let X_1 and X_2 be subspaces of the topological space X. Denote $i_\nu : X_\nu \longrightarrow X$ as the injection. $\nu = 1,2$.

(X_1, X_2) is said to be an excisive couple of subspaces if the inclusion chain map

$$C_q(X_1,G) + C_q(X_2,G) \longrightarrow C_q(X_1 \cup X_2, G)$$

induces an isomorphism of homology.

For given topological pairs (X,Y), (X',Y'), we define their product $(X,Y) \times (X',Y')$ to be the pair $(X \times X', X \times Y' \cup Y \times X')$.

If G is a field, and if $\{X \times Y', Y \times X'\}$ is an excisive couple in $X \times X'$, then the cross product is an isomorphism:

$$H_*(X,Y;G) \otimes H_*(X',Y';G) \cong H_*((X,Y) \times (X',Y');G),$$

i.e.,

$$H_q(X \times X', X \times Y' \cup Y \times X';G) \cong \bigoplus_{p=0}^{q} H_q(X,Y;G) H_{q-p}(X',Y';G),$$

$\forall\, q = 0,1,2,\dots .$

In the case where $G =$ a field Q,

$$\text{rank } H_q(X,Y;Q) = \dim H_q(X,Y;Q),$$

we write

$$\chi(X,Y;Q) = \sum_{q=0}^{\infty} (-1)^q \dim H_q(X,Y;Q),$$

and call it the Euler characteristic of (X, Y).

The following homology groups are often used.

(1) $$H_q(S^n, G) \cong \begin{cases} 0 & q \neq n, \text{ when } q, n \geq 1 \\ G & q = n \geq 1, \text{ and } q = 0, n \geq 1, \\ G^2 & q = n = 0. \end{cases}$$

(2) $$H_q(B^n, S^{n-1}, G) \cong \begin{cases} 0 & q \neq n, \\ G & q = n, \end{cases}$$

where B^n is the n-ball, and $S^{n-1} = \partial B^n$.

(3) $$H_q(T^n, G) \cong \begin{cases} G^{C_q^n} & 0 \leq q \leq n, \\ 0 & q > n, \end{cases}$$

where $T^n = S^1 \times \cdots \times S^1$ is the n-torus.

(4) $$H_q(P^n, Z_2) \cong \begin{cases} 0 & q > n \\ Z_2 & q \leq n, \end{cases}$$

where P^n is the real n-projective space.

(5) $$H_q(CP^n, G) \cong \begin{cases} 0 & q > 2n \text{ or } q \text{ odd}, \\ G & q \text{ even such that } 0 \leq q \leq 2n, \end{cases}$$

where CP^n is the complex n-projective space, and $G = Q$, the rational field, or $\mathbb{Z}$.

Now we turn our study to singular cohomology. The singular q-cochain is defined to be the homomorphism $c : C_q(X, G) \longrightarrow G$:

$$[\sigma_1 + \sigma_2, c] = [\sigma_1, c] + [\sigma_2, c], \ \forall\, \sigma_1, \sigma_2 \in C_q(X, G),$$
$$[g \cdot \sigma, c] = g \cdot [\sigma, c] \ \forall\, g \in G, \ \forall\, \sigma \in C_q(X, G).$$

The set of all singular q-cochains $\mathrm{Hom}(C_q(X, G), G)$ is denoted by $C^q(X, G)$. $C^q(X, G)$ is a module:

$$[\sigma, c_1 + c_2] = [\sigma, c_1] + [\sigma, c_2] \ \forall\, c_1, c_2 \in C^q(X, G), \ \forall\, \sigma \in C_q(X, G),$$
$$[\sigma, g \cdot c] = g \cdot [\sigma, c], \ \forall\, g \in G, \ \forall\, \sigma \in C_q(X, G), \ \forall\, c \in C^q(X, G).$$

Thus the duality $[\ ,\ ]$ is a bilinear form on $C_q(X, G) \times C^q(X, G)$.

The dual operator of the boundary operator ∂ with respect to $[\ ,\ ]$ is called the coboundary operator and is denoted by δ:

$$[\partial\sigma, c] = [\sigma, \delta c] \ \forall\, \sigma \in C_q(X, G), \ \forall\, c \in C^{q-1}(X, G).$$

Hence, $\delta : C^{q-1}(X,G) \longrightarrow C^q(X,G)$ is a homomorphism, and $\partial^2 = 0$ implies

$$\delta^2 c = 0 \; \forall \; c \in C^q(X,G).$$

Singular cohomology is defined as follows: For a topological pair (X,Y), let

$$\overline{C}^q(X,Y;G) = \operatorname{Hom}(C_q(X,G)/C_q(Y,G),G),$$

and let

$$\overline{\delta} : \overline{C}^{q-1}(X,Y) \longrightarrow \overline{C}^q(X,Y)$$

be the dual operator of the boundary operator $\overline{\partial} : C_q(X,Y;G) \longrightarrow C_{q-1}(X,Y;G)$. Then define

$$H^q(X,Y;G) = \ker(\overline{\delta}) \,/\, \operatorname{Im}(\overline{\delta}).$$

It is easily seen that $\overline{C}^q(X,Y;G)$ is isomorphic to

$$C^q(X,Y;G) = \{c \in C^q(X,G) \mid [\sigma,c] = 0 \qquad \forall \, \sigma \in C_q(Y,G)\}\,.$$

The isomorphism is realized by the dual homomorphism

$$P^* : \overline{C}^q(X,Y;G) \longrightarrow \overline{C}^q(X,G)$$

of the homomorphism

$$P : C_q(X,G) \longrightarrow C_q(X,Y;G).$$

Therefore

$$Z^q(X,Y;G) := \ker(\overline{\delta}) = \{c \in C^q(X,G) | [\sigma,c] = 0 \; \forall \; \sigma \in B_q(X,Y;G)\}\,,$$
$$B^q(X,Y;G) := \operatorname{Im}(\overline{\delta}) = \{c \in C^q(X,G) | [\sigma,c] = 0 \; \forall \; \sigma \in Z_q(X,Y;G)\}\,.$$

In general, we have a canonical homomorphism:

$$\alpha : H^q(X,Y;G) \longrightarrow H_q(X,Y;G)^*.$$

In the case where G is a field, α is surjective.

The properties of cohomology are very similar to those of homology. The important difference is as follows: Singular homology is a covariant functor of topological pairs, but singular cohomology is a contravariant functor.

(1′) If $f : (X,Y) \longrightarrow (X',Y')$ is continuous, then

$$f^* : H^*(X',Y',G) \longrightarrow H^*(X,Y;G)$$

We have

(a) If $f = \text{id}$, then $f^* = \text{id}$.
(b) If $g: (X', Y') \longrightarrow (X'', Y'')$ is continuous, then $(gf)^* = f^* g^*$.
(c) $\overline{\delta} f^* = f^* \overline{\delta}$.

(2′) If, $f, g: (X, Y) \longrightarrow (X', Y')$ are homotopic, then $f^* = g^*$. If $(X, Y) \sim (X', Y')$, then $H^*(X, Y; G) \cong H^*(X', Y'; G)$.
(3′) (Excision) $H^*(X \backslash U, Y \backslash U; G) \cong H^*(X, Y; G)$, if $\overline{U} \subset \text{int}(Y)$.
(4′) (Exactness) If $Z \subset Y \subset X$, then the sequence

$$\cdots \leftarrow H^q(Y, Z; G) \overset{i^*}{\leftarrow} H^q(X, Z; G) \overset{j^*}{\leftarrow} H^q(X, Y; G) \overset{\overline{\delta}}{\leftarrow} H^{q-1}(Y, Z; G) \leftarrow \cdots$$

is exact.

(5′) $H^q(\{p\}, G) = \begin{cases} G & q = 0 \\ 0 & q \neq 0. \end{cases}$

(6′) If $(X \times Y', Y \times X')$ is an excisive couple in $X \times X'$, and $H^*(X, Y; G)$ is of finite type, i.e., $H^q(X, Y; G)$ is finitely generated for each q, and G is a field, then

$$H^*(X, Y; G) \otimes H^*(X', Y'; G) \cong H^*((X, Y) \times (X', Y'); G).$$

We can define a product on singular cohomolgy groups such that the singular cohomology groups become graded algebras.

We denote $C^*(X, G) = \oplus_{q=0}^{\infty} C^q(X, G)$, and define a cup product as follows: $\forall\, c \in C^p(X, G)$, $\forall\, d \in C^q(X, G)$, $\forall\, \sigma \in C_{p+q}(X, G)$, we consider affine maps

$$\lambda_p : \Delta_p \longrightarrow \Delta_{p+q}$$
$$\rho_q : \Delta_q \longrightarrow \Delta_{p+q}$$

to be

$$\lambda_p = (e_0, \ldots, e_p), \quad \rho_q = (e_p, e_{p+1}, \ldots, e_{p+q}).$$

and then define

$$[\sigma, c \cup d] = [\sigma \lambda_p, c] \cdot [\sigma \rho_q, d].$$

The cup product is bilinear, associative, and possesses the unit, i.e., the 0-cochain 1, which is defined by $[x, 1] = e \;\forall\, x \in X$.

We may easily prove that

$$\delta(c \cup d) = \delta c \cup d + (-1)^p c \cup \delta d \quad \forall\, c \in C^p(X, G),\ \forall\, d \in C^q(X, G).$$

Hence, $Z^*(X, G)$ is a subalgebra of $C^*(X, G)$ and $B^*(X, G)$ is an ideal of $Z^*(X, G)$. The cup product $\cup$ is well defined on $H^*(X, G)$, and makes it a graded algebra. Furthermore, if $f: X \longrightarrow Y$ is continuous, then $f^*: H^*(Y) \longrightarrow H^*(X)$ is a ring homomorphism: $f^*(c \cup d) = f^*(c) \cup f^*(d)$, which satisfies $f^* \delta = \delta f^*$.

The cap product is defined as the dual operator of the cup product, i.e., $\cap: C_{p+q}(X) \times C^p(X) \longrightarrow C_q(X)$,

$$\forall\, c \in C^p(X),\ \forall\, d \in C^q(X),\ \forall\, \sigma \in C_{p+q}(X),$$
$$[\sigma \cap c, d] = [\sigma, c \cup d],$$

or, equivalently,

$$\sigma \cap c = [\sigma\lambda_p, c]\sigma\rho_q.$$

The boundary operators relate the cap product as follows:

$$\partial(\sigma \cap c) = (-1)^p[(\partial\sigma) \cap c - \sigma \cap \delta c].$$

$\forall\, \sigma \in C_{p+q}(X)$, $\forall\, c \in C^p(X)$.

If $f : X \longrightarrow Y$ is continuous, then we have

$$f_*[\sigma \cap f^*(c)] = f_*(\sigma) \cap c.$$

Since $\forall\ \sigma\ \in\ Z_{p+q}(X)$, $\forall\ c\ \in\ Z^p(X)$, we have $\sigma \cap c \in Z_q(X)$, and $\forall\, \sigma \in B_{p+q}(X)$, $\forall\, c \in Z^p(X)$, we have $\sigma \cap c \in B_q(X)$, the cap product is well-defined on homology groups:

$$\cap : H_{p+q}(X) \times H^p(X) \longrightarrow H_q(X).$$

The definition of cup product and cap product can be extended to topological pairs. In fact, we have

$$\cap : H_{p+q}(X, Y; G) \times H^p(X, Y; G) \longrightarrow H_q(X, G)$$
$$\cap : H_{p+q}(X, Y; G) \times H^p(X, G) \longrightarrow H_q(X, Y; G),$$

and

$$\cup : H^p(X, Y_1; G) \times H^q(X, Y_2 : G) \longrightarrow H^{p+q}(X, Y_1 \cup Y_2; G),$$

if (Y_1, Y_2) is an excisive couple in X.

The cup length of a topological space X is defined as

$$CL(X) = \max\{l \in Z_+ \mid \exists\, c_1, ..., c_l \in H^*(X, G),$$
$$\dim(c_i) > 0, i = 1, \ldots, l, \text{ such that } c_1 \cup \cdots \cup c_l \neq 0\}.$$

This is a topological invariant which is very useful in critical point theory.

More generally, we define the cup length for a topological pair (X, Y).

$$CL(X, Y) = \max\{l \in Z_+ \mid \exists\, c_0 \in H^*(X, Y),\ \exists\, c_1, c_2, \ldots, c_l \in H^*(X),$$

with $\dim(c_i) > 0$, $i = 1, 2, \ldots, l$, such that $c_0 \cup c_1 \cup \cdots \cup c_l \neq 0\}$.

In the case where $Y = \emptyset$, we just take $c_0 \in H^0(X)$. These two definitions are the same.

We may characterize $CL(X,Y)$ by its dual.

Definition 1.1. Let (X,Y) be a pair of topological spaces and $Y \subset X$. For two nontrivial singular homology classes $[\sigma_1], [\sigma_2] \in H_*(X,Y)$, we say that $[\sigma_1]$ is subordinate to $[\sigma_2]$, denoted by $[\sigma_1] < [\sigma_2]$, if there exists $c \in H^*(X)$, with $\dim c > 0$ such that

$$[\sigma_1] = [\sigma_2] \cap c,$$

where $\cap$ is the cap product.

Let us define

$$L(X,Y) = \max\{l \in Z_+ \mid \exists \text{ nontrivial classes } [\sigma_j] \in H_*(X,Y), \\ 1 \le j \le l, \text{ such that } [\sigma_1] < [\sigma_2] < \cdots < [\sigma_l]\}.$$

Theorem 1.1. *$L(X,Y) = CL(X,Y) + 1$.*

Proof. For $L(X,Y) = l+1$ if and only if $\exists$ nontrivial classes $[\sigma_0] < [\sigma_1] < \cdots < [\sigma_l]$ in $H_*(X,Y)$, i.e., $\exists$ $c_i \in H^*(X)$, dim $c_i > 0, 1 \le i \le l$, such that

$$[\sigma_{i+1}] = [\sigma_i] \cap c_i, \; i = 1, 2, \ldots, l.$$

However, $\exists$ $c_0 \in H^*(X,Y)$ such that $[\,[\sigma_0], c_0\,] \ne 0$ is equivalent to the nontriviality of $[\sigma_0]$. And since

$$[[\sigma_l], c_l \cup c_{l-1} \cup \cdots \cup c_0] = [[\sigma_{l-1}], c_{l-1} \cup c_{l-2} \cup \cdots \cup c_0] \\ = \cdots = [[\sigma_0], c_0],$$

$\therefore$ $L(X,Y) = l+1$ if and only if $CL(X,Y) = l$. □

The homotopy group is another important topological invariant. Let us recall some basic definitions and properties in homotopy theory.

Let X be a topological space and p be a point in X. We call (X,p) a pointed space with base point p. A topological pair (X,Y), in which Y is a subspace of X that contains p, is called a pointed pair (often written (X,Y,p)).

A map f from pointed space (X,p) to a pointed space (X',p'), $f : X \to X'$, with $f(p) = p'$, is called a pointed map. Similarly, we define a pointed pair map, pointed homotopy, pointed pair homotopy, and so forth.

Let I^n denote the n-dimensional unit cube, $n \ge 1$, $I^{n-1} \subset I^n$ the bottom space $(t = (t_1, t_2, \ldots, t_n) \in I^n$, if $0 \le t_i \le 1$, $i = 1, 2, \ldots, n$, and $t \in I^{n-1}$

if, further, $t_n = 0$). Set $J^{n-1} = \partial I^n \backslash I^{n-1}$, i.e., the union of all the other faces. We denote by $\Omega_n(X,Y,p)$ the set of all continuous maps

$$\phi : \left(I^n, \partial I^{n-1}, J^{n-1}\right) \to (X,Y,p),$$

i.e.,

$$\phi : I^n \to X, \phi(\partial I^{n-1}) \subset Y \text{ and } \phi(J^{n-1}) = p.$$

(Denote by $\Omega_n(X,p)$ the set of all continuous maps $\phi : (I^n, \partial I^{n-1}) \to (X,p)$).

The n-relative homotopy group $\pi_n(X,Y;p)$ is defined as the set of all components of $\Omega_n(X,Y;p)$ (the n homotopy group $\pi_n(X,p)$ for those of $\Omega_n(X,p)$). In fact, $\pi_n(X,p) = \pi_n(X,p;p)$.

One may define a multiplication on $\Omega_n(X,Y,p)$: For $n \geq 2$ (and, on $\Omega_n(X,p)$, for $n \geq 1$) : $\forall\, \phi, \psi \in \Omega_n(X,Y,p)$,

$$(\phi * \psi)(t) = \begin{cases} \phi(2t_1, t_2, \dots, t_n) & 0 \leq t_1 \leq \frac{1}{2} \\ \psi(2t_1 - 1, t_2, \dots, t_n) & \frac{1}{2} \leq t_1 \leq 1 \end{cases} \qquad \forall\, t \in I^n.$$

The homotopy class $[\phi * \psi]$ clearly depends only on $[\phi]$ and $[\psi]$. Hence we may define a multiplication on $\pi_n(X,Y,p)$ by taking

$$[\phi] * [\psi] = [\phi * \psi].$$

According to the multiplication, which is obviously associative, the identity element $[e]$ is the class which contains the unique constant map

$$e : X \to p.$$

An inverse element of $[\phi]$ is the class $[\phi \circ \theta]$, where $\theta : I^n \to I^n$ denotes the map defined by

$$\theta(t) = (1 - t_1, t_2, \dots, t_n)$$

for every $t \in I^n$.

With the multiplication structure, $\pi_n(X,Y,p)$ is a group for $n \geq 2$, and $\pi_n(X,p)$ is a group for $n \geq 1$.

For $n = 0$, we define $\pi_0(X,p)$ to be the set of path-connected components of X with the path component of p as a distinguished element. So, $\pi_0(X,p)$ is only a set, without group structure, as is $\pi_1(X,y,p)$.

Moreover, one can show that $\pi_n(X,p)$, for $n \geq 3$, and $\pi_n(X,Y,p)$, for $n \geq 2$, are abelian groups.

There are alternative definitions of homotopy groups.

Since $J^{n-1} = \partial I^n \backslash I^{n-1}$ is contractible, let $z_0 = \theta$ in I^n, and the inclusion map

$$(I^n, \partial I^n, z_0) \to \left(I^n, \partial I^n, J^{n-1}\right)$$

is a homotopy equivalence. Hence $\pi_n(X, Y, p)$ is defined as equivalent to the set of homotopy classes of maps:

$$\phi : (I^n, \partial I^n, z_0) \to (X, Y, p).$$

Again, let $x_0 = (1, 0, \dots, 0) \in \mathbb{R}^n$, $D^n =$ the unit ball in $\mathbb{R}^n$, and $S^{n-1} = \partial D^n$, $\pi_n(X, Y, p)$ as the set of homotopy classes of maps

$$\phi : \left(D^n, S^{n-1}, x_0\right) \to (X, Y, p).$$

The homotopy groups enjoy analogues of all the axioms of homology groups except the excision axiom.

(1) Suppose $f : (X, Y, p) \to (X', Y', p')$ is a continuous pointed pair map, then there is a reduced homomorphism

$$f_* : \pi_k(X, Y, p) \to \pi_k(X', Y', p').$$

(a) If $f =$ id, then $f_* =$ id.
(b) If $g : (X', Y', p') \to (X'', Y'', p'')$ is another continuous pointed pair map, then $(gf)_* = g_* \cdot f_*$.
(c) Let $\partial : \pi_k(X, Y, p) \to \pi_{k-1}(Y, p)$ be defined by restriction, i.e., for a given $\phi : (I^n, \partial I^n, \partial I^n \backslash I^{n-1}) \to (X, Y, p)$, we define

$$\partial[\phi] = [\phi|_{\partial I^n}]\,.$$

We have the following commutative diagram:

$$\begin{array}{ccc}
\pi_k(X, Y, p) & \xrightarrow{\partial} & \pi_{k-1}(Y, p) \\
\downarrow f_* & & \downarrow (f|_Y)_* \\
\pi_k\left(X', Y', p'\right) & \xrightarrow{\partial} & \pi_{k-1}\left(Y', p'\right).
\end{array}$$

(2) (Homotopy invariance). Suppose that f is pointed pair homotopic to g, where $f, g : (X, Y, p) \to (X', Y', p')$. Then

$$f_* = g_*.$$

Thus, if (X, Y, p) and (X', Y', p') are pointed pair homotopic equivalent, then

$$\pi_k(X, Y, p) = \pi_k\left(X', Y', p'\right).$$

(3) (Exactness). Let $p \in Z \subset Y \subset X$, then the following sequence is exact:

$$\dots \to \pi_{k+1}(X, Y, p) \xrightarrow{\partial'} \pi_k(Y, Z, p)$$
$$\xrightarrow{i_*} \pi_k(X, Z, p) \xrightarrow{j_*} \pi_k(X, Y, p) \to \dots$$

where $i : (Y,Z) \to (X,Z)$, $j : (X,Z) \to (X,Y)$ are inclusions, and define $\partial' : \pi_{k+1}(X,Y,p) \to \pi_k(Y,Z,p)$, $k \geq 1$, to equal the composition $\pi_{k+1}(X,Y,p) \xrightarrow{\partial} \pi_k(Y,p) \xrightarrow{j'_*} \pi_k(Y,Z,p)$, where $j' : (Y,p) \to (Y,Z)$ is an inclusion.

(4) If Y is path-connected, then $\pi_k(X,Y,p)$ does not depend on $p \in Y$. In this case, we abbreviate $\pi_k(X,Y)$.

(5) If $X = \{p\}$, then $\pi_k(X,p) = 0 \ \forall\, k \geq 1$.

Thus, if X is contractible, i.e., X is homotopic to a point, then $\pi_k(X,p) = 0 \ \forall\, k \geq 1$.

(6) If (X,p), (Y,q) are pointed spaces, then

$$\pi_k(X \times Y, (p,q)) \cong \pi_k(X,p) \oplus \pi_k(Y,q).$$

(7) If (X,p) is a pointed space, we define its loop space $\Omega(X,p)$ to be the function space consisting of all continuous maps $\phi : (I^1, \partial I^1) \mapsto (X,p)$ endowed with compact open topology. Then

$$\pi_{k-1}(\Omega,\omega) = \pi_k(X,p) \quad \text{for } k \geq 2,$$

where $\Omega = \Omega(X,p)$, and ω is the constant map $\omega(t) = p \ \ \forall\, t \in I$.

(8) If $0 < k < n$, then $\pi_k(S^n) = 0$, and $\pi_n(S^n) \neq 0$,

$$\begin{aligned} \pi_k\left(S^n\right) &\cong \pi_k\left(\mathbb{R}P^n\right) \qquad && \forall\, k \geq 2, \\ \pi_k\left(S^{2n+1}\right) &\cong \pi_k\left(\mathbb{C}P^n\right) \qquad && \forall\, k \geq 3. \end{aligned}$$

The following important property establishes the relationship between homology and homotopy.

(9) **Theorem 1.2.** (Hurewicz Isomorphism Theorem). *Let (X,Y,p) be a pointed pair, and assume that Y and X are simply connected. If there is a $k \geq 2$ such that $H_q(X,Y) \cong 0 \ \ \forall\, q < k$, then $\pi_q(X,Y,p) \cong 0 \ \ \forall\, q < k$, and there is an isomorphism*

$$\phi : \pi_k(X,Y,p) \xrightarrow[\cong]{} H_k(X,Y).$$

For (absolute) homotopy groups, the same conclusion holds for $k \geq 1$ and $Y = \{p\}$, where these homology groups are of integral coefficients.

(10) Let (X,Y) be a (pointed) pair (with base point p) such that Y is a retract of X. Then

$$\begin{aligned} & H_*(X) \cong H_*(Y) \oplus H_*(X,Y) \\ (\text{and} \qquad & \pi_k(X,p) \cong \pi_k(Y,p) \oplus \pi_k(X,Y,p) \ \ \forall\, k \geq 2). \end{aligned}$$

Claim. Let $i : Y \to X$ and $j : (X,\emptyset) \to (X,Y)$ (or $(X,p) \to (X,Y)$, where p is the base point of the pair (X,Y)), be the inclusions, and r

is the retraction: $X \to Y$. It follows that r_*i_* is the identity map on $H_*(Y)$ (or $\pi_k(Y,p)$ $\forall$ $k \geq 1$). This implies that i_* is a monomorphism and r_* is an epimorphism, i.e., $H_*(X) \cong H_*(Y) \oplus \ker r_*$ (or $\pi_k(X,p) \cong \pi_k(Y,p) \oplus \ker r_*$, $\forall$ $k \geq 1$).

From the exactness of the homology (homotopy) sequence

$$\begin{aligned} &\to H_{k+1}(X,Y) \xrightarrow{\partial_*} H_k(Y) \xrightarrow{i_*} H_k(X) \\ &\xrightarrow{j_*} H_k(X,Y) \to \\ &\Big(\to \pi_{k+1}(X,Y,p) \xrightarrow{\partial_*} \pi_k(Y,p) \xrightarrow{i_*} \pi_k(X,p) \\ &\xrightarrow{j_*} \pi_k(X,Y,p) \to \Big), \end{aligned}$$

because $\ker i_* = 0$, ∂_* is a trivial map, and therefore j_* is an epimorphism. Since $\ker j_* = \operatorname{Im} i_*$, j_* induces an isomorphism from $\ker r_*$ onto $H_*(X,Y)$ (onto $\pi_k(X,Y,p)$ if $k \geq 2$, resp.).

(11) The following theorem is useful in the sequel.

Theorem 1.3. (Palais). *Let V be a Banach space and E be a dense subspace (linear) of V with finite topology.*

Given $\mathcal{O}$, an open set in V, let $\widetilde{\mathcal{O}} = \mathcal{O} \cap E$ be a subspace of E. Then the inclusion map $i : \widetilde{\mathcal{O}} \to \mathcal{O}$ is a homotopy equivalence.

Corollary 1.1. *Let V_1 and V_2 be Banach spaces, and let $f : V_1 \to V_2$ be a linear continuous map of V_1 onto a dense linear subspace of V_2. Given $\mathcal{O}$ open in V_2, let $\widetilde{\mathcal{O}} = f^{-1}(\mathcal{O})$ and $\tilde{f} = f|_{\widetilde{\mathcal{O}}}$. Then $\tilde{f} : \widetilde{\mathcal{O}} \to \mathcal{O}$ is a homotopy equivalence.* (cf. Palais [Pal 2, Theorems 16, 17].)

Thus, if X is a Banach space embedded continuously into a Hilbert space H as a dense linear subspace, then we have

$$H_*(A,B;G) = H_*\left(A|_X, B|_X; G\right),$$

for any pair of open sets (A,B) in H, where $(A|_X, B|_X)$ is the restriction of (A,B) on X.

2. A Review of the Banach-Finsler Manifold

Most manifolds interested in analysis are infinite dimensional.

Definition 2.1. Let X be a Banach space, and let M be a connected Hausdorff space. We say that M is a Banach C^r manifold, $r \geq 1$ (integer), modeled on X, if

(1) $\exists$ a family of open coverings $\{U_i \mid i \in \Lambda\}$,

(2) $\exists$ a family of coordinates $\{\phi_i \mid : U_i \to \phi_i(U_i) \subset X$, homeomorphism, $i \in \Lambda\}$,

(3) $\phi_i \circ \phi_{i'}^{-1} : \phi_{i'}(U_i \cap U_{i'}) \to \phi_i(U_i \cap U_{i'})$ is a C^r-diffeomorphism, $\forall\, i, i' \in \Lambda$.

Each pair (ϕ_i, U_i) is called a chart. The set $\{(\phi_i, U_i) \mid i \in \Lambda\}$ is called an atlas.

In a similar way, we define $C^r(C^{r-0})$ maps between two C^r Banach manifolds, and vector bundles over Banach manifolds (we omit the definitions and basic properties), in particular, the tangent bundle $T(M)$ and cotangent bundle $T^*(M)$.

Let $\Xi = (E, \pi, M)$ be a vector bundle. $\xi : M \to E$ is called a section, if $\pi \circ \xi = \mathrm{id}_M$. A section ξ is called C^r (or C^{r-0}) continuous, if it is a C^r (or C^{r-0}) map from M to E.

A section of the tangent bundle is called a vector field, and a section of the cotangent bundle is called a co-vector field.

For a given C^{1-0} vector field ξ, and a given point $p \in M$, there exists an unique maximal semiflow $\alpha : [0, T) \to M$ satisfying

$$\begin{aligned} \dot{\alpha}(t) &= \xi(\alpha(t)), \\ \alpha(0) &= p. \end{aligned}$$

A Riemannian manifold (M, g) is metrizable. The metric d is defined by the arclength of the geodesics, and thus it is defined by the Riemannian metric g:

$$d(x,y) = \inf \Big\{ \int_0^1 g(\dot{\sigma}(t), \dot{\sigma}(t))^{\frac{1}{2}}\, dt \\ \qquad\qquad \mid \sigma \in C^1([0,1], M) \quad \sigma(0) = x, \sigma(1) = y \Big\}.$$

As a metric space (M, d), the topology coincides with (is equivalent to) the topology of the manifold.

Generally speaking, in order to define the distance (or metric d), the inner product g is not necessary. A norm is enough. We may introduce a metric d on a Banach manifold.

Since the Riemannian metric is defined on $T(M)$ globally, we shall introduce a Finsler structure on a Banach manifold in a similar manner.

Definition 2.2. Let $\pi : E \to M$ be a Banach vector bundle. $\|\cdot\| : E \to R^1_+$ is called a Finsler structure if

(1) $\|\cdot\|$ is continuous,

(2) $\forall\, p \in M, \|\cdot\|_p := \|\cdot\| \,|_{E_p}$ is an equivalent norm on $E_p := \pi^{-1}(p)$,

(3) $\forall\, p_0 \in M$, for any neighborhood U of p_0 which trivializes the vector bundle E, i.e., $E \mid U = \pi^{-1}(U) \approx U \times E_{p_0}; \forall\, k > 1,\ \exists$ a neighborhood V of M with $V \supset U$ such that

$$\frac{1}{k}\|\cdot\|_p \le \|\cdot\|_{p_0} \le k\|\cdot\|_p \quad \forall\, p \in V.$$

Example 1. If E is a trivial bundle $U \times X$, then

$$\|(p,x)\| = \|x\|_X$$

is a Finsler structure.

Example 2. Suppose that $\pi : E \to M$ is a Hilbert bundle, with $g : C^\infty(E) \times C^\infty(E) \to C^\infty(M)$, a bilinear symmetric positive definite C^∞ map (Riemannian metric). Then

$$\|(p,x)\| = \sqrt{g(\sigma,\sigma)_p} \quad (\text{where } \sigma(p) = x)$$

is a Finsler structure.

Claim. $\forall\, p \in M$, $\exists$ a trivialized neighborhood U_i and the corresponding map $\tau_i : \pi^{-1}(U_i) \to U_i \times H_i$, where H_i is a Hilbert space. It follows that $\exists\, A_i(\cdot) : U_i \to L(H_i, H_i)$ such that $A_i(p)$ is positive definite and satisfies

$$g(\sigma,\sigma)_p = (A_i(p)x, x)_i.$$

(1) is obvious, (2) follows from the positive definiteness of $A_i(p)$. And the continuity of $p \mapsto A_i(p)$, as well as of $p \mapsto A_i^{\pm\frac{1}{2}}(p)$ implies (3).

Lemma 2.1. *Suppose that M is a paracompact C^1-Banach manifold and that $\mathcal{O} = \{(\phi_i, U_i) | i \in \Lambda\}$ is an atlas, then there exists a C^{1-0}-partition of unity subordinate to $\mathcal{O}$.*

Proof. Since M is paracompact, there is a locally finite refinement $\{o_\beta \mid \beta \in B\}$ of $\mathcal{O}$, i.e., $\forall\, \beta \in B, \exists\, \alpha = \alpha(\beta) \in \Lambda$ such that $\bar{o}_\beta \subset U_{\alpha(\beta)}$.

Let $\tilde{V}_\beta = \phi_{\alpha(\beta)}(o_\beta) \subset X$, and let

$$h_\beta(x) = \inf\left\{\|x - y\| \mid y \notin \tilde{V}_\beta\right\},$$

then $h_\beta \in C^{1-0}(X, R^1_+)$ and $\tilde{V}_\beta = \{x \in X \mid h_\beta(x) > 0\}$.

Claim. $\forall\, x_1, x_2 \in X$, $\forall\, \epsilon > 0$, $\exists\, y \in \tilde{V}_\beta$ such that

$$\|x_2 - y\| \le h_\beta(x_2) + \epsilon,$$

and so

$$\begin{aligned} h_\beta(x_1) &\le \|x_1 - y\| \le \|x_1 - x_2\| + \|x_2 - y\| \\ &\le \|x_1 - x_2\| + h_\beta(x_2) + \epsilon, \end{aligned}$$

i.e., $h_\beta(x_1) - h_\beta(x_2) \le \|x_1 - x_2\| + \epsilon$. We may change the positions of x_1 and x_2, and since $\epsilon > 0$ is arbitrary, we obtain

$$|h_\beta(x_1) - h_\beta(x_2)| \le \|x_1 - x_2\|.$$

Define

$$f_\beta = h_\beta \circ \phi_{\alpha(\beta)},$$

and define

$$\eta_\beta(p) = \frac{f_\beta(p)}{\sum\limits_{\beta \in B} f_\beta(p)}.$$

Then we have

$$\sup \eta_\beta \subset \bar{o}_\beta \subset U_{\alpha(\beta)}, \ \forall\, \beta \in B,$$

and

$$\sum_{\beta \in B} \eta_\beta(p) = 1.$$

□

Theorem 2.1. *Suppose that M is a paracompact Banach manifold and that $\pi : E \to M$ is a Banach vector bundle, then there exists a Finsler structure on the vector bundle.*

Proof. Choose an open covering $\{(\tau_i, U_i) \mid i \in \Lambda\}$ which trivializes the vector bundle. On each $\pi^{-1}(U_i)$, we have a natural Finsler structure $\|\cdot\|_i$. Since M is paracompact, there is a locally finite refinement $\{o_\beta \mid \beta \in B\}$ of the above open covering and the corresponding C^{1-0}-partition of unity $\{o_\beta, \eta_\beta) \mid \beta \in B\}$, say, $\forall\, \beta \in B$, $\exists\, \alpha = \alpha(\beta) \in \Lambda$ such that $\bar{o}_\beta \subset U_{\alpha(\beta)}$. Let

$$|||\cdot||| = \sum_{\beta \in B} \eta_\beta \circ \pi(\cdot)\|\cdot\|_{\alpha(\beta)}.$$

This is a Finsler structure on E.

Claim. The continuity of $|||\cdot|||$ is trivial. $\forall\, p \in M, \exists$ finitely many $\beta \in B$ such that $\eta_\beta(p) \neq 0$, since $\|x\|_{p,\alpha(\beta)} = \|(p,x)\,|_{\alpha(\beta)}$ is a norm $\forall\, \beta \in B$. Therefore

$$|||x|||_p = \sum_{\beta \in B} \eta_\beta \circ \pi(x)\|(p,x)\|_{\alpha(\beta)} = \sum_{\beta \in B} \eta_\beta(p)\|x\|_{p,\alpha(\beta)}$$

is also a norm, which is an equivalent norm on E_p.

As to (3), $\forall\, p_0 \in M$, $\forall\, k > 1$, by definition, $\exists$ a neighborhood U of p_0 such that there are at most finitely many $\{\beta_i\}_1^n$ such that $\eta_\beta(p) \equiv 0$, $\forall\, \beta \notin \{\beta_i\}_1^n$, $\forall\, p \in U$. Since $\|\cdot\|_i$ is a Finsler structure, $\exists\, V \subset U$, a smaller neighborhood of p_0 such that

$$k^{-\frac{1}{2}}\|x\|_{p,\alpha(\beta_i)} \leq \|x\|_{p_0,\alpha(\beta_i)} \leq k^{\frac{1}{2}}\|x\|_{p,\alpha(\beta_i)}$$

and

$$k^{-\frac{1}{2}}\eta_{\beta_i}(p) \leq \eta_{\beta_i}(p_0) \leq k^{\frac{1}{2}}\eta_{\beta_i}(p).$$

$\forall\, p \in U$, $i = 1, 2, \dots, n$, we obtain

$$k^{-1}\|x\|_p \le \|x\|_{p_0} \le k\|x\|_p, \ \forall\, p \in V.$$

□

Definition 2.3. A regular C^1-Banach manifold M, together with a Finsler structure on its tangent bundle $T(M)$, is called a Finsler manifold.

Example 3. Any paracompact Banach manifold possesses a Finsler structure on its tangent bundle, making it to be a Finsler manifold.

Example 4. If M is a Finsler manifold, with Finsler structure $\|\cdot\|$, then we can define

$$\|(p, x^*)\| = \sup\{\langle x^*, x\rangle \,|\, \|x\|_p \le 1 \ \forall\, x \in T_p(M)\}.$$

$\forall\, x^* \in T_p(M)^*$, it is a Finsler structure on $T^*(M)$.

Particularly, if $f : M \xrightarrow{C^1} R^1$, where M is a paracompact Banach manifold, then $\|(p,\, df(p))\|$ is well defined and $p \mapsto \|(p,\, df(p))\|$ is a continuous function.

In the following, we omit p in $(p,\, df(p))$, and denote it by $\|df(p)\|$.

For a Finsler manifold M, we may define a metric d as follows:

$$d(x, y) = \inf\left\{L(\sigma) | \sigma \in C^1([0,1], M),\ \sigma(0) = x,\ \sigma(1) = y\right\},$$

where

$$L(\sigma) = \int_0^1 \|(\sigma(t), \sigma'(t))\|\, dt.$$

One can prove the following:

Theorem 2.2. *Suppose that M is a Finsler manifold, and that the function d is defined above. Then d is a metric on M, and the reduced topology is equivalent to the topology on the manifold.*

For a proof we refer to Palais [Pal4].

Finally, we come to the main subject of our theory.

Definition 2.4. Let M be a C^1-Finsler manifold. Let $f \in C^1(M, \mathbb{R}^1)$; a point $p \in M$ is called a *critical point* of f if $df(p) = \theta$. The set $K = \{p \in M | df(p) = \theta\}$ is called the *critical set*. A real number c is called a *critical value* if $f^{-1}(c) \cap K \ne \emptyset$.

A real number, if it is not a critical value of f, is called a *regular value*. The complement set of K, i.e., $M^* = M \backslash K$, is called *regular set*. A point in M^* is called a *regular point*.

3. A Pseudo Gradient Vector Field and Deformation Theorems

The basic idea in critical point theory is to investigate the variations of topological structures of the level sets of a given function f. Certain flows, depending on the gradient of the function f, are used to deform these level sets. In finite dimensional manifolds, or more generally, in Hilbert Riemannian manifolds, gradient flow is a natural candidate. However, there are two disadvantages to gradient flow: (1) It needs more smoothness, say C^{2-0}, and (2) it only works on Hilbert Riemannian manifolds, because, for a Banach Finsler manifold $M, f : M \to R^1, df \in T^*(M)$ rather than $T(M)$, the gradient flow is not well-defined.

We are introduced to

Definition 3.1. (Pseudo gradient vector field). Let M be a Finsler manifold, and let $f : M \to R^1$ be differentiable at $p \in M$. $X \in T_p(M)$ is called a pseudo gradient if

$$\begin{aligned} &(1) \quad \|X\| \leq 2\|df(p)\|, \\ &(2) \quad \langle df(p), X\rangle \geq \|df(p)\,|^2, \end{aligned}$$

where $\langle\ ,\ \rangle_p$ is the duality on $T_p(M)$, and $\|\quad\|$ is the Finsler structure.

Let $S \subset M$ be a subset, and, if f is differentiable on S, X is called a pseudo gradient vector field on S (p.g.v.f. in short), if $\forall\ p \in S, X_p$ is a pseudo gradient of f at p.

Lemma 3.1. *Suppose that M is a Finsler manifold, and that $f : M \xrightarrow{C^1} R^1$. Let $M^* = M\backslash K$, where $K = \{p \in M \mid df(p) = 0\}$ is the critical set; then $\forall\ p \in M^*$, there exists a pseudo gradient of f at p.*

Proof. Since $\|df(p)\| \neq 0\ \forall\ p \in M^*$, by definition, $\exists\ X \in T_p(M)$ such that $\|X\| = 1$ and $\langle df(p), X\rangle > \frac{2}{3}\|df(p)\|$. Let $Y = \frac{3}{2}\|df(p)\|X$, then $\|Y\| = \frac{3}{2}\|df(p)\| < 2\|df(p)\|$, and $\langle df(p), Y\rangle_p > \|df(p)\|^2$. □

Theorem 3.1. *Suppose that M is a C^2-Finsler manifold and that $f : M \xrightarrow{C^1} R^1$. There exists a C^{1-0} p.g.v.f. of f on M^*.*

Proof. According to the lemma, $\forall\ p_0 \in M^*, \exists\ X_{p_0} \in T_p(M)$ such that

$$\begin{aligned} &\|X_{p_0}\| < 2\|df(p_0)\|, \\ &\langle df(p_0), X_{p_0}\rangle_{p_0} > \|df(p_0)\|^2. \end{aligned}$$

The continuity of the Finsler structure and the continuity of $df(p)$ imply that there is a neighborhood V_{p_0} of $p_0, V_{p_0} \subset M^*$, such that

$$\|X_{p_0}\| < 2\|df(p)\|,$$

and

$$\langle df(p), X_{p_0}\rangle_p > \|df(p)\|^2, \forall\, p \in V_{p_0}.$$

Since M^* is metrizable, it is paracompact. There is a locally finite C^{1-0} partition of unity $\{\eta_\beta \mid \beta \in B\}$, with supp $\eta_\beta \subset V_{p_0}$, for some $p_0 = p_0(\beta) \in M^*$. Let

$$X = X(p) = \sum_{\beta\in B} \eta_\beta(p) X_{p_0(\beta)}.$$

This is the p.g.v.f we need.

Claim.

$$\|X(p)\| \le \sum \eta_\beta(p)\|x_{p_0(\beta)}\| < 2\|\, df(p)\|,$$
$$\langle\, df(p), X(p)\rangle = \sum \eta_\beta(p)\langle df(p), X_{p_0(\beta)}\rangle > \|df(p)\|^2.$$

The local finiteness of the supports of $\{\eta_\beta \mid \beta \in B\}$ implies the C^{1-0} smoothness.

Once we have a p.g.v.f. of a function f, we get a decreasing flow by solving the following ordinary differential equation (ODE):

$$\dot\sigma(t) = -X(\sigma(t))$$
$$\sigma(0) = x_0 \in M^*.$$

The equation is locally solvable. Along the flow, the function $t \mapsto f\circ\sigma(t)$ is decreasing:

$$\begin{aligned}\frac{d}{dt} f\circ\sigma(t) &= \langle f'(\sigma(t)), \dot\sigma(t)\rangle\\ &= -\langle f'(\sigma(t)), X(\sigma(t))\rangle\\ &\le -\|f'(\sigma(t))\|^2 .\end{aligned}$$

Our first goal is to prove that if $f^{-1}[a,b]\cap K = \emptyset$, i.e., f has no critical value in the interval $[a,b]$, then f_a is a deformation retract of f_b. The deformation will be realized by the flow. Now the problem is that the pseudo gradient flow so far is only defined locally; we do not know if it could exist as long as it arrives at f_a.

A condition on the function f is needed.

Definition 3.2. Given $f : M \xrightarrow{C^1} R^1$ and $c \in R^1$, we say that f satisfies the $(\mathrm{PS})_c$ condition if any sequence $\{x_n\} \subset M$ along which $f(x_n) \to c$ and $df(x_n) \to \theta$ (strongly) possesses a convergent subsequence.

We say that f satisfies the (PS) condition if it satisfies $(\mathrm{PS})_c$ for all $c \in R^1$. We have the following facts:

(1) If $F : M \xrightarrow{C^1} R^1$ satisfies $(\mathrm{PS})_c\ \forall\, c \in [a,b]$, and if $K \cap f^{-1}[a,b] = \emptyset$, then $\exists\, \epsilon_0, \delta_0 > 0$ such that

$$\|df(x)\| \geq \epsilon_0\ \forall\, x \in f^{-1}[a-\delta_0, b+\delta_0].$$

Claim. If not, $\exists\, x_n \in f^{-1}[a-\frac{1}{n}, b+\frac{1}{n}], n = 1,2,\ldots$, satisfying $df(x_n) \to \theta$. According to the $(\mathrm{PS})_c\ \forall\, c \in [a,b]$, there exists a convergent subsequence $x_{n_i} \to x^*$, which implies $x^* \in K \cap f^{-1}[a,b]$. This is a contradiction.

(2) If $f : M \xrightarrow{C^1} R^1$ satisfies $(\mathrm{PS})_c$, then $K_c := K \cap f^{-1}(c)$ is compact.

Claim. If $\{x_n\} \subset K_c$, then $f(x_n) = c$ and $df(x_n) = \theta$. The $(\mathrm{PS})_c$ condition implies a convergent subsequence.

Lemma 3.2. (Deformation). *If $f \in C^1(M, R^1)$ satisfies $(PS)_c$, $\forall\, c \in [a,b]$, and if $K \cap f^{-1}(a,b] = \emptyset$, then f_a is a strong deformation retract of f_b.*

Proof. 1. We consider the pseudo-gradient (semi) flow on $f^{-1}[a,b]$:

$$\begin{cases} \dot{\sigma}(t) &= -X(\sigma(t))/\|X(\sigma(t))\|^2 \\ \sigma(0) &= x_0 \in f^{-1}[a,b]. \end{cases}$$

We want to show that the maximal solvable half interval $[0, T_{x_0})$ satisfies

(1) $$T_{x_0} < +\infty,$$
(2) $$f(\sigma(T_{x_0} - 0)) = a.$$

Claim. Since

$$f(\sigma(t)) - f(x_0) = -\int_0^t < df(\sigma(\tau)),\ \dot{\sigma}(\tau) > d\tau\ < -\frac{t}{4}.$$

each initial point $x_0 \in f^{-1}[a,b]$ arrives at f_a along the flow in a finite time, i.e., $T_{x_0} < +\infty$. Noticing

$$\| \int_t^{t'} \dot{\sigma}(\tau) d\tau \| \leq \int_t^{t'} \frac{d\tau}{\| X(\sigma(\tau)) \|} \leq \frac{1}{\epsilon_0}|t'-t|,$$

this implies $\sigma^* = \lim_{t \to T_{x_0}-0} \sigma(t)$ exists. If $f(\sigma^*) > a$, then by fact(1), the flow $\sigma(t)$ can be extended beyond T_{x_0}. This is in contradiction with maximality.

We call T_{x_0} the arriving time.

2. The arriving time function $x \to T_x : f^{-1}[a,b] \to R^1$ is continuous.

Claim. $t = T_{x_0}$ is the solution of the equation

$$f(\sigma(t, x_0)) = a,$$

where we write $\sigma(t, x_0)$ the flow $\sigma(t)$ emanating from x_0.

Since

$$\frac{\partial}{\partial t} f \circ \sigma(t, x_0)|_{t=T_{x_0}} = \langle df(\sigma(T_{x_0}, x_0)), \dot{\sigma}(T_{x_0}, x_0)\rangle < -\frac{1}{4},$$

the implicit function theorem is applied.

3. Define

$$\eta(t, x_0) = \begin{cases} x_0 & \text{if } x_0 \in f_a \\ \sigma(T_{x_0} t, x_0) & \text{if } x_0 \in f_b \backslash f_a, \end{cases}$$

then $\eta : [0,1] \times f_b \to f_a$ satisfies

$$\begin{aligned} &\eta(0, \cdot) = \text{id}, \\ &\eta(1, f_b) \subset f_a, \\ &\eta(t, \cdot)|_{f_a} = \text{id}_{f_a} \ \ \forall\, t \in [0,1]. \end{aligned}$$

We only want to verify the continuity of η. The verification is divided into three cases:

(1) on $[0,1] \times \overset{\circ}{f_a}$, $\eta = \text{id}$,
(2) on $[0,1] \times (f_b \backslash f_a)$, the continuity of η follows from the theory of ODE,
(3) on $[0,1] \times f^{-1}(a)$.

$\forall \epsilon > 0$ we want to find $\delta > 0$ such that

$$f(x) = a \quad \| x - y \| < \delta \Longrightarrow \| x - \eta(t, y) \| < \epsilon \ \ \forall\, t \in [0,1]$$

If $y \in f_a$, then $\delta = \epsilon$.
If $y \notin f_a$, then

$$\begin{aligned} \|\eta(t, y) - x\| &= \|\sigma(T_y t, y) - x\| \\ &\le \|x - y\| + \int_0^{T_y t} \|\dot{\sigma}(s, y)\| ds \\ &\le \delta + \frac{1}{\epsilon_0} T_y = \delta + \frac{1}{\epsilon_0}(T_y - T_x). \end{aligned}$$

By the continuity of the function T_x, $\exists\, \delta_0 > 0$ such that if $\|y - x\| < \delta_0$, we have

$$T_y < T_x + \frac{1}{2}\epsilon_0 \epsilon.$$

Therefore, $\|\eta(t, y) - x\| < \epsilon$, when $\|x - y\| < \delta = \min\left\{\frac{\epsilon}{2}, \delta_0\right\}.$

Theorem 3.2. (Second deformation lemma). *Let M be a C^2 Finsler manifold. Suppose that $f \in C^1(M, \mathbb{R}^1)$ satisfies the $(PS)_c$ condition $\forall\, c \in [a,b]$ and that a is the only critical value of f in $[a,b)$. Assume that the connected components of K_a are only isolated points. Then f_a is a strong deformation retract of $f_b \backslash K_b$.*

Before giving the proof, we need a fact from point set topology (cf. Kelley [Kel1]).

Lemma 3.3. *Suppose that K is a metric compact space and that F_1, F_2 are two compact subsets. Then either $\exists$ a connected component of K, joining F_1 and F_2, or $\exists$ compact subsets K_1, K_2 such that $K_1 \cup K_2 = K$, $K_1 \cap K_2 = \emptyset$, and $F_i \subset K_i$, $i = 1, 2$.*

Proof. This is an improvement of Lemma 3.2, so we define the pseudo-gradient flow as before:

$$\begin{cases} \dot\sigma(t,x) &= -\frac{X(\sigma(t,x))}{\|X(\sigma(t,x))\|^2} \\ \sigma(0,x) &= x \in f^{-1}[a,b]\backslash K_b. \end{cases}$$

As we have shown that $\forall\, x \in f^{-1}(a,b]\backslash K_b$, there exists the arriving time $T_x > 0$ such that $\lim\limits_{t\to T_x - 0} f\circ\sigma(t,x) = a$.

1. We shall prove that the $\lim\limits_{t\to T_x-0} \sigma(t,x)$ does exist and then

$$f(\sigma(T_x - 0, x)) = a.$$

Claim. $(\mathrm{PS})_a$ implies K_a is compact. Either one of the following cases occurs:

$$\text{(a)} \quad \inf_{t\in[0,T_x)} \operatorname{dist}(\sigma(t,x), K_a) > 0,$$
$$\text{(b)} \quad \inf_{t\in[0,T_x)} \operatorname{dist}(\sigma(t,x), K_a) = 0.$$

In case (a), again by $(\mathrm{PS})_c$ $\forall\, c \in [a,b]$, $\exists\, \alpha > 0$ such that

$$\inf_{t\in[0,T_x)} \|df(\sigma(t,x))\| \geq \alpha.$$

Thus

$$\begin{aligned} \operatorname{dist}(\sigma(t_2,x),\ \sigma(t_1,x)) &\leq \int_{t_1}^{t_2} \left\| \frac{d\sigma}{dt} \right\| dt \\ &\leq \int_{t_1}^{t_2} \frac{dt}{\|X(\sigma(t,x))\|} \leq \int_{t_1}^{t_2} \frac{dt}{\|df(\sigma(t,x))\|} \leq \frac{|t_2 - t_1|}{\alpha}. \end{aligned}$$

Since T_x is finite, $\lim_{t\to T_x-0}\sigma(t,x)$ exists.

In case (b), we shall prove that $\exists\, z\in K_a$ such that

$$\lim_{t\to T_x-0}\sigma(t,x)=z.$$

First, we assert that

$$\lim_{t\to T_x-0}\operatorname{dist}(\sigma(t,x),K_a)=0.$$

If not, $\exists\, t_i\to T_x-0$, such that

$$\operatorname{dist}(\sigma(t_i,x),\ K_a)\geq\epsilon_0>0.$$

But case (b) implies that $\exists\, t_i'\to T_x-0$ such that

$$\lim_{i\to\infty}\operatorname{dist}(\sigma(t_i',x),\ K_a)=0.$$

Thus we have two sequences $t_i^*<t_i^{**}$, both converging to T_x, such that

$$\begin{aligned}&\operatorname{dist}(\sigma(t_i^*,x),\ K_a)=\frac{\epsilon_0}{2},\\&\operatorname{dist}(\sigma(t_i^{**},x),\ K_a)=\epsilon_0,\end{aligned}$$

and

$$\sigma(t,x)\in(\bar K_a)_{\epsilon_0}\backslash(K_a)_{\frac{\epsilon_0}{2}}\quad\forall\, t\in[t_i^*,t_i^{**}]$$

where $(K_a)_\delta$ denotes the δ-neighborhood of K_a.
Again by $(\mathrm{PS})_c$ $\forall\, c\in[a,b]$, we have

$$\inf_{t\in[t_i^*,t_i^{**}]}\|df(\sigma(t,x))\|\geq\alpha>0.$$

Therefore

$$\begin{aligned}\frac{\epsilon_0}{2}&\leq\operatorname{dist}(\sigma(t_i^{**},x),\ \sigma(t_i^*,x))\\&\leq\int_{t_i^*}^{t_i^{**}}\left\|\frac{d\sigma}{dt}\right\|dt\\&\leq\frac{1}{\alpha}|t_i^{**}-t_i^*|\to0.\end{aligned}$$

This is a contradiction.

It follows that

$$\lim_{t\to T_x-0}df(\sigma(t,x))=\theta,$$

provided by the compactness of K_a.

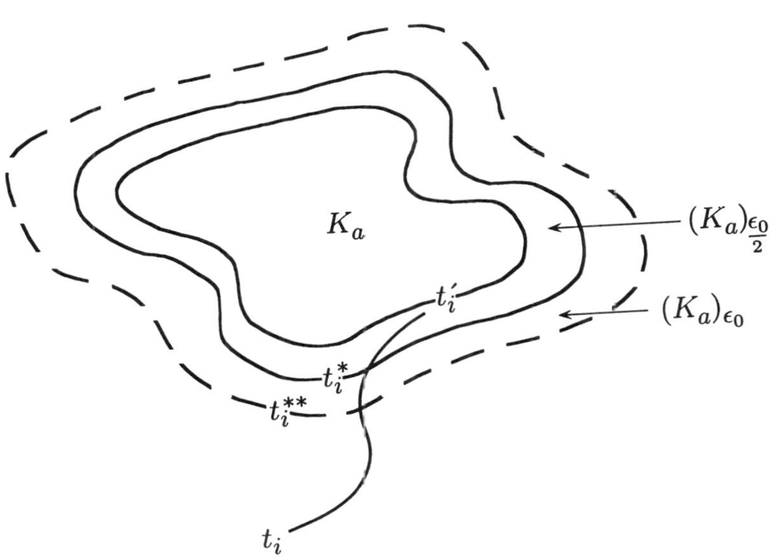

$(\text{PS})_a$ condition then implies that the limit set A of the orbit $\{\sigma(t,x)|t \in [0,T_x)\}$ is nonempty and that, for each sequence $t_i \to t_{x-0}$, there exists a subsequence $\tilde{t}_i$ such that $\sigma(\tilde{t}_i, x)$ is convergent.

Next we prove that A is a compact connected subset of K_a. The compactness is obvious. We only want to prove the connectedness. If not, $\exists$ open subsets O and O' such that

$$O \cap O' = \emptyset,\ A = (O \cap A) \cup (O' \cap A)$$

and $O \cap A \neq \emptyset$, $O' \cap A \neq \emptyset$.

Choosing $z \in O \cap A$, $z' \in O' \cap A$, $\exists\, t_i \to T_x - 0$, $t_i' \to T_x - 0$ satisfying $\sigma(t_i, x) \to z$, $\sigma(t_i', x) \to z'$. For large i, we have

$$\sigma(t_i, x) \in O,\ \sigma(t_i', x) \in O'$$

so $\exists\, t_i^* \in [t_i, t_i']$ (or $[t_i', t_i]$) such that

$$\sigma(t_i^*, x) \notin O \cup O'.$$

Denote the limit of $\sigma(t_i^*, x)$ by z^*. (It exists because of $(\text{PS})_a$.) So $z^* \in A$, but $a^* \notin O \cup O'$.

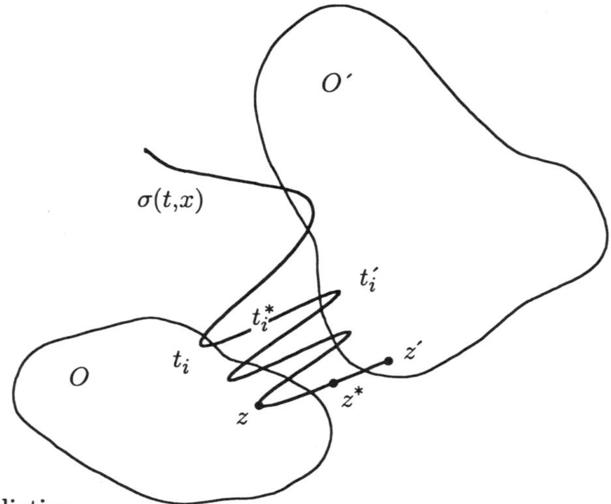

This is a contradiction.

According to the assumption that the connected components of K_a are isolated points, we conclude that $A =$ a point $z \in K_a$.

2. We shall prove the continuity of the function T_x.

As shown in Lemma 3.2, if $\sigma(T_{x_0} - 0, x_0) \notin K_a$, then the function T_x is continuous at x_0. So we restrict ourselves to the case $z = \sigma(T_{x_0} - 0, x_0) \in K_a$.

If T_x is not continuous at such a x_0, then $\exists\, x_n \to x_0$ such that

$$|T_{x_n} - T_{x_0}| \geq \epsilon_0 > 0,$$

so either (subsequence)

$$T_{x_n} \leq T_{x_0} - \epsilon_0, \text{ or } T_{x_n} \geq T_{x_0} + \epsilon_0.$$

Since

$$\begin{aligned} f \circ \sigma(T_x - \epsilon, x) - f \circ \sigma(t, x) &= \int_t^{T_x - \epsilon} \frac{df(\sigma(t,x))}{dt} dt \\ &\leq -\frac{1}{4}(T_x - \epsilon - t), \end{aligned}$$

we have

$$f \circ (\sigma(t,x)) \geq a + \frac{1}{4}(T_x - t).$$

But for any fixed $\epsilon > 0$, according to ODE,

$$\text{dist}(\sigma(T_{x_0} - \epsilon, x_n),\ (\sigma(T_{x_0} - \epsilon, x_0)) \to 0,$$

so that

$$\begin{aligned} f \circ \sigma(T_{x_0} - \epsilon, x_0) &= \lim_{n\to\infty} f \circ \sigma(T_{x_0} - \epsilon, x_n) \\ &\geq \lim_{n\to\infty} \left[a + \frac{1}{4}(T_{x_n} - T_{x_0} + \epsilon) \right] \\ &\geq a + \frac{1}{4}(\epsilon_0 + \epsilon), \text{ if } T_{x_n} \geq T_{x_0} + \epsilon_0. \end{aligned}$$

Letting $\epsilon \to 0$, we obtain

$$a \geq a + \frac{\epsilon_0}{4}.$$

This is a contradiction.

Similarly, we prove that $T_{x_n} \leq T_{x_0} - \epsilon_0$ is impossible.

3. Finally, we define the deformation retract as before

$$\eta(t,x) = \begin{cases} x & \text{if } (t,x) \in [0,1] \times f_a, \\ \sigma(T_x t, x) & \text{if } (t,x) \in [0,1) \times (f_b \backslash (f_a \cup K_b)), \\ \sigma(T_x - 0, x) & \text{if } (t,x) \in \{1\} \times (f_b \backslash (f_a \cup K_b)). \end{cases}$$

Claim. Only the continuity of η has to be verified. Four cases are distinguished:

(a) $(t,x) \in [0,1] \times \overset{\circ}{f}_a$,
(b) $(t,x) \in [0,1) \times (f^{-1}(a,b] \backslash K_b)$,
(c) $(t,x) \in \{1\} \times (f^{-1}(a,b] \backslash K_b)$,
(d) $(t,x) \in [0,1] \times f^{-1}(a)$.

We only want to verify cases (c) and (d). Since their proofs are similar, we only give the verification for (c).

If η is discontinuous at $(1, x_0)$, then $\exists\, \epsilon > 0$ and $t_n \to T_{x_0} - 0$, $x_n \to x_0$ such that

$$\text{dist}(\sigma(t_n, x_n), \sigma(T_{x_0} - 0, x_0)) \geq \epsilon.$$

Let $z = \sigma(T_{x_0} - 0, x_0) (\in K_a)$, and let

$$F_1 = \{z\}, \; F_2 = (M \backslash \overset{\circ}{B}(z,\epsilon)) \cap K_a.$$

F_1 and F_2 are compact subsets of K_a. Provided by the assumption of K_a, and the lemma, we have compact subsets K_1, $K_2 \subset K_a$ such that $K_1 \cap K_2 = \emptyset$, $F_i \subset K_i$, $i = 1, 2$, and $K_1 \cup K_2 = K_a$. Obviously, we may take $K_1 \subset \overset{\circ}{B}(z,\epsilon)$, and $N = K_2 \cup (M \backslash \overset{\circ}{B}(z,\epsilon))$, then

$$\alpha = \text{dist}(N, K_1) > 0.$$

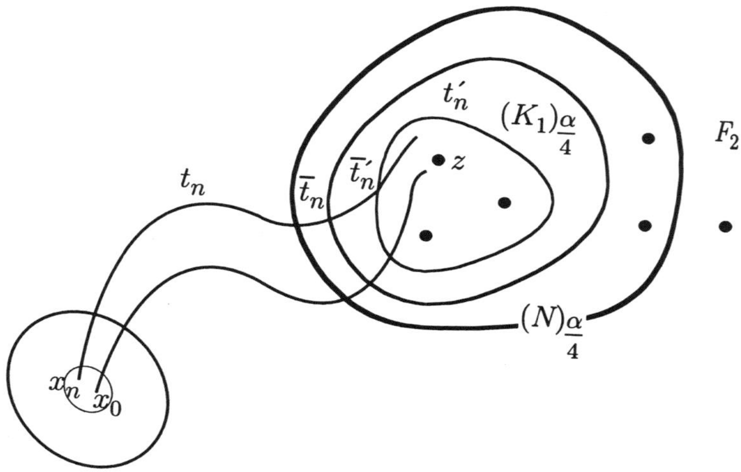

K_1 • $K_2\, x$

The continuity of the flow as well as of the arriving time T_x implies that $\exists\, \delta > 0$ such that

$$\operatorname{dist}(\sigma(t, x_0),\ \sigma(T_{x_0} - 0, x_0)) < \frac{\alpha}{8}$$

if $t \in [T_{x_0} - \delta, T_{x_0})$, and $\exists\, \delta_1 > 0$ such that

$$T_x > T_{x_0} - \delta$$

if $x \in B(x_0, \delta_1)$.

For $t \in [T_{x_0} - \delta, T_{x_0}) \cap [T_{x_0} - \delta, T_x)$, $x \in B(x_0, \delta_1)$, $\exists\, \delta_t \in (0, \delta)$ such that

$$\operatorname{dist}(\sigma(t, x),\ \sigma(t, x_0)) < \frac{\alpha}{8}.$$

In summary, for such a t, and for any $x \in B(x_0, \delta_t)$,

$$\sigma(t, x) \in (K_1)_{\frac{\alpha}{4}},\ \text{the } \frac{\alpha}{4} \text{ neighborhood of } K_1.$$

For large n, $x_n \in B(x_0, \delta_t)$, denote $t_n' =$ such a t, satisfying

$$\sigma(t_n', x_n) \in (K_1)_{\frac{\alpha}{4}}.$$

Reducing δ, and repeating the above procedure, we obtain a subsequence $\{x_n, t_n', t_n\}$ such that

$$\begin{cases} t_n', t_n \to T_{x_0} - 0 \\ \sigma(t_n', x_n) \in (K_1)_{\frac{\alpha}{4}} \\ \sigma(t_n, x_n) \notin B(z, \epsilon). \end{cases}$$

We may assume $t'_n < t_n$; then we have $\bar{t}'_n, \bar{t}_n$ such that

$$t'_n \leq \bar{t}'_n < \bar{t}_n \leq t_n$$

and

$$\begin{aligned} \sigma(\bar{t}'_n, x_n) &\in \partial[(K_1)_{\frac{\alpha}{4}}] \\ \sigma(\bar{t}'_n, x_n) &\in \partial[(N)_{\frac{\alpha}{4}}] \\ \sigma(t, x_n) &\notin (K_1)_{\frac{\alpha}{4}} \cup (N)_{\frac{\alpha}{4}} \ \forall t \in [\bar{t}'_n, \bar{t}_n], \end{aligned}$$

$n = 1, 2, \ldots.$

According to $(\mathrm{PS})_c$, $\forall\, c \in [a, b]$

$$\beta = \inf\left\{\|df(x)\| \,|\, x \in f^{-1}[a,b) \backslash \left((K_1)_{\frac{\alpha}{4}} \cup (N)_{\frac{\alpha}{4}}\right)\right\} > 0.$$

Therefore

$$\frac{\alpha}{2} \leq \mathrm{dist}(\sigma(\bar{t}'_n, x_n),\ \sigma(\bar{t}_n, x_n)) \leq \int_{\bar{t}'_n}^{\bar{t}_n} \|\frac{d\sigma}{dt}\| dt \leq \frac{1}{\beta}|\bar{t}_n - \bar{t}'_n| \to 0.$$

This is a contradiction. The proof is completed. □

Remark 3.1. Theorem 3.2 is due to E. Rothe [Rot1], K.C. Chang [Cha1] and Z.Q. Wang [WaZ4].

Another deformation lemma is also often used.

Theorem 3.3. (First deformation lemma). *Let M be a C^2 Finsler manifold. Suppose that $f \in C^1(M, \mathbb{R}^1)$ satisfies the $(PS)_c$ condition. Assume that N is a closed neighborhood of $K_c = K \cap f^{-1}(c)$. Then there exist a continuous map $\eta : [0,1] \times M \to M$ and constants $\bar{\epsilon} > \epsilon > 0$ such that*

(1) $\eta(0, \cdot) = id$,

(2) $\eta(t, \cdot)|_{Cf^{-1}[c-\bar{\epsilon}, c+\bar{\epsilon}]} = id|_{Cf^{-1}[c-\bar{\epsilon}, c+\bar{\epsilon}]}$,

(3) $\eta(t, \cdot) : M \to M$ *is a homeomorphism* $\forall\, t \in [0,1]$,

(4) $\eta(1, f_{c+\epsilon} \backslash N) \subset f_{c-\epsilon}$,

(5) $f \circ \eta(t, x)$ *is nonincreasing in* t $\forall (t, x) \in [0,1] \times M$.

We would rather prove a more general version of Theorem 3.3, which indicates the fact that $\bar{\epsilon}$ depends on constants derived from the Palais-Smale condition. Namely,

Theorem 3.4. *Let M be a C^2-Finsler manifold. Suppose $f \in C^1(M, \mathbb{R}^1)$. Let $N' \subset N$ be two closed neighborhoods satisfying*

$$dist(N', \partial N) \geq \frac{7}{8}\delta,\ \delta > 0.$$

Suppose that there are constants b and $\bar{\epsilon}$ positive, such that

$$\|df(x)\| \geq b\,\forall\, x \in f_{c+\epsilon}\backslash(f_{c-\epsilon} \cup N'),$$
$$0 < \bar{\epsilon} < Min\left\{\frac{1}{4}\delta b^2,\ \frac{1}{8}\delta b\right\}.$$

Then for any $0 < \epsilon < \frac{\bar{\epsilon}}{2}$, there exists $\eta \in C([0,1] \times M,\ M)$ satisfying (1)-(5) in Theorem 3.3.

We assume Theorem 3.4 at this point; then we give the

Proof of Theorem 3.3. According to $(\text{PS})_c$, K_c is compact. Hence, for $\delta > 0$ sufficiently small, $N(\delta) = \{x \in M \,|\, \text{dist}(x, K_c) < \delta\} \subset N$. Again, by $(\text{PS})_c$, there exist constants b, $\bar{\epsilon} > 0$, such that

$$\|df(x)\| \geq b \quad \forall\, x \in f_{c+\bar{\epsilon}}\backslash(f_{c-\bar{\epsilon}} \cup N(\frac{\delta}{8})). \tag{3.1}$$

Since (3.1) remains valid, if $\bar{\epsilon}$ is decreased, we may assume

$$0 < \bar{\epsilon} < \text{Min}\left\{\frac{1}{4}\delta b^2,\ \frac{1}{8}\delta b\right\}.$$

Theorem 3.3 follows from Theorem 3.4, if we take $N' = N(\frac{\delta}{8})$.

Proof of Theorem 3.4. Define a smooth function:

$$p(s) = \begin{cases} 0 & \text{for } s \notin [c - \bar{\epsilon},\ c + \bar{\epsilon}], \\ 1 & \text{for } s \in [c - \epsilon,\ c + \epsilon], \end{cases}$$

with $0 \leq p(s) \leq 1$. Let $A = \overline{M\backslash(N')_{\frac{\delta}{8}}}$, where $(N')_\delta = \{x \in M | \text{dist}(x, N') \leq \delta\}$, and $B = N'$ be two closed subsets. Let

$$g(x) = \frac{\text{dist}(x, B)}{\text{dist}(x, A) + \text{dist}(x, B)}.$$

We see $0 \leq g \leq 1$, $g = 0$ on N' and $g = 1$ outside $(N')_{\frac{\delta}{8}}$. Define

$$q(s) = \begin{cases} 1 & 0 \leq s \leq 1, \\ \frac{1}{s} & s \geq 1. \end{cases}$$

Letting X be a p.g.v.f. of f, we define

$$V(x) = -g(x)\,p(f(x))\,q(\|X(x)\|)X(s).$$

Thus $V \in C^{1-0}$, and $\|V(x)\| \le 1$. Consider the ODE

$$\begin{cases} \dot\sigma(t) & = V(\sigma(t)) \\ \sigma(0) & = x_0 \ \ \forall x_0 \in M. \end{cases}$$

Since V is bounded, the global existence and uniqueness of the flow $\sigma(t)$ on $\mathbb{R}^1$ are known. Let

$$\eta(t,x) = \sigma(t), \text{ with } \sigma(0) = x.$$

Then $\eta \in C([0,1]\times M,\ M)$ satisfies (1), (2), (3), and (5). It remains for us to verify (4). From (5), one needs only to verify (4) for $x \in f_{c+\epsilon}\backslash(f_{c-\epsilon}\cup N)$. More precisely, we shall prove:

$$f \circ \eta(\frac{3}{4}\delta, x) \le c - \epsilon \ \ \forall x \in f_{c+\epsilon}\backslash(f_{c-\epsilon}\cup N).$$

Indeed, if not, $\exists\, x \in f^{-1}[c-\epsilon, c+\epsilon]\backslash N$ such that

$$c-\epsilon < f\circ\eta(t,x) \le c+\epsilon \quad \forall\, t \in (0, \frac{3}{4}\delta).$$

It follows

$$p(f\circ\eta(t,x)) = 1.$$

Noticing that $\|\eta(t,x) - \eta(0,x)\| \le t$, we have

$$\begin{aligned} \text{dist}(\eta(t,x), (N')_{\frac{\delta}{8}}) &\ge \text{dist}(\eta(0,x), (N')_{\frac{\delta}{8}}) - t \\ &> (\frac{7}{8} - \frac{1}{8} - \frac{3}{4})\delta = 0, \end{aligned}$$

so $g\circ\eta(t,x) = 1$. Now,

$$\begin{aligned} \frac{d}{dt} f\circ\eta(t,x) &= \langle df(\eta(t,x)), \dot\eta(t,x)\rangle \\ &= -q(\|V(\eta(t,x))\|)\langle df(\eta(t,x)), V(\eta(t,x))\rangle \\ &\le -\frac{1}{2} q(\|V(\eta(t,x))\|)\|df(\eta(t,x))\|^2. \end{aligned}$$

If $\|V(\eta(t,x))\| \le 1$, then we have

$$\frac{d}{dt} f\circ\eta(t,x) \le -\frac{1}{2}\|df(\eta(t,x))\|^2 \le -\frac{1}{2}b^2.$$

Otherwise,

$$\begin{aligned}\frac{d}{dt}f\circ\eta(t,x) &\le -\frac{1}{2}\|df(\eta(t,x))\|^2/\|V(\eta(t,x))\|\\ &\le -\frac{1}{4}\|df(\eta(t,x))\| < -\frac{1}{4}b.\end{aligned}$$

In summary,

$$\frac{d}{dt}f\circ\eta(t,x) \le -\mathrm{Min}\{\frac{1}{2}b^2, \frac{1}{4}b\}.$$

Thus,

$$\begin{aligned}f\circ\eta(t,x) &\le f\circ\eta(0,x) - \frac{1}{2}\delta\mathrm{Min}\{\frac{1}{2}b^2,\frac{1}{4}b\}\\ &\le c+\epsilon-\delta\mathrm{Min}\{\frac{1}{4}b^2,\frac{1}{8}b\} < c-\epsilon.\end{aligned}$$

This is a contradiction.

Remark 3.2. Theorem 3.3 is due to Palais [Pal1]; see also P.H. Rabinowitz [Rab1]. For Theorem 3.4, cf. J.Q. Liu [Liu2], and K.C. Chang, J.Q. Liu [ChL2].

4. Critical Groups and Morse Type Numbers

Basic Morse theory is set up in two steps:

(1) Locally, we define a sequence of groups, which we call *critical groups*, to describe the local behavior of a function f near its critical point.

(2) Globally, we define a sequence of numbers, which count the critical points in accordance with the critical groups. These numbers are called *Morse type numbers.* We study the relationship between these Morse type numbers and the deformation property of the underlying manifold via the deformation property.

Definition 4.1. Let p be an isolated critical point of f, and let $c = f(p)$. We call

$$C_q(f,p) = H_q\left(f_c\cap U_p, (f_c\backslash\{p\})\cap U_p, G\right)$$

the qth critical group, with coefficient group G of f at $p, q = 0,1,2,\ldots,$ where U_p is a neighborhood of p such that $K\cap(f_c\cap U_p) = \{p\}$, and $H_*(X,Y;G)$ stands for the singular relative homology groups with the abelian coefficient group G.

According to the excision property of the singular homology theory, the critical groups are well-defined; i.e., they do not depend on a special choice of the neighborhood U_p.

Now we give some examples.

Example 1. Letting p be an isolated local minimum of $f \in C(M, R^1)$, then

$$C_q(f,p) = \begin{cases} G & q = 0 \\ 0 & q \neq 0. \end{cases}$$

In a finite dimensional manifold M^n, if p is an isolated local maximum of $f \in C(M^n, R^1)$, then

$$C_q(f,p) = \begin{cases} G & q = n \\ 0 & q \neq n. \end{cases}$$

Example 2. Let M be a 1-dimensional manifold. If p is an isolated critical point of $f \in C(M, R^1)$, which is neither a local maximum nor a local minimum, then we have

$$C_0(f,p) = C_1(f,p) = 0.$$

Example 3. (Monkey saddle.) Let f be the function $x^3 - 3xy^2$ defined on R^2. Then we have

$$C_0(f,\theta) = C_2(f,\theta) = 0, \quad \text{and} \quad C_1(f,\theta) = G \oplus G.$$

Let M be a Hilbert-Riemannian manifold. $f \in C^2(M, R^1)$, $p \in K$ is called a *nondegenerate critical point*, if $d^2 f(p)$ has a bounded inverse. Since $A = d^2 f(p)$ is a self-adjoint operator which possesses a resolution of identity, we call the dimension of the negative space corresponding to the spectral decomposing, the *Morse index* of p, denoted by $\text{ind}(f,p)$. (It can be ∞).

Now we are in a position to compute the critical groups of a nondegenerate critical point via its Morse index.

The following Morse lemma is a cornerstone in studying the local behavior of a nondegenerate critical point. The proof is postponed to the next section in which a general splitting lemma will be proved.

Morse Lemma 4.1. *Suppose that $f \in C^2(M, R^1)$ and that p is a nondegenerate critical point; then there exists a neighborhood U_p of p and a local diffeomorphism $\Phi : U_p \to T_p(M)$ with $\Phi(p) = \theta$, such that*

$$f \circ \Phi^{-1}(\xi) = f(p) + \frac{1}{2}\left(d^2 f(p)\xi, \xi\right), \ \forall\, \xi \in \Phi\,(U_p),$$

where $(\ ,\ )$ is the inner product of the Hilbert space H, on which the Riemannian manifold M is modeled.

Theorem 4.1. *Suppose that $f \in C^2(M, R^1)$ and p is a nondegenerate critical point of f with Morse index j; then*

$$C_q(f,p) = \begin{cases} G & g = j \\ 0 & q \neq j. \end{cases}$$

Proof. According to the Morse Lemma, we may restrict ourselves to a special case where f is a quadratic function on the Hilbert space H : $f(x) = \frac{1}{2}(Ax, x)$, where A is a bounded, invertible, self-adjoint operator. Let $P_\pm$ be the orthogonal projection onto the positive (negative) subspace $H_\pm$ with respect to the spectral decomposition of A. We have

$$f(x) = \frac{1}{2}\left(\| (AP_+)^{\frac{1}{2}} x\|^2 - \| (-AP_-)^{\frac{1}{2}} x\|^2\right).$$

On an ϵ-ball B_ϵ centered at θ, it is easily seen that

$$B_\epsilon \cap f_0 = \left\{x \in H \mid \|x\| \leq \epsilon,\ \|y_+\| \leq \|y_-\|, y_\pm = (\pm AP_\pm)^{\frac{1}{2}} x\right\}.$$

We define a deformation

$$\eta(t, x) = y_- + ty_+ \qquad \forall\ (t, x) \in [0, 1] \times (B_\epsilon \cap f_0).$$

It is a strong deformation retract from $(B_\epsilon \cap f_0,\ B_\epsilon \cap (f_0 \backslash \{\theta\})$ to $(H_- \cap B_\epsilon, H_- \backslash \{\theta\}) \cap B_\epsilon)$. Thus

$$\begin{aligned} C_q(f,p) &\cong H_q\left(f_0 \cap B_\epsilon, (f_0 \backslash \{\theta\}) \cap B_\epsilon\right) \\ &\cong H_q\left(H_- \cap B_\epsilon, (H_- \backslash \{\theta\}) \cap B_\epsilon\right) \\ &\cong H_q\left(B^j, S^{(j-1)}\right) \cong \begin{cases} G & q = j \\ 0 & q \neq j \end{cases} \end{aligned}$$

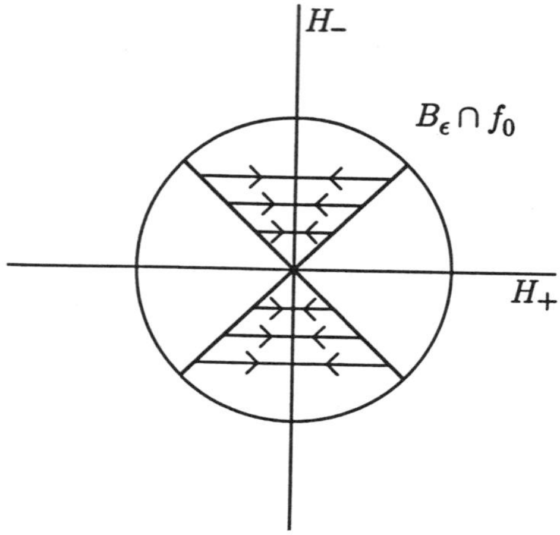

if j is finite.

Nevertheless, for $j = +\infty$, since S^∞ is contractible, we always have

$$C_q(f,p) \cong 0.$$

The conclusion is proved. □

Let M be a C^2-Finsler manifold, and let $f \in C^1(M, \mathbb{R}^1)$. Suppose that f has only isolated critical values and that each of them corresponds to a finite number of critical points; say

$$\ldots < c_{-2} < c_{-1} < c_0 < c_1 < c_2 < \ldots$$

are critical values with

$$K_{c_i} = \{z_j^i\}_{j=1}^{m_i}, \quad i = 0, \pm 1, \pm 2, \ldots .$$

We choose

$$0 < \epsilon_i < \min\{c_{i+1} - c_i, c_i - c_{i-1}\}, \quad i = 0, \pm 1, \pm 2, \ldots .$$

Definition 4.2. For a pair of regular values $a < b$, we call

$$M_q(a,b) = \sum_{a<c_i<b} \operatorname{rank} H_q\left(f_{c_i+\epsilon_i}, f_{c_i-\epsilon_i}, G\right)$$

the qth Morse type number of the function f with respect to (a,b), $q = 0, 1, 2, \ldots$.

For functions f satisfying the (PS) condition, Morse type numbers are well-defined according to the deformation lemma, i.e., they are independent of the special choice of $\{\epsilon_i\}$. In the case where $f \in C^1(M, R^1)$, assume that c is an isolated critical value and $K_c = \{z_j\}_1^m$, then for sufficiently small $\epsilon > 0$, we have

Theorem 4.2.

$$H_*\left(f_{c+\epsilon}, f_{c-\epsilon}; G\right) \cong H_*\left(f_c, f_c \backslash K_c; G\right) \cong \bigoplus_{j=1}^{m} C_*\left(f, z_j\right).$$

Proof. By the deformation theorem and the homotopy invariance of singular homology groups, we have

$$H_q\left(f_{c+\epsilon}, f_{c-\epsilon}, G\right) \cong H_q\left(f_c, f_{c-\epsilon}; G\right)$$

and

$$H_q\left(f_c \backslash K_c, f_{c-\epsilon}; G\right) \cong H_q\left(f_{c-\epsilon}, f_{c-\epsilon}; G\right) \cong 0.$$

Applying the exactness of singular homology groups to the triple $(f_c, f_c\backslash K_c, f_{c-\epsilon})$:

$$\ldots \to H_q\left(f_c\backslash K_c, f_{c-\epsilon}\right) \to H_q\left(f_c, f_{c-\epsilon}\right) \\ \to H_q\left(f_c, f_c\backslash K_c\right) \to H_{q-1}\left(f_c\backslash K_c, f_{c-\epsilon}\right) \to \ldots$$

we find

$$0 \to H_q\left(f_c, f_{c-\epsilon}\right) \to H_q\left(f_c, f_c\backslash K_c\right) \to 0,$$

i.e.,

$$H_q\left(f_c, f_{c-\epsilon}\right) \cong H_q\left(f_c, f_c\backslash K_c\right).$$

By using the excision property, we may decompose the relative singular homology groups into critical groups

$$H_*\left(f_c, f_c\backslash K_c\right) \cong H_*\left(f_c \cap \bigcup_{j=1}^{m} B\left(z_j, \epsilon\right), f_c \cap \bigcup_{j=1}^{m}\left(B\left(z_j, \epsilon\right)\backslash\{z_j\}\right)\right) \\ \cong \bigoplus_{j=1}^{m} C_*\left(f, z_j\right),$$

for $\epsilon > 0$ small enough, where $B(x, \epsilon)$ is the ball centered at x with radius ϵ. □

Corollary 4.1. $M_q(a,b) = \sum_{a<c_i<b}\sum_{j=1}^{m_i} \text{rank}\, C_q(f, z_j^i)$ $q = 0, 1, 2, \ldots$.

This is the connection between critical groups and Morse type numbers. Now we formulate the Morse inequalities. Let us define

$$\beta_q = \beta_q(a,b) = \text{ rank } H_q\left(f_b, f_a, G\right),$$

$q = 0, 1, \ldots$, where $a < b$ are regular values of f.

Theorem 4.3. *Suppose that M is a C^2-Finsler Banach manifold and that $f \in C^1(M, \mathbb{R}^1)$ satisfies the $(PS)_c$ condition $\forall\ c \in [a,b]$, where a, b are regular values of f. Assume $K \cap f^{-1}[a,b] = \{z_1, z_2, \ldots, z_\ell\}$. Then we have*

$$\sum_{q=0}^{\infty} M_q t^q = \sum_{q=0}^{\infty} \beta_q t^q + (1+t)Q(t), \tag{4.1}$$

where Q is a formal series with nonnegative coefficients, $M_q = M_q(a,b) = \sum_{j=1}^{\ell}$ rank $C_q(f, z_j)$, and $\beta_q = \beta_q(a,b) =$ rank $H_q(f_b, f_a)$, $q = 0, 1, 2, \ldots$.

Note. (4.1) is formal: It means

$$\sum_{j=0}^{q}(-1)^{q-j}M_j \geq \sum_{j=0}^{q}(-1)^{q-j}\beta_j, \quad q=0,1,2,\dots,$$

and

$$\sum_{q=0}^{\infty}(-1)^{q}M_q = \sum_{q=0}^{\infty}(-1)^{q}\beta_q,$$

if all M_q, β_q, $q=0,1,2,\dots$, are finite and the series converge.

Proof. (4.1) follows from the subadditivity of the functions

$$S_q(X,Y) = \sum_{j=0}^{q}(-1)^{q-j} \operatorname{rank} H_j(X,Y).$$

More precisely, for a triple $X \supset Y \supset Z$, we have

$$S_q(X,Z) \leq S_q(X,Y) + S_q(Y,Z), \qquad q=0,1,2,\dots. \tag{4.2}$$

Indeed, let $c_1 < c_2 < \dots < c_n$ be the critical values of f in (a,b). We choose $a_0, a_1, \dots, a_n$ such that

$$a = a_0 < c_1 < a_1 < c_2 < \dots < c_n < a_n = b$$

If M_q and β_q are finite, then by Corollary 4.1

$$\operatorname{rank} H_q\left(f_{a_i}, f_{a_{i-1}}\right), \quad i=1,2,\dots,n, \quad q=0,1,\dots$$

are all finite. It follows from (4.2),

$$\sum_{i=1}^{n}\sum_{j=0}^{q}(-1)^{q-j} \operatorname{rank} H_j\left(f_{a_i}, f_{a_{i-1}}\right) \geq \sum_{j=0}^{q}(-1)^{q-j} \operatorname{rank} H_j\left(f_{a_n}, f_{a_0}\right).$$

Consequently, by Corollary 4.1 and Theorem 4.2

$$\sum_{j=0}^{q}(-1)^{q-j}M_j \geq \sum_{j=0}^{q}(-1)^{q-j}\beta_j.$$

Since we have assumed the convergence of the series, there exists q_0 such that

$$M_q = \beta_q = 0 \quad \forall\, q \geq q_0.$$

The identity

$$\sum_{q=0}^{\infty}(-1)^q M_q = \sum_{q=0}^{\infty}(-1)^q \beta_q$$

is nothing but the two consecutive inequalities above.

Now we prove (4.2). In an exact sequence,

$$\dots \xrightarrow{i} A \xrightarrow{j} B \longrightarrow \dots .$$

Since $\ker j = \operatorname{Im} i$,

$$\text{rank } A = \text{ rank Im } j + \text{ rank Im } i.$$

Let us denote rank Im i by $\epsilon(A)$, then

$$\text{rank } A = \epsilon(B) + \epsilon(A).$$

For a triple $Z \subset Y \subset X$, we have the exact sequence

$$\dots \to H_q(Y,Z) \to H_q(X,Z) \to H_q(X,Y) \to H_{q-1}(Y,Z) \to \dots \to 0,$$

thus

$$\begin{aligned}
\text{rank } H_q(X,Y) &= \epsilon_q(X,Y) + \epsilon_{q-1}(Y,Z), \\
\text{rank } H_q(X,Z) &= \epsilon_q(X,Z) + \epsilon_q(X,Y), \\
\text{rank } H_q(Y,Z) &= \epsilon_q(Y,Z) + \epsilon_q(X,Z).
\end{aligned}$$

It follows that

$$\begin{aligned}
&\text{rank } H_q(X,Y) + \text{ rank } H_q(Y,Z) - \text{ rank } (X,Z) \\
&\quad = \epsilon_q(Y,Z) + \epsilon_{q-1}(Y,Z).
\end{aligned}$$

Consequently, we obtain

$$S_q(Y,Z) + S_q(X,Y) - S_q(X,Z) = \epsilon_q(Y,Z) \geq 0, \qquad q = 0,1,2,\dots .$$

(4.2) is proved. □

The geometric interpretation of these Morse inequalities is as follows: If f is a nondegenerate function on a compact smooth manifold M, and if we choose $b > \max_{x\in M} f(x)$, $a < \min_{x\in M} f(x)$, then M_q is the number of critical points with Morse index q, and β_q is the Betti number of M.

The following handle body theorem describes a clear picture of the changes in topological structure as the level sets pass through a critical value on which there are only nondegenerate critical points.

Theorem 4.4. *Suppose that $f \in C^2(M, \mathbb{R}^1)$ satisfies the (PS) condition, where M is a C^2-Hilbert Riemannian manifold, and that c is an isolated critical value. Assume that $K_c = \{z_1, z_2, \dots, z_\ell\}$ consists of only nondegenerate critical points with Morse indices $\{m_1, m_2, \dots, m_\ell\}$ resp. Then there exist $\epsilon > 0$ and homeomorphisms $h_i : B^{m_i} \to M$ (into), $i = 1, 2, \dots, \ell$, such that*

$$f_{c+\epsilon} \simeq f_{c-\epsilon} \cup \bigcup_{i=1}^{\ell} h_i \left(B^{m_i}\right),$$

satisfying

$$(4.3) \quad f_{c-\epsilon} \cap h_i \left(B^{m_i}\right) = f^{-1}(c-\epsilon) \cap h_i \left(B^{m_i}\right) = h_i \left(\partial B^{m_i}\right), \quad i = 1, 2, \dots, \ell,$$

where B^{m_i} is the m_i dimensional ball, and the notation $\simeq$ means a deformation retract.

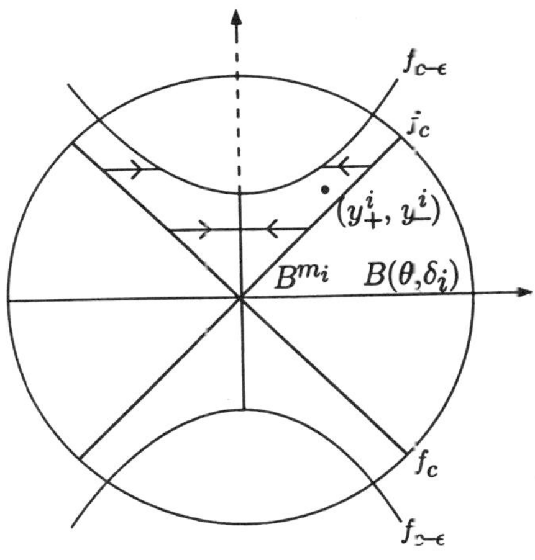

Proof. From the second deformation lemma, we have $\epsilon > 0$ such that $f_{c+\epsilon} \simeq f_c$.

We choose $\delta > 0$ small enough such that $B(z_i, \delta) \cap B(z_j, \delta) = \emptyset \; \forall \, i \neq j$, and that the Morse lemma in $B(z_i, \delta)$ is applicable, which yields a local homeomorphism

$$\Phi_i : B\left(\theta, \delta_i\right) \to B\left(z_i, \delta\right) \text{ (into)}$$

satisfying

$$f \circ \Phi_i \left(y^i\right) = \frac{1}{2}\left(\|y^i_+\|^2 - \|y^i_-\|^2\right) + c,$$

where $y^i \in B(\theta, \delta_i)$, the ball centered at θ, with radius $\delta_i > 0$, in the tangent space $T_{z_i}(M)$, and $y^i = y^i_+ + y^i_-$ is the orthogonal decomposition according to $d^2 f(z_i)$, $i, j, = 1, 2, \dots, \ell$.

Let $N = \cup_{i=1}^{\ell} \Phi_i B(\theta, \delta_i)$.

We apply the first deformation theorem to $\overline{N}$, yielding a deformation retract $\eta : \eta(f_{c+\epsilon} \backslash \overline{N}) \subset f_{c-\epsilon}$.

Since η decreases the value of f,

$$\eta\left(f_c\right) \subset \left(f_c \cap \overline{N}\right) \cup f_{c-\epsilon}.$$

It suffices to show $(f_c \cap \overline{N}) \cup f_{c-\epsilon} \simeq \cup_{i=1}^{\ell} h_i(B^{m_i}) \cup f_{c-\epsilon}$.

We choose $\epsilon_1 \in (0, \epsilon)$ satisfying

$$\sqrt{2\epsilon_1} < \min\left\{\delta_i, \quad i = 1, 2, \ldots, \ell\right\},$$

and define

$$\xi(t,x) = \begin{cases} x & \text{if } x \in f_{c-\epsilon} \\ \Phi\left(y_-^i + \left(1 + t\left(\frac{1}{\|y_+^i\|}\sqrt{\max\left\{\|y_-^i\|^2 - 2\epsilon_1, 0\right\}} - 1\right)\right) y_+^i\right), \end{cases}$$

where $t \in [0,1]$, $y^i = \Phi^{-1}(x)$ as $x \in B(z_i, \delta)$, and $y^i = y_+^i + y_-^i$ according to the orthogonal decomposition, $i = 1, 2, \ldots, \ell$. Thus

$$f_{c-\epsilon_1} \cup \left(f_c \cap \bigcup_{i=1}^{\ell} \Phi\left(B\left(\theta, \delta_i\right)\right)\right) \simeq f_{c-\epsilon_1} \cup \bigcup_{i=1}^{\ell} \Phi\left(B^{m_i}\right).$$

The restriction $\Phi|_{B(\theta, \sqrt{2\epsilon_1}) \cap P_i^- T_{z_i}(M)}$ yields the homeomorphism h_i, where P_i^- is the orthogonal projection onto the negative eigenspace of $d^2 f(z_i)$, $i = 1, 2, \ldots, \ell$.

Combining η with ξ, we obtain the deformation retract. (4.3) is easily verified.

This completes the proof. □

The references for Theorems 4.1 through 4.4 are Milnor [Mil1], Schwartz [ScJ1], Rothe [Rot1], Palais [Pal1], Pitcher [Pit1] and Marino Prodi [MaP1]. In Theorem 4.4, the handle body theorem is established on Hilbert Riemannian manifolds, where the Morse Lemma holds, and the local behavior of a nondegenerate critical point is quite clear. In order to extend this theorem to Finsler manifolds, or to Banach spaces, new difficulties arise in two ways. One way is conceptual: The above definition of nondegeneracy does not make sense in Banach spaces because the Hessian $d^2 f(x)$ is a bounded linear operator from the space to its dual, so that one cannot say that $d^2 f(x)$ has a bounded inverse except when the space is isomorphic to its dual. The other difficulty is technical. The Morse lemma is no longer useful because it is not compatible with the Palais-Smale condition, e.g., the quadratic functional $f(u) = \int_0^1 u^2(t)\, dt$ does not satisfy the (PS) condition on the space $L^p[0,1]$, for $p > 2$.

The theory can be extended as follows.

An operator $L \in B(X)$, the Banach algebra of all bounded linear operators from a Banach space X into itself, is said to be hyperbolic, if the spectrum $\sigma(L)$ is contained in two compact subsets separated by the imaginary axis.

Definition 4.3. Let M be a C^2 Finsler manifold, $f \in C^2(M, \mathbb{R}^1)$. Let p_0 be an isolated critical point. We say that p_0 is nondegenerate if there is a neighborhood U of p_0 on which $T(M)$ is trivialized to a be $U \times X$, such that there is a hyperbolic operator $L \in B(X)$ satisfying

(1) $d^2f(p_0)(Lx, y) = d^2f(p_0)(x, Ly), \quad \forall\, x, y \in X,$

(2) $d^2f(p_0)(Lx, x) > 0 \quad \forall\, x \in X\backslash\{\theta\},$

(3) $\langle df(p), Lx\rangle > 0, \quad \forall\, p \in U,\ p = p_0 + x$ in the local coordinates.

The dimension of the negative invariant subspace of L is called the index of p.

According to the new definition (which coincides with the old if M is Hilbert-Riemannian by taking $L = d^2f(p_o)^{-1}$), Theorem 4.4 does hold for Finsler manifolds modeled on Banach spaces with differentiable norms. The reader is referred to Chang [Cha2] and K. Uhlenbeck [Uhl1]. A different statement was also given by T. Tromba [Tro3].

Remark 4.1. The above theorems as well as the corollary can be extended to functions defined on a manifold M with boundary ∂M. The same proof works and the same conclusion holds for functions f under the following assumptions:

(1) $K \cap \partial M = \emptyset;$

(2) $\exists$ a p.g.v.f. V of f such that $-V|_{\partial M\backslash f^{-1}(a)}$

direct inward (i.e., the negative pseudo-gradient flow $\eta(t, x) \in \overset{\circ}{M},\ \forall\, x \in \partial M\backslash f^{-1}(a),\ \forall\, t > 0$ small). For a more general boundary condition, see Section 6.1.

Now we are interested in defining a reparameterized flow in order to make the flow have infinite arriving time if it passes through a critical point.

Let

$$\begin{cases} \dot{\rho}(\tau) = -X(\rho(\tau)) \\ \rho(0) = x_0 \in f^{-1}(a, b]\backslash K_b. \end{cases}$$

Then $\rho = \rho(\tau, x_0)$ is a reparametrization of the p.g. flow σ defined in Theorem 3.2.

We shall prove

Theorem 4.5. *If $f \in C^{2-0}(M, \mathbb{R}^1)$, and if $\sigma(T_{x_0} - 0, x_0) = z \in K_a$, then $\rho(+\infty, x_0) = \lim_{t\to+\infty} \rho(\tau, x_0) = z$.*

Proof. Let

$$s = f \circ \rho(\tau, x_0)\,. \tag{4.4}$$

We have

$$\frac{ds}{d\tau} = -\,\langle df, X\rangle \circ \rho(\tau, x_0)\,.$$

Suppose that $\widehat{T}$ is the arriving time in the new parameter τ, and assume that $\widehat{T} < +\infty$; we obtain

$$\begin{aligned}\widehat{T} &= \lim_{\delta\to+0}\int_{f(x_0)}^{a+\delta}\frac{d\tau}{ds}ds\\ &= \lim_{\delta\to+0}\int_{a+\delta}^{f(x_0)}\frac{ds}{\langle df, X\rangle \circ \rho(\tau, x_0)}.\end{aligned}$$

Let $\phi(s) = \langle df, X\rangle \circ \rho(\tau)$, where s and τ are related in (4.4). We have

$$\int_a^{f(x_0)}\frac{ds}{\phi(s)} < \infty.$$

From

$$\begin{aligned}\phi(s) &\le \|X\circ\rho(\tau)\|^2\\ &\le 4\,\|df\circ\rho(\tau,x_0) - df(z)\|^2\\ &\le 4C\left\|\int_\tau^{\hat{T}}\dot{\rho}(\tau',x_0)\,d\tau'\right\|^2\\ &\le 4C\left(\int_a^s\left\|\dot{\rho}(\tau',x_0)\right\|\left|\frac{d\tau'}{ds'}\right|ds'\right)^2\\ &\le 4C\left(\int_a^s\|X\circ\rho(\tau',x_0)\|^2\frac{ds'}{|\phi(s')|^2}\right)(s-a)\\ &\le 4C'\int_a^s\frac{ds'}{\phi(s')}(s-a) \le M(s-a)\end{aligned}$$

for some constants, C, C' and $M > 0$, it follows

$$\int_{a+\delta}^{f(x_0)}\frac{ds}{\phi(s)} \ge \frac{1}{M}\int_{a+\delta}^{f(x_0)}\frac{ds}{s-a} = +\frac{1}{M}\left(\ell n\,(f\,(x_0) - a) - \ell n\delta\right) \to +\infty$$

as $\delta \to +0$. This is a contradiction.

The new reparametrized flow $\rho(t, x_0)$ is well-defined for all $(t, x_0) \in R^1 \times (M\backslash K)$.

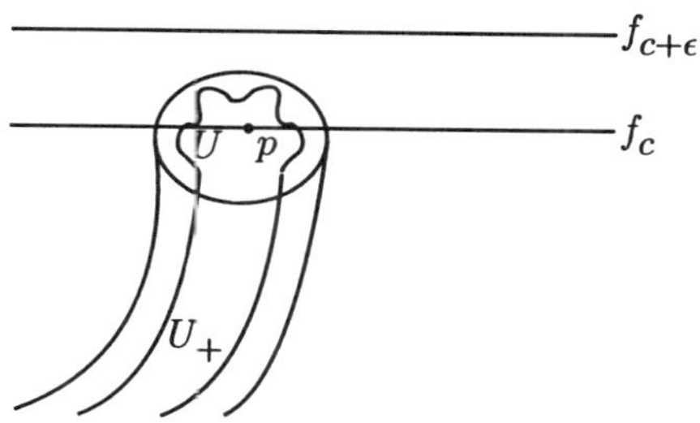

Theorem 4.6. *If $f \in C^{2-0}(M, \mathbb{R}^1)$ satisfies (PS) and if p is an isolated critical point of f, which is not a local minimum, then*

$$C_0(f,p) = 0.$$

Proof. We choose a path-connected neighborhood U of p such that $K \cap U = \{p\}$, and $U \subset f_{c+\epsilon}$ for some $\epsilon > 0$ where $c = f(p)$.

Let $U_+ = \cup_{t \geq 0} \rho(t, U)$, where ρ is the reparametrized flow defined above. Then we have

$$\begin{aligned} C_*(f,p) &\cong H_*\left(f_c \cap U_+, (f_c \backslash \{p\}) \cap U_+\right) \\ &\cong H_*\left(f_{c+\epsilon} \cap U_+, (f_c \backslash \{p\}) \cap U_+\right) \end{aligned}$$

because $f_c \cap U_+$ is a strong deformation retract of $f_{c+\epsilon} \cap U_+$ (Deformation Theorem). Since $f_{c+\epsilon} \cap U_+$ is path connected, and $(f_c \backslash \{p\}) \cap U_+ \neq \emptyset$.

$$\therefore C_0(f,p) = 0.$$

As an extension of Example 2, we have

Example 4. If $f \in C^{2-0}(M^n, R^n)$, and if p is an isolated critical point of f, which is neither a local maximum nor a local minimum, then

$$C_0(f,p) = C_n(f,p) = 0.$$

5. Gromoll-Meyer Theory

The contributions of Gromoll and Meyer to isolated critical point theory are threefold: (1) a splitting lemma, which is a generalization of the Morse lemma (cf. the previous section); (2) an alternative definition of critical groups; (3) a shifting theorem which reduces the critical groups of a degenerate isolated point to the critical groups of the function restricted to its degenerate sub-manifold. In this section, we shall rewrite their theory with slight improvements and prove the equivalence of the two definitions of critical groups.

Theorem 5.1. *Suppose that U is a neighborhood of θ in a Hilbert space H and that $f \in C^2(U, \mathbb{R}^1)$. Assume that θ is the only critical point of f and that $A = d^2 f(\theta)$ with kernel N. If 0 is either an isolated point of the spectrum $\sigma(A)$ or not in $\sigma(A)$, then there exist a ball B_δ, $\delta > 0$, centered at θ, an origin-preserving local homeomorphism ϕ defined on B_δ, and a C^1 mapping $h : B_\delta \cap N \to N^\perp$ such that*

$$f \circ \phi(z+y) = \frac{1}{2}(Az, z) + f(h(y) + y), \quad \forall\, x \in B_\delta, \tag{5.1}$$

where $y = P_N x$, $z = P_{N^\perp} x$, and P_N is the orthogonal projection onto the subspace N.

Proof. 1. Decomposing the space H into $N \oplus N^\perp$, we have

$$d_z f\,(\theta_1 + \theta_2) = \theta_1 \qquad \left(\theta_1 = P_{N^\perp}\theta,\ \theta_2 = P_N \theta\right),$$

and

$$d_z^2 f\,(\theta_1 + \theta_2) = A|_{N^\perp}.$$

Because of the implicit function theorem, there is a function $h : B_\delta \cap N \to N^\perp$, $\delta > 0$, such that

$$d_z f(y + h(y)) = \theta_1.$$

Let $u = z - h(y)$, and let

$$F(u, y) = f(z+y) - f(h(y) + y). \tag{5.2}$$

$$F_2(u) = \frac{1}{2}(Au, u). \tag{5.3}$$

Then we obtain

$$\begin{aligned} F\,(\theta_1, y) &= 0 \\ d_u F\,(\theta_1, y) &= d_z f(h(y) + y) = \theta_1, \\ d_u^2 f\,(\theta_1, \theta_2) &= d_z^2 f(\theta) = A|_{N^\perp}. \end{aligned}$$

2. Define $\xi : (u, y) \mapsto u_0 \in F_2^{-1} \circ F(u, y) \cap \{\eta(t, u) \mid |t| < \|u\|\}$, where η is the flow defined by the following ODE

$$\begin{aligned} \dot{\eta}(s) &= -\frac{A\eta(s)}{\|A\eta(s)\|} \\ \eta(0) &= u. \end{aligned}$$

Claim. η is well-defined for $|t| < \|u\|$. Since

$$\|\eta(s) - u\| \le |s|,$$

we have $\|\eta(t,u)\| \ge \|u\| - |t|$. From this, together with $\eta(t,u) \in N^\perp$, it follows that the denominator of the vector field is not zero for $|t| < \|u\|$.

Claim. ξ is well-defined on $B_\delta^N \times B_\delta^\perp$, where $B_\delta^N = B_\delta \cap N$, and $B_\delta^\perp = B_\delta \cap N^\perp$ for some $\delta > 0$. In fact, the following inequalities hold:

(a) $\forall\, \epsilon > 0$, $\exists\, \delta_1 = \delta(\epsilon) > 0$ such that

$$\begin{aligned}|F(u,y) - F_2(u)| &= |F(u,y) - F(\theta_1, y) - (d_u F(\theta_1, y), u) - F_2(u)| \\ &= \left|\int_0^1 (1-t)\left(\left(d_u^2 F(tu,y) - d_u^2 F(\theta_1,\theta_2)\right) u, u\right) dt\right| \\ &< \epsilon \|u\|^2,\end{aligned}$$

where $(u,y) \in B_{\delta_1}^N \times B_{\delta_1}^\perp$;

(b)

$$\begin{aligned}|F_2(\eta(t,u)) - F_2(u)| &= \left|\int_0^t \frac{d}{ds} F_2(\eta(s,u))\, ds\right| \\ &= \left|\int_0^t \left(dF_2(\eta), \dot{\eta}\right) ds\right| \\ &= \int_0^{|t|} \|A\eta(s,u)\|\, ds \\ &\ge C \int_0^{|t|} \|\eta(s,u)\|\, ds \\ &\ge C\left(\|u\|\,|t| - \frac{t^2}{2}\right),\end{aligned}$$

where $C > 0$ is a constant determined by the spectrum of A. We conclude that

(c) $F_2(\eta(t,u))$, as a function of t, is strictly decreasing on $(-\|u\|, \|u\|)$;

(d) $F_2(\eta(-t,u)) > F(u,y) > F_2(\eta(t,u))$ holds for

$$\left(1 - \sqrt{1 - \frac{2\epsilon}{C}}\right) \|u\| \le t \le \|u\|. \tag{5.4}$$

Therefore, there exists a unique $\bar{t}(u,y)$ with

$$|\bar{t}(u,y)| \le \left(1 - \sqrt{1 - \frac{2\epsilon}{C}}\right) \|u\| \tag{5.5}$$

such that

$$F_2\left(\eta\left(\bar{t}(u,y),u\right)\right)=F(u,y). \tag{5.6}$$

Thus the function ξ is of the form:

$$\xi(u,y)=\begin{cases}\theta_1 & u=\theta_1\\ \eta\left(\bar{t}(u,y),u\right) & u\neq\theta_1.\end{cases}$$

3. Define a map $\psi:(u,y)\to(\xi(u,y),y)$. We shall verify that ψ is a local homeomorphism. That $\bar{t}(u,y)$ is continuous easily follows from the implicit function theorem for $u=u_0\neq\theta$, provided

$$\frac{\partial}{\partial t}F_2\circ\eta\left(\bar{t},u_0\right)=-\|A\eta\left(\bar{t},u_0\right)\|\neq 0; \tag{5.7}$$

and for $u=\theta$, provided by (5.5).

We have used the path $\eta(t,u)$ to carry a point (u,y) to the point $(\xi(u,y),y)$; the same path can be used for the opposite purpose, i.e., to define the inverse map $\Phi=\psi^{-1}$. The same reason is provided to verify the continuity of Φ. Therefore Φ is a homeomorphism.

The equality (5.1) follows directly from (5.6). □

Note. The function $y\to f(y+h(y))$ is C^2.

In the case where $N=\{\theta\}$, the Morse Lemma is a consequence of this theorem except the conclusion is that Φ is a diffeomorphism.

Proof of the Morse Lemma. We have proved that Φ is a homeomorphism. Now we shall prove it is a diffeomorphism. That $\bar{t}(u)$, and then $\xi(u)$, is continuously differentiable for $u\in B_\delta\backslash\{\theta\}$, follows from the implicit function theorem and (5.7). It is also easily obtained that $d\xi(\theta)=\text{id}$, by using

$$\|\eta\left(\bar{t}(u),u\right)-u\|\leq|\bar{t}(u)|\leq\left(1-\sqrt{1-\frac{2\epsilon}{C}}\right)\|u\|=o(\|u\|) \tag{5.8}$$

(i.e., $\eta(\bar{t}(u),u)=u+o(\|u\|)$ as $u\to\theta$). Hence $\|d\xi(u)-\text{id}\|=o(1)$ remains to be proved. Since

$$\eta(t,u)=u-\int_0^t\frac{A\eta(s,u)}{\|A\eta(s,u)\|}ds\quad\text{for}\quad|t|<\|u\|,$$

we write $\eta(t)=\eta(t,u)$, $\eta_u(t)=d_u\eta(t,u)$ for simplicity,

$$\eta_u(t)=\text{id}-\int_0^t\left(\frac{A\eta_u(s)}{\|A\eta(s)\|}-\frac{A\eta(s)\otimes\left((A\eta_u(s))^*A\eta(s)\right)}{\|A\eta(s)\|^3}\right)ds, \tag{5.9}$$

that is

$$\|\eta_u(t)\| \le 1 + C_1 \int_0^1 \frac{1}{\|\eta(s)\|} \|\eta_u(s)\| \, ds,$$

where $C_1 > 0$ is a constant depending on $\sigma(A)$. But

$$\sqrt{1 - \frac{2\epsilon}{C}} \|u\| \le \|\eta(s)\| \le 2\|u\| \quad \text{as} \quad |s| \le \bar{t}(u)| . \tag{5.10}$$

Applying the Gronwall inequality,

$$\|\eta_u(t)\| \le 1 + e^{\frac{2C_1}{\|u\|} t} \le C_2 \quad \text{for} \quad 0 < |t| \le \bar{t}(u).$$

Thus

$$\|\eta_u\left(\bar{t}(u), u\right) - \text{id}\| \le C_3 \frac{|\bar{t}(u)|}{\|u\|} = o(1) \quad \text{as} \quad \|u\| \to 0.$$

Since

$$\|d\xi(u) - \text{id}\| = \|\eta_u\left(\bar{t}(u), u\right) - \text{id} + d_u\bar{t}(u) \cdot \eta_t'\left(\bar{t}(u), u\right)|$$

and by (5.6) and (5.10), we have

$$\begin{aligned} \|d_u\bar{t}(u)\| &= \frac{\|d_uF(u) - \eta_u\left(\bar{t}(u), u\right) A\eta\left(\bar{t}(u), u\right)\|}{\|A\eta\left(\bar{t}(u), u\right)\|} \\ &= \frac{\|Au + o(\|u\|) - (\text{id} + o(1))A(u + o(\|u\|))\|}{\|A(u + o(\|u\|))\|} = o(1) \\ &\qquad \text{as} \quad u \to \theta. \end{aligned}$$

This proves that $\|d\xi(u) - \text{id}\| \to 0$ as $\|u\| \to 0$. The implicit function theorem is employed, and ξ is a local diffeomorphism. □

Remark 5.1 There is no difficulty extending the Morse Lemma as well as Theorem 5.1 to the case where M is a Hilbert Riemannian manifold. The theorem was first obtained by Gromoll and Meyer [GrM1] under the condition $f \in C^3$. However, the C^2-Morse Lemma was proved by N. H. Kuiper [Kui1], from whom the idea of the above proof was taken. The C^3-Morse Lemma was given by Palais [Pal2], Nirenberg [Ni1] and Schwartz [Sch1] (cf. Appendix).

As to Theorem 5.1, for a different proof see J. Mawhin and M. Willem [MaW2] based on [Cam1], in which the condition that 0 is either isolated or not in $\sigma(d^2f(\theta))$ is replaced by one in which $d^2f(\theta)$ is a Fredholm operator with index 0.

However, our condition seems weaker, because a self-adjoint Fredholm operator A must have a finitely dimensional kernel $N(A)$ and a closed range $R(A)$. In the case where $N(A) \ne \{\theta\}$, the induced operator

$$\hat{A} : H/N(A) \to R(A),$$

defined by

$$\hat{A}[x] = Ax \quad \forall\, x \in [x],\ [x] \in H/N(A),$$

is invertible. Because of the Banach theorem, $\hat{A}^{-1}$ is bounded, which implies that 0 is isolated in $\sigma(A)$.

On the other hand, the condition $\dim N(A) < \infty$ is not assumed in Theorem 5.1.

Let M be a C^2 Finsler manifold, and let $f \in C^1(M, \mathbb{R}^1)$ be a function satisfying the (PS) condition. Suppose that $V : M \backslash K \to T(M)$ is a p.g.v.f. of f.

Definition 5.1. Let p be an isolated critical point of f, $c = f(p)$. A pair of topological spaces (W, W_-) is called a Gromoll-Meyer pair with respect to V, if

(1) W is a closed neighborhood of p possessing the mean value property, i.e., $\forall\ t_1 < t_2$, $\eta(t_i) \in W$, $i = 1, 2$, implies $\eta(t) \in W$ for all $t \in [t_1, t_2]$, where $\eta(t)$ is the decreasing flow with respect to V. And there exists $\epsilon > 0$ such that $W \cap f_{c-\epsilon} = f^{-1}[c - \epsilon, c) \cap K = \emptyset$, $W \cap K = \{p\}$;

(2) $W_- := \{x \in W \mid \eta(t, x) \notin W\ \forall\, t > 0\}$ is closed in W;

(3) W_- is a piecewise submanifold, and the flow η is transversal to W_-, i.e., $\eta \pitchfork W_-$.

At this point, the existence of a Gromoll-Meyer pair is assumed. The following theorem claims the motivation of the definition.

Theorem 5.2. *Let (W, W_-) be a Gromoll-Meyer pair with respect to a p.g.v.f. V of an isolated critical point p of the function $f \in C^1(M, \mathbb{R}^1)$ satisfying the (P, S) condition. Then we have*

$$H_*(W, W_-; G) \simeq C_*(f, p).$$

Proof. We now introduce two sets $U_+ = \cup_{0<t<\infty}\eta(t, W)$ and $\tilde{U}_+ = \cup_{0<t<\infty}\eta(t, W_-)$, and we define a continuous function $\gamma(x)$ on $\tilde{U}_+$ such that $\eta(\gamma(x), x) \in f_{c-\epsilon} \cap \tilde{U}_+$, but $\eta(t, x) \notin f_{c-\epsilon} \cap \tilde{U}_+$ if $t < \gamma(x)$. Thus

$$\sigma(t, x) = \eta(t\gamma(x), x) \quad t \in [0, 1],\ x \in \tilde{U}_+$$

defines a strong deformation retract $\tilde{U}_+ \to f_{c-\epsilon} \cap \tilde{U}_+ = f_{c-\epsilon} \cap U_+$, which reduces to

$$H_*\left(f_{c+\epsilon} \cap U_+, f_{c-\epsilon} \cap U_+; G\right) \simeq H_*\left(U_+, \tilde{U}_+; G\right).$$

Noticing that $\eta \pitchfork W_-$, $\forall\ y \in \tilde{U}_+$, there are unique $x \in W_-$, $t \geq 0$, such that $y = \eta(t, x)$ and $t = t(y)$, $x = x(y)$ continuously depend on y. The mean value property implies that $U_+ = W \cup \cup_{0<t<\infty}\eta(t, W_-)$.

Let $\tilde{U}_\delta = \cup_{\delta<t<\infty}\eta(t, W_-)$, for $\delta > 0$, i.e., $\tilde{U}_\delta = \{y \in U_+ \mid \hat{t}(y) > \delta\}$, where we define $\hat{t}(y) = 0 \;\forall\, y \in W$, and $\hat{t}(y) = t(y) \;\forall\, y \in \tilde{U}_+$. Hence $\tilde{U}_\delta$ is open,

$$\bar{\tilde{U}}_\delta = \bigcup_{\delta\leq t<\infty} \eta(t, W_-) \subset \bigcup_{0<t<\infty} \eta(t, W_-) = \operatorname{int}\left(\tilde{U}_+\right).$$

By the excision property, we have

$$H_*\left(U_+, \tilde{U}_+; G\right) \simeq H_*\left(U_+\backslash\tilde{U}_\delta, \tilde{U}_+\backslash\tilde{U}_\delta; G\right).$$

Again, there is a strong deformation retract, defined by the reversed flow $\eta(-t) : ((U_+\backslash\tilde{U}_\delta),\ (\tilde{U}_+\backslash\tilde{U}_\delta)) \to (W, W_-)$, provided $\eta \pitchfork W_-$, we obtain

$$H_*\left(U_+\backslash\tilde{U}_\delta, \tilde{U}_+\backslash\tilde{U}_\delta; G\right) = H_*(W, W_-; G).$$

On the other hand, by the Deformation Theorem 3.2,

$$\begin{aligned} &H_*\left(f_{c+\epsilon} \cap U_+, f_{c-\epsilon} \cap U_+; G\right) \\ &\qquad\simeq H_*\left(f_c \cap U_+, (f_c\backslash\{p\}) \cap U_+; G\right) \simeq C_*(f, p). \end{aligned}$$

Combining all these isomorphisms we obtain $H_*(W, W_-; G) \cong C_*(f, p)$. □

We now turn our attention to constructing a Gromoll-Meyer pair for an isolated critical point. For simplicity, we may assume that M is a Hilbert space, and $p = \theta$, $f(\theta) = 0$. Choose $\epsilon > 0$, $\delta > 0$ such that 0 is the unique critical value in $[-\epsilon, \epsilon]$, and θ is the unique critical point in the ball B_δ centered at θ. Define a function

$$g(x) = \lambda f(x) + \|x\|^2,$$

a neighborhood

$$W = f^{-1}[\gamma, \gamma] \cap g_\mu \tag{5.11}$$

and a subset of W,

$$W_- = f^{-1}(-\gamma) \cap W, \tag{5.12}$$

where λ, μ and γ are positive numbers to be determined by the following conditions:

$$B_{\delta/2} \cap f^{-1}[-\gamma, \gamma] \subset W \subset B_\delta \cap f^{-1}[-\epsilon, \epsilon] \tag{5.13}$$

$$f^{-1}[-\gamma, \gamma] \cap g^{-1}(\mu) \subset B_\delta\backslash B_{\delta/2} \tag{5.14}$$

$$(dg(x), df(x)) > 0 \;\;\forall\, x \in B_\delta/\mathring{B}_{\delta/2}. \tag{5.15}$$

Due to the (PS) condition, there exists $\beta = \inf_{x\in B_\delta/B_{\delta/2}} \|df(x)\| > 0$. Let $m = \sup_{x\in B_\delta} \|df(x)\|$. λ, μ and γ are then determined consecutively: $\lambda > \frac{2\delta m}{\beta^2}$, $0 < \gamma < \min\{\epsilon, \frac{3\delta^2}{8\lambda}\}$, and $\frac{\delta^2}{4} + \lambda\gamma \le \mu \le \delta^2 - \lambda\gamma$.

Theorem 5.3. *The pair (5.11), (5.12) is a Gromoll-Meyer pair with respect to the negative gradient vector field $-df(x)$.*

Proof. We only want to verify

(1) the mean value property,

(2) $W_- = \{x \in W \mid \eta(t,x) \notin W, \ \forall\, t > 0$, for the negative gradient flow η.

Claim (1). For simplicity, we may assume $\eta(0), \eta(t) \in W$. We wish to prove $\eta(s) \in W \ \forall\, s \in [0,t]$. Define $T = \sup\{s \in [0,t] \mid \eta(s') \in W \ \forall\, s' \le s\}$. If $T < t$; then $\eta(T) \notin \overset{\circ}{B}_{\delta\backslash 2}$. But

$$\gamma \ge f\circ\eta(0) \ge f\circ\eta(T) > f\circ\eta(t) \ge -\gamma$$
$$(g\circ\eta)'(T) = -(dg, df)|_{\eta(T)} < 0,$$

provided by (5.13) and (5.15). This contradicts the fact that T is a supremum. Therefore $T = t$.

Claim (2). We write $W^- = \{x \in W \mid \eta(t,x) \notin W, \ \forall\, t > 0\}$. Obviously we see $W_- \subset W^-$. Now we prove $W^- \subset W_-$. By definition, $W^- \subset \partial W$ and $\partial W = W_- \cup (f^{-1}(\gamma) \cap \overset{\circ}{g}_\mu) \cup (g^{-1}(\mu) \cap (W\backslash W_-))$. If $x \in f^{-1}(\gamma) \cap \overset{\circ}{g}_\mu$, then $x \notin W^-$. For $x \in g^{-1}(\mu) \cap (W\backslash W_-)$, according to (5.14) and (5.15), we have $(g\circ\eta)'(0,x) < 0$ and $f(x) > -\gamma$. These imply that $\exists\, \tau > 0$ such that $g\circ\eta(\tau,x) \le \mu$ and $|(f\circ\eta)(\tau,x)| \le \gamma$, i.e., $x \notin W^-$. We have proved $W_- = W^-$. □

We intend to compute the critical groups of an isolated critical point p, which may be degenerate; the splitting theorem is employed. Let Φ be a local parametrization of M defined in some open neighborhood U of θ in $T_p(M) \cong H$ with $\Phi(\theta) = p$, such that $f\circ\Phi(z,y) = \frac{1}{2}(Az,z) + f_0(y)$, where $A = d^2f(p)$, 0 is either an isolated point of the spectrum $\sigma(A)$ or not in $\sigma(A)$, and f_0 is a function defined on N, the null space of A. We call $N = \Phi(U\cap N)$ the characteristic submanifold of M for f at p with respect to the parametrization Φ. The following theorem sets up the relationship between the critical groups of f and those of $\tilde f := f\,|_N$.

Theorem 5.4. (Shifting theorem). *Assume that the Morse index of f at p is j, then we have*

$$C_q(f,p) = C_{q-j}\left(\tilde f, p\right), \ q = 0, 1, \dots .$$

First we need:

Lemma 5.1. *Suppose that $H = H_1 \oplus H_2$, $g_i \in C^2(H_i, \mathbb{R}^1)$, θ_i is an isolated critical point of g_i, $i = 1, 2$, where H_1, H_2 are Hilbert spaces. Assume that (W_i, W_{i-}) is a Gromoll-Meyer pair of θ_i with respect to the gradient vector field of g_i, $i = 1, 2$; then $(W_1 \times W_2, (W_{1-} \times W_2) \cup (W_1 \times W_{2-}))$ is a Gromoll-Meyer pair of the function $f = g_1 + g_2$ at $\theta = \theta_1 + \theta_2$ with respect to the gradient vector field df, if θ is an isolated critical point of f.*

This is easy to check. We omit the proof.

Theorem 5.5. *Under the assumptions of* Lemma 5.1, *we have*

$$C_*(f, \theta) = C_*(g_1, \theta_1) \otimes C_*(g_2, \theta_2).$$

Proof. This is a combination of Theorem 5.2, Lemma 5.1 and the Künneth formula.

Proof of Theorem 5.4. This is a combination of Theorems 5.1, 5.5.

Remark 5.2. Theorem 5.5 was conjectured by Gromoll and Meyer [GrM1] and was solved by G. Tian [Tia1]. In [Dan1], Dancer independently proved the conjecture in the finite dimensional case.

In order to verify that the pair $((W_{1-} \times W_2), (W_1 \times W_{2-}))$ is an excision couple, we use the notations: U_+, $\tilde{U}_+$, V_+, $\tilde{V}_+$, defined as in Theorem 5.2 where $U_+ = \cup_{t \geq 0} \eta_1(t, W_1)$, $\tilde{U}_+ = \cup_{t>0} \eta_1(t, W_{1-})$, $V_+ = \cup_{t \geq 0} \eta_2(t, W_2)$, $\tilde{V}_+ = \cup_{t>0} \eta_2(T, W_{2-})$, and η_1, η_2 are the decreasing flows with respect to dg_1 and dg_2 respectively. Also $\tilde{U}_\delta$, $\tilde{V}_\delta$, $\delta > 0$, stand for the same meaning. Thus

$$\begin{aligned}
&H_*\left((W_{1-} \times W_2) \cup (W_1 \times W_{2-}), W_1 \times W_{2-}\right) \\
&\cong H_*\left(\left(\tilde{U}_+ \times V_+\right) \cup \left(U_+ \times \tilde{V}_+\right), U_+ \times \tilde{V}_+\right) \\
&\cong H_*\left(\tilde{U}_\delta \times V_+, \tilde{U}_\delta \times \tilde{V}_+\right) \qquad \text{(excision)} \\
&\cong H_*\left(W_{1-} \times W_2, W_{1-} \times W_{2-}\right).
\end{aligned}$$

Corollary 5.1. *Suppose that N is finite dimensional with dimension k and p is*

(1) *a local minimum of $\tilde{f}$, then*

$$C_q(f, p) = \delta_{qj} G,$$

(2) *a local maximum of $\tilde{f}$, then*

$$C_q(f, p) = \delta_{q,j+k} G,$$

(3) *neither a local maximum nor a local minimum of* $\tilde{f}$, *then*

$$C_q(f,p) = 0 \quad \text{for} \quad q \le j, \quad \text{and} \quad q \ge j + k.$$

Finally, we prove the homotopy invariance of the critical groups. It is similar to the Leray Schauder index for isolated zeroes of vector fields. The perturbation should preserve the isolatedness of the critical points.

Lemma 5.2. *Let* (W, W_-) *be a Gromoll-Meyer pair of an isolated critical point* p *of a* C^{2-0} *function* f, *defined on a Hilbert space* H, *satisfying the (PS) condition, with respect to* $-df(x)$. *Then there exists* $\epsilon > 0$ *such that* (W, W_-) *is also a Gromoll-Meyer pair of* g *satisfying the (PS) condition with respect to certain p.g.v.f. of* g, *provided that* g *has a unique critical point* q *in* W, *and* $\|f - g\|_{C^1(W)} < \epsilon$.

Proof. $\exists\, r > 0$ such that $B(p, r) \subset \overset{\circ}{W}$. Due to (PS)

$$\beta = \inf \left\{ \|df(x)\| \mid x \in W \backslash B\left(p, \frac{r}{2}\right) \right\} > 0.$$

Define $\rho \in C^{1-0}(H, \mathbb{R}^1)$, satisfying

$$\rho(x) = \begin{cases} 1 & x \in B\left(p, \frac{r}{2}\right), \\ 0 & x \notin B(p, r). \end{cases}$$

with $0 \le \rho(x) \le 1$, and a vector field

$$V(x) = \frac{3}{2}[\rho(x)dg(x) + (1 - \rho(x))df(x)].$$

Choosing $0 < \epsilon < \frac{\beta}{4}$, we obtain

$$\|V(x)\| \le 2\|dg(x)\|,$$

and

$$(V(x), dg(x)) \ge \|dg(x)\|^2 \qquad \forall\, x \in W,$$

for $\|g - f\|_{c^1(W)} < \epsilon$.

Claim. $\forall\, x \notin B(p, \frac{r}{2})$

$$\|dg(x)\| \ge \|df(x)\| - \epsilon \ge \beta - \epsilon > 3\epsilon.$$

We have

$$\begin{aligned}
(V(x), dg(x)) &\ge \frac{3}{2}\left[\|dg(x)\|^2 - \epsilon\|dg(x)\|\right] \\
&\ge \frac{3}{2}\left[\|dg(x)\|^2 - \frac{1}{3}\|dg(x)\|^2\right] = \|dg(x)\|^2 \\
\|V(x)\| &\le \frac{3}{2}(\|dg(x)\| + \epsilon) \le 2\|dg(x)\|.
\end{aligned}$$

Since $\forall\, x \in B(p, \frac{r}{2})$, $V(x) = dg(x)$, the verification is trivial.

Notice that $V(x) = -df(x)$ outside a ball $B(p,r) \subset \overset{\circ}{W}$, and q is the only critical point of g in W. It is not difficult to verify that (W, W_-) is a GM pair of g with respect to V. (The mean value property holds because the flow is the same negative gradient flow of f outside $B(p,r)$, particularly in a neighborhood of ∂W. Similarly, for other properties assumed on W_-.)

Theorem 5.6. *Suppose that $\{f_\sigma \in C^2(H, \mathbb{R}^1) \mid \sigma \in [0,1]\}$ is a family of functions satisfying the (PS) condition. Suppose that there exists an open set N such that f_σ has a unique critical point p_σ in N, $\forall\, \sigma \in [0,1]$, and that $\sigma \to f_\sigma$ is continuous in $C^1(\overline{N})$ topology. Then*

$$C_*(f_\sigma, p_\sigma) \quad \text{is independent of} \quad \sigma.$$

Proof. Due to the (PS) condition, $\sigma \to p_\sigma$ is continuous. Applying Theorem 5.3, we may construct a GM pair $(W_\sigma, W_{\sigma -})$ for p_σ, $\forall\, \sigma \in [0,1]$. Combining Theorem 5.2, Lemma 5.2, and the finite covering, we obtain our conclusion directly.

Theorem 5.7. (Marino Prodi [MaP2]) *Suppose that $f \in C^2(M, \mathbb{R}^1)$ has a critical value c with $K_c = \{p_1, p_2, \ldots, p_\ell\}$. Assume that $d^2 f(p_i)$ are Fredholm operators, $i = 1, 2, \ldots, \ell$. Then $\forall\, \epsilon > 0$, there exists a function $g \in C^2(M, \mathbb{R}^1)$ such that*

(1) $g = f$ in $M \setminus \cup_{j=1}^{\ell} B(p_j, \epsilon)$,
(2) $\|g - f\|_{C^2(M)} \leq \epsilon$,
(3) *g has only nondegenerate critical points, all concentrated in $\cup_{j=1}^{\ell} B(p_j, \epsilon)$.*
(4) *Let $\mathrm{ind}(f, p_j) = m_j$, and $\dim \ker d^2 f(p_j) = n_j$; then the Morse indices of those nondegenerate critical points of g in $B(p_j, \epsilon)$ are in $[m_j, m_j + n_j]$, $j = 1, 2, \ldots, \ell$. Furthermore, if f satisfies the (PS) condition, then the same is true for g.*

Proof. One may choose $\epsilon > 0$ so small that all balls $B(p_j, \epsilon)$, $j = 1, 2, \ldots, \ell$ are disjoint. We only change f in small neighborhoods of p_j, $j = 1, 2, \ldots, \ell$, so the function g can be constructed consecutively. With no loss of generality, we assume $K_c = \{p\}$ to be a single point, with Morse index m and nullity n.

Because of the splitting theorem, we choose $\epsilon > 0$ so small that under suitable coordinates $\phi : B(p, \epsilon) \to H$, the Hilbert space on which M is modeled. The function f is written as

$$f \circ \phi^{-1}(x) = c + \frac{1}{2}\left(\|x_+\|^2 - \|x_-\|^2\right) + h(y),$$

where $x = (x_+, x_-, y) \in (H_+ \oplus H_- \oplus H_0) \cap B(\theta, \delta)$, for some $\delta > 0$, and $h \in C^2$, possesses θ as the unique critical point with $h(y) = o\left(\|y\|^2\right)$.

Let us define a smooth function $\rho : \mathbb{R}^1_+ \to \mathbb{R}^1$, $0 \le \rho \le 1$, $\rho(t) = 1$ for $t \le \delta/2$, $\rho(t) = 0$ for $t \ge \delta$, and $|\rho'(t)| < 4/\delta$. $\forall\, b \in \mathbb{R}^n$, set

$$h_b(y) = h(y) + (b, y)\rho(\|y\|),$$

where (,) is the inner product on $\mathbb{R}^n$.

Since $dh(y) \neq \theta \quad \forall\, y \in B(\theta,\delta)\backslash B(\theta, \delta/2)$, one finds that $c > 0$ such that

$$\|dh(y)\| \ge c \quad \forall\, y \in B(\theta,\delta)\backslash B\left(\theta, \delta/2\right).$$

Thus

$$\begin{aligned}\|dh_b(y)\| &= \left\| dh(y) + b\rho(\|y\|) + (b,y)\rho'(\|y\|)\frac{y}{\|y\|}\right\| \\ &\ge c - 5b > 0, \quad \text{if} \quad b < c/5.\end{aligned}$$

Therefore h_b has no critical point in $B(\theta,\delta)\backslash B(\theta,\delta/2)$.

In $B(\theta, \delta/2)$, however, $h_b(y) = h(y) + (b, y)$. Because of Sard's Theorem, $h_b(y)$ has only nondegenerate critical points for a.e. $b \in \mathbb{R}^n$; we may choose b small.

Fixing such a $b \in \mathbb{R}^n$, define

$$g \circ \phi^{-1}(x) = c + \frac{1}{2}\left(\|x_+\|^2 - \|x_-\|^2\right) + h_b(y),$$

and

$$g = f \quad \text{outside} \quad \phi^{-1}B(\theta,\delta) \subset B(p,\epsilon).$$

This is what we need. In fact (1), (3), and (4) follow directly from the construction. (2) holds if $\|b\|$ is small.

Furthermore, g differs from f only in the neighborhood $\phi^{-1}B(\theta,\delta)$. Since $H_0 \cap \overline{B(\theta,\delta)}$ is compact ($\dim H_0 = \dim\ker d^2f(p) < +\infty$) and $d^2 g \circ \phi^{-1} \mid_{(H_+ \oplus H_-)\cap B(\theta,\delta)}$ is invertible, we conclude that g satisfies the (PS) condition if f does.

6. Extensions of Morse Theory

In previous sections, Morse theory was studied on a manifold without boundary, except in Remark 4.1, in which we assumed the negative gradient points inward at the boundary. The purpose of this section is to extend the previous study in two aspects:

(1) under more general boundary conditions, and
(2) when the underlying space is not a manifold but a locally convex closed set.

6.1. Morse Theory Under General Boundary Conditions

Let M be a Hilbert Riemannian manifold with boundary $\partial M = \Sigma$, which is a smooth oriented submanifold with codimension 1. Thus the outward normal unit $n(x)$ of Σ at $x \in \Sigma$ is well-defined.

Given two real numbers a and b and a function $f \in C^1(M, \mathbb{R}^1)$, set $M[a,b] = M \cap f^{-1}[a,b], \Sigma[a,b] = \Sigma \cap f^{-1}[a,b]$. In the following, if there is no confusion, we shall simply write M and Σ in place of $M[a,b]$ and $\Sigma[a,b]$ respectively.

Definition 6.1. $f \in C^1(M, \mathbb{R}^1)$ is said to be satisfying the general boundary condition (GBC in short), if

(1) $K \cap \Sigma = \emptyset$, i.e., it has no critical point on Σ;

(2) the restriction $\hat{f} = f\,|_\Sigma$, as a function on Σ, has only isolated critical points.

Let $\Sigma\text{-} = \{x \in \Sigma \mid \langle f'(x), n(x)\rangle \leq 0\}$, where $\langle\ ,\ \rangle$ is the inner product of the Hilbert space H.

Suppose that f has only isolated critical points $\{x_1, x_2, \dots, x_n, \dots\}$; we define the Morse type numbers as follows:

$$M_q = \Sigma_{n=1}^{\infty} \operatorname{rank} C_q\,(x_n, f)\,, \qquad q = 0,1,2,\dots .$$

Let $\{m_0, m_1, \dots, m_n, \dots\}$ and $\{\mu_0, \mu_1, \dots, \mu_n, \dots\}$ be the Morse type numbers of the functions f and $\hat{f}\,|_{\Sigma_-}$ on M and $\Sigma\text{-}$ respectively. We define the augmented Morse type numbers:

$$M_n = m_n + \mu_n \qquad n = 0,1,2,\dots .$$

Our main results are the following:

Theorem 6.1. *Suppose that both the functions f and $\hat{f}$ satisfy the (PS) condition on M and Σ respectively, and that f has only isolated critical points and satisfies the GBC. Then we have the following compact version of Morse inequalities:*

$$\Sigma_{n\geq 0} M_n t^n = \Sigma_{n \geq 0} \beta_n t^n + (1+t)Q(t)$$

where β_n is the nth Betti number of the manifold $M[a,b]$, $n = 0,1,2,\dots$, and $Q(t)$ is a formal power series with nonnegative coefficients.

Theorem 6.2. *Suppose that $f \in C^2$, $K \cap f^{-1}[a,b] = \emptyset$, and that $\hat{f}$ is a Morse function on $\Sigma\text{-}$ which has the critical set $\{y_1, y_2, \dots, y_\iota\}$ with Morse indices $\{m_1, m_2, \dots, m_\iota\}$. Assume that both f and $\hat{f}$ satisfy the (PS) condition. Then*

$$f_b \cong f_a \cup h_1\,(B^{m_1}) \cup \cdots \cup h_\iota\,(B^{m_\iota})\,,$$

where B^m is the m ball, and h is a homeomorphism.

It is well-known that if the negative gradient vector field $-f'(x)$ satisfies the inward condition at the boundary Σ, i.e., $\Sigma- = \emptyset$, then the above theorems hold. The main idea in proving our theorems is as follows:

(1) We first extend the known theorems from $\Sigma- = \emptyset$, to $K- := \{x \in \Sigma- \mid \hat{f}'(x) = \theta\} = \emptyset$.

(2) We then define a new function f_1 which satisfies the following conditions:

a. f_1 coincides with f except for a neighborhood of the boundary Σ,
b. the set $K-$ for f_1 is empty, and
c. the critical points of f_1, in excess of those of f, correspond biuniquely to the critical points of $f|_{\Sigma-}$ with preservation of the Morse indices.

Lemma 6.1. *Suppose that $f \in C^1(M, \mathbb{R}^1)$, and that $\hat{f}$ satisfies the (PS) condition. If $K- = \emptyset$, then there exist a vector field X on $M^* := M\backslash K$, $K = \{x \in M \mid f'(x) = \theta\}$, and a positive number d, such that*

$$\begin{aligned} &(1)\ \|X(x)\| \le 1, \\ &(2)\ \langle f'(x), X(x)\rangle \ge \text{ Min } \{d, 1/2\|f'(x)\|\}, \qquad \forall\, x \in M, \\ &(3)\ \langle n(x), X(x)\rangle > 0 \qquad \forall\, x \in \Sigma. \end{aligned}$$

Proof. The proof is an extension of the standard proof of the existence of a pseudo gradient vector field for manifolds without boundaries. We shall pay more attention to the construction of such a vector field near the boundary Σ. First, we note that $\exists\, d > 0$, such that $\|\hat{f}'(x)\| \ge 2d \quad \forall\, x \in \Sigma-$.

Claim. If not, $\exists\, x_n \in \Sigma-$ such that $\hat{f}'(x_n) \to \theta$. According to the (PS) condition for the function $\hat{f}$, $\exists\, x^* \in \Sigma-$, with $\hat{f}'(x^*) = \theta$, i.e., $x^* \in K-$; this is a contradiction.

Next, let $\{U_\alpha \mid \alpha \in \Lambda\}$ be an open covering of M^* satisfying $T(M)\,|_{U_\alpha} \cong U_\alpha \times H$, $\forall\, \alpha \in \Lambda$. If either $x \in \text{int}(M^*)$, or $x \in \Sigma_+ := \{x \in \Sigma \mid \langle f'(x), n(x)\rangle > 0\}$, then we choose $V_x = f'(x)/\|f'(x)\| \in T_x(M) \cong H$, and a neighborhood $0_x \subset U_\alpha$ of x, for some α, such that $\langle V_x, f'(y)\rangle > 1/2\ \|f'(y)\|\ \forall\, y \in 0_x$, and

$$\begin{aligned} &0_x \cap \Sigma = \emptyset && \text{if } \ x \in \text{ int } (M^*), \\ &\langle n(y), V_x\rangle > 0 \qquad \forall\, y \in 0_x \cap \Sigma && \text{if } x \in \Sigma_+. \end{aligned}$$

In the remaining case, i.e., $x \in \Sigma-$, one may choose $W_x \in T_x(\Sigma)$, i.e., $\langle n(x), W_x\rangle = 0$, satisfying $\|W_x\| = 1$, and $\langle W_x, f'(x)\rangle > d$, provided by

$\|\hat{f}'(x)\| \geq 2d$. After a small perturbation, we have $V_x \in T_x(M)$, satisfying $\|V_x\| = 1$, $\langle V_x, f'(x)\rangle > d$, and $\langle n(x), V_x\rangle > 0$. Again, we have a small neighborhood $0_x \subset U_\alpha$ for some α, such that

$$\begin{aligned} \langle V_x, f'(y)\rangle &> d \qquad \forall\, y \in 0_x, \text{ and} \\ \langle n(y), V_x\rangle &> 0 \qquad \forall\, y \in 0_x \subset \Sigma. \end{aligned}$$

Finally, to the open covering $\{0_x \mid x \in M\}$, we have a locally finite refinement covering $\{C_\beta \mid \beta \in B\}$, and an associated C^{1-0} partition of unity $\{\phi_\beta \mid \beta \in B\}$, i.e., supp $\phi_\beta \subset C_\beta$, $0 \leq \phi_\beta \leq 1$, and $\sum_{\beta\in B}\phi_\beta(x) = 1$, $\forall\, x \in M$. For each β, $\exists\, x = x(\beta) \in M^*$ such that supp $\phi_\beta \subset 0_x$.

Let us define $X(x) = \sum_{\beta\in B}\phi_\beta(x)V_{x(\beta)}$. This is just what we need. In fact, (1) is trivial. By the definition,

$$\begin{aligned} \langle X(x), f'(x)\rangle &= \sum_{\beta\in B}\phi_\beta(x)\langle V_{x(\beta)}, f'(x)\rangle \\ &\geq \text{Min}\,\{d, 1/2\,\|f'(x)\|\} \qquad \forall\, x \in M^*, \end{aligned}$$

(2) follows. And $\langle n(x), X(x)\rangle = \sum_{\beta\in B}\langle n(x), V_{x(\beta)}\rangle \cdot \phi_\beta(x) > 0$, $\forall\, x \in \Sigma$, because $\forall\, x \in \Sigma$, $x \in$ Supp $\phi_\beta \subset 0_{x(\beta)}$ implies $x(\beta) \in \Sigma$. □

In the case $K_- = \emptyset$, both the Morse inequalities and the Morse handle body theorem hold under the following assumption:

(6.1) $\qquad f \in C^1(M, \mathbb{R}^1)$, and both the functions f and $\hat{f}$ satisfy the (PS) conditions on M and Σ respectively.

Now, we are going to reduce the GBC problem to case $K_- = \emptyset$, by perturbations.

Assuming that $\{y\} \in K_-$, we choose a chart (U, ϕ), $\phi : U \to H_- := \{\xi \in H \mid \langle \xi, e\rangle \leq 0\}$ for some fixed $e \in H\backslash\{\theta\}$, satisfying $K \cap U = \emptyset$, $K_- \cap U = \{y\}$, $\phi(y) = \theta$, $n(y) = e$, and $\Sigma \cap U = \phi^{-1}(\phi(U) \cap H_0)$, where $H_0 = \text{span}\,\{e\}^\perp$. In other words, if we write $\phi(x) = te + z$, then $t \leq 0$, and $t = 0$ if and only if $x \in \Sigma$. Define

$$w = f \circ \phi^{-1}(te + z) - f \circ \phi^{-1}(z) \tag{6.2}$$

then

$$f' \circ \phi^{-1}(\theta) \cdot \left(\phi^{-1}\right)_t(\theta) = \langle f'(y), n(y)\rangle < 0,$$

because $\hat{f}'(y) = \theta$, and f has no critical point on Σ, so $f'(y) \neq \theta$. According to the Implicit Function Theorem, $\exists\, \tau, \epsilon > 0$ and a C^1 function $t = t(w, z)$, which solves (6.2) in $[-\tau, 0] \times (B(\theta, \epsilon) \cap H_\sigma) \subset \phi(U)$, with $t(0, z) = 0$. Now, (w, z) is regarded to be a new local coordinate of $x = \psi(w, z) =$

$\phi^{-1}(t(w,z)e+z)$, for $(w,z) \in [0,\tau_1] \times (B(\theta,\epsilon) \cap H_0)$, where $\tau_1 > 0$ depends on τ. Let us define

$$\rho(w) = \begin{cases} 0 & w > \delta, \\ (\delta - w)^3/\delta^2 & \delta \geq w \geq 0, \end{cases}$$

where $0 < \delta < \tau_1$ is to be determined later; and let

$$\chi(s) = \begin{cases} 0 & s \geq \epsilon, \\ 1 & \epsilon/2 \geq s \geq 0, \end{cases}$$

where $\chi \in C^\infty$, $0 \leq \chi \leq 1$, and $|\chi'| \leq 3/\epsilon$. We define

$$\zeta(x) = \rho(w) \cdot \chi(|z|) \quad \text{for} \quad x \in U_1 := \psi\left([0,\tau_1] \times (B(\theta,\epsilon) \cap H_0)\right)$$

and equal to zero elsewhere. Then we define

$$f_1(x) = f(x) + \zeta(x).$$

Lemma 6.2. *Suppose the assumption* (6.1), $K \cap U = \emptyset$, *and* $K- \cap U = \{y\}$. *Then we have* $y^* \in U$, *and the following:*

(1) K_1, *the critical set of* f_1, *equals* $K \cup \{y^*\}$, (K_1)-, *the set* K- *for the function* f_1, *equals* $K- \backslash \{y\}$,

(2) f_1 *and* $\hat{f}_1$ *possess the (PS) condition;*

(3) $C_*(f_1, y^*) = C_*(f, y)$.

wherever $\delta > 0$ *small.*

Proof. (1) Both functions f_1 and $\hat{f}_1$ have no critical point in $\epsilon/2 \leq |z| \leq \epsilon$. In fact, provided by the (PS) condition, $\exists\ d > 0$ such that $\|\hat{f}'(x)\| > d\, \forall\, x \in U \cap \Sigma$, with $\epsilon/2 \leq |z| \leq \epsilon$. Let $g(z) = f \circ \phi^{-1}(z)$; then

$$\|g'(z)\| > d \qquad \text{for} \quad \epsilon/2 \leq |z| \leq \epsilon.$$

Thus,

$$\begin{aligned} \|\partial_z f_1(x)\| &= \|g'(z) + \rho(w)\chi'(|z|)z/|z|\,\| \\ &\geq d - 3\delta/\epsilon > d/2 \end{aligned}$$

$\forall\ (w,z) \in [0,\delta] \times [(B(\theta,\epsilon) \backslash B(\theta,\epsilon/2)) \cap H_0)$, if we choose $\delta < d\epsilon/6$.

(2) f_1 has a unique critical point $y^* = \psi(w_0,\theta)$, in U. In fact, we only want to restrict ourselves to the neighborhood $U_0 = \psi([0,\delta] \times (B(\theta,\epsilon/2) \cap H_0)$. Note that

$$f_1(x) = f(x) + \rho(w)\chi(z) = g(z) + w + \rho(w) \tag{6.3}$$

in this neighborhood, and that

$$\partial_z f_1(x) = \theta \quad \text{if and only if} \quad z = \theta;$$
$$\partial_w f_1(x) = \theta \quad \text{if and only if} \quad 1 + \rho'(w) = 0,$$
$$\text{i.e.,} \quad w_0 = (1 - \frac{1}{\sqrt{3}})\delta.$$

(3) If $\{x_n\}$ is a sequence, along which $f_1'(x_n)$ converges to θ and $f_1(x_n)$ is bounded, then we want to show that there is a convergent subsequence. In fact, if it has a subsequence outside U_1, then there must be a convergent subsequence, provided by the (PS) condition for f; otherwise, according to paragraph (1), there must be a subsequence in U_0. Since f_1 is of separate variables in U_0, $f_1 = g(z) + w + \rho(w)$, $g(z) = \hat{f}(x)$, and since $\hat{f}$ satisfies the (PS) condition, again, we obtain a convergent subsequence.

Analogously, we verify the (PS) condition for $\hat{f}_1$.($\hat{f}_1 = g(z)+\delta$ in $U_0 \cap \Sigma$).

(4) We shall verify (K_1)- $\cap U = \emptyset$. As before, we are only concerned with $\hat{f}_1$ in the neighborhood $U_0 \cap \Sigma$, so the critical point there is uniquely $\{y\}$. Since

$$\begin{aligned}\langle f_1'(y), n(y)\rangle &= (1 + \rho'(w)) \circ \partial_t w \mid_{(t,z)=(0,\theta)} \\ &= -2\langle f'(y), n(y)\rangle > 0\end{aligned}$$

thus, $y \notin (K_1)$-.

(5) Since f_1 is of separate variables in a neighborhood of y^*, and w_0 is a minimum of $w + \rho(w)$ we have

$$\begin{aligned}C_*(f_1, y^*) &= C_*\left(\hat{f}, y\right) \otimes C_*(\text{id} + \rho, w_0) \\ &= C_*\left(\hat{f}, y\right),\end{aligned}$$

provided by Theorem 5.5. The lemma is proved. □

Proof of Theorem 6.1. According to the isolatedness of the critical points of f and $\hat{f}$, and the (PS) conditions, we conclude that both f and $\hat{f}$ have only finitely many critical points, say, $\{z_1, z_2, \dots, z_k\}$ and $\{y_1, y_2, \dots, y_l\}$ respectively. Now we apply Lemma 6.2 l times, and the new function f_1 satisfies the condition (K_1)- $= \emptyset$. Both $f_1, \hat{f}_1$ possess the (PS) condition, and f_1 has the critical set $\{z_1, z_2, \dots, z_k, y_1^*, y_2^*, \dots, y_l^*\}$. The Morse inequalities are applied for f_1.

Since a, b are regular values, $\exists\ \alpha > 0$, such that $K \cap f^{-1}[a - 2\alpha, a] = K \cap f^{-1}[b, b + 2\alpha] = \emptyset$. If we choose $\epsilon, \delta > 0$ so small that

$$a - \alpha < f\left(\psi\left([0, \delta] \times (B(\theta, \epsilon) \cap H_0)\right)\right) < b + \alpha,$$

for each neighborhood of y_i, $i = 1, 2, \ldots, l$ and $\delta < \alpha$; then

$$f_1^{-1}[a - 2\alpha, b + 2\alpha] = f^{-1}[a - 2\alpha, b + 2\alpha].$$

Let β_* denote the Betti numbers of the manifold $M \cap f_1^{-1}[a - 2\alpha, b + 2\alpha]$. Hence,

$$\sum_{n \geq 0} M_n t^n = \sum_{n \geq 0} \beta'_n t^n + (1 + t)Q(t).$$

However, $f^{-1}[a - 2\alpha, b + 2\alpha] \cong f^{-1}[a, b]$, thus

$$\begin{aligned} \beta'_* &= \text{rank } H_* \left(f_1^{-1}[a - 2\alpha, b + 2\alpha]\right) \\ &= \text{rank } H_* \left(f^{-1}[a - 2\alpha, b + 2\alpha]\right) \\ &= \text{rank } H_* \left(f^{-1}[a, b]\right) = \beta_*. \end{aligned}$$

The Morse inequalities are proved. □

Proof of Theorem 6.2. Similar to the proof of Theorem 6.1., but apply the handle body theorem instead. □

Remark 6.1. Theorems 6.1 and 6.2 generalize the results due to Morse Van Schaach, cf. M. Morse, S. S. Cairns [MoC1], and E. Rothe, cf. [Rot2,4]. They were obtained in Chang, Liu [ChL1].

6.2. Morse Theory on a Locally Convex Closed Set

The variational inequality problem is a different kind of variational problem arising from mechanics and physics. For example, let C be a closed convex set in a Banach space X, and let $f : C \to \mathbb{R}^1$ be a restriction of a C^1 function. We shall find a point $x_0 \in C$ such that

$$\langle df(x_0), x - x_0 \rangle \geq 0 \qquad \forall\, x \in C,$$

where $\langle\ ,\ \rangle$ is the duality between X^* and X.

We may even extend the concept of convexity from Banach spaces to Banach manifolds, and study the variational inequality problems via the minimax methods and the Morse theory on the extended convex sets.

Definition 6.2. Let M be a C^1 Banach manifold modeled on a Banach space X. A subset $S \subset M$ is called locally convex if, $\forall\, x \in S$, $\exists$ a chart $\phi : U \to X$ such that $\phi(U \cap S)$ is convex in X.

The local convexity depends on the special choice of the atlas $\mathcal{A} = \{(U_x, \phi_x) \mid x \in S\}$. This is an additional structure on S. In the following, when we say that S is locally convex, it always means that there is a structure $(M, \mathcal{A})$, such that S is locally convex with respect to $(M, \mathcal{A})$.

Definition 6.3. Let $S \subset M$ be a closed subset. A vector $v \in T_xM$ is called tangent to S at $x \in S$, if $\exists$ a chart $\phi : U \to X$ at x such that

$$\lim_{h \downarrow +0} h^{-1} d\left(\phi(x) + h\phi'(x)v, \phi(U \cap S)\right) = 0,$$

where $d(\cdot, \cdot)$ is the distance on X.

A vector field V defined on a neighborhood Ω of S is said to be tangent to S if $V(x)$ is tangent to S, $\forall\, x \in S$.

Let us denote $T_x(S)$ as the set of all tangent vectors at $x \in S$. By definition, $T_x(S)$ is a nonempty closed cone, if $S \neq \emptyset$. In particular, if C is a convex set in a Banach space X, and

$$\widehat{T}_x(C) = \{v \in X \mid \exists\, \epsilon > 0 \quad \text{such that} \quad x + \epsilon v \in C\}, \tag{6.4}$$

then

$$\overline{\widehat{T}_x(C)} = T_x(C), \ \forall\, x \in C$$

Claim. $\widehat{T}_x(C) \subset T_x(C)$ is trivial, and since $T_x(C)$ is closed, $\overline{\widehat{T}_x(C)} \subset T_x(C)$. It remains to verify that $\forall\, v \in T_x(C)\ \exists\, \epsilon_n > 0$, $v_n \in X$ such that $x + \epsilon_n v_n \in C$, and $v_n \to v$.

By Definition 6.3, we have $y_n \in C$, $h_n \downarrow 0$ such that

$$h_n^{-1} \left\| x + h_n v - y_n \right\| \to 0.$$

Set $v_n = h_n^{-1}(y_n - x)$, and $\epsilon_n = h_n$, $n = 1, 2, \ldots$; these are what we need.

Lemma 6.3. *Assume that S is a locally convex closed set with respect to $(M, \mathcal{A})$. Then $T_x(S)$ is a closed convex cone in T_xM $\forall\, x \in S$.*

Claim. In fact, if (V, ϕ) is a chart in $\mathcal{A}$ at $x, \phi : U \to X$, then

$$T_x(S) = \phi'(x)^{-1} \cdot T_{\phi(x)}(\phi(U \cap S)).$$

According to (6.4), $T_{\phi(x)}(\phi(U \cap S))$ is convex, and since $\phi'(x)$ is linear, the convexity of the cone $T_x(S)$ follows from (6.4). Therefore $T_x(S)$ is convex. By Definition 6.3, if $v \in T_x(S)$, then $tv \in T_x(S)$, so that $T_x(S)$ is a cone.

Definition 6.4. Let S be a locally convex closed set with respect to $(M, \mathcal{A})$, and let $f \in C^1(\Omega, \mathbb{R}^1)$ where Ω is an open neighborhood of S. We say that $x_0 \in S$ is a critical point of f with respect to S if

$$\langle df(x_0), v\rangle \geq 0 \qquad \forall\, v \in T_{x_0}(S),$$

where $\langle\ ,\ \rangle$ is the duality between $T^*_{x_0}M$ and $T_{x_0}M$. Or, equivalently, we say that x_0 is critical with respect to S if

$$\left\langle df(x_0), \phi'(x_0)^{-1}\left(y - \phi\left(x_0\right)\right)\right\rangle \geq 0 \quad \forall\, y \in \phi(U \cap S).$$

Now, $\forall\, x^* \in T^*_{x_0}M$, let

$$|x^*|_{x_0} = \operatorname{Sup}\left\{\langle x^*, v\rangle \mid v \in T_{x_0}(S) \quad \text{with} \quad \|\phi'(x_0)v\|_X \leq 1\right\}.$$

Therefore:

$$\begin{aligned} &x_0 \in S \quad \text{is a critical point of} \quad f \quad \text{with respect to} \quad S \\ &\text{if and only if} \quad |-df\,(x_0)\,|_{x_0} = 0. \end{aligned}$$

The Palais-Smale condition $(\mathrm{PS})_c$ with respect to S is extended as follows:

$$\begin{aligned} &\text{Any sequence } \{x_n\} \subset S, \text{ along which} \\ &f\,(x_n) \to c \quad \text{and} \quad |-df\,(x_n)\,|_{x_n} \to 0, \\ &\text{implies a convergent subsequence.} \end{aligned}$$

Applying the same argument employed in Theorem 3.3, we have

Lemma 6.4. *Suppose that $f \in C^1(\Omega, \mathbb{R}^1)$ satisfies $(PS)_c$ with respect to a locally convex closed set S. Then the critical set $K_c = K \cap f^{-1}(c)$ with respect to S is compact, and for any closed neighborhood N (if $K_c = \emptyset$, then $N = \emptyset$) of K_c, $\exists$ constants $b, \epsilon > 0$ such that*

$$|-df(x)|_x \geq b, \quad \forall\, x \in S \cap f^{-1}[c - \epsilon, c + \epsilon].$$

Now we are in a position to establish both the critical point theory as well as the Morse theory with respect to S as in Sections 3 and 4. Since the idea is the same, it suffices to outline the main steps and the necessary modifications.

1. *The construction of a p.g.v.f.*

From now on, we assume that M is paracompact. First we point out that the function $x \mapsto |-df(x)|_x$ is continuous on S.

Claim. According to Lemmas 6.3 and (6.4),

$$|-df(x)|_x = \text{Sup}\,\{\langle -df(x), \phi'(x)^{-1}(y-\phi(x))\rangle \mid y \in \phi(U\cap S) \text{ with } \|y-\phi(x)\|_X \le 1\}.$$

The conclusion is obvious.

Second, $\forall\, x_0 \in S^* := S\backslash K$, where K is the critical set with respect to S, we choose $v_0 \in \widehat{T}_{x_0}(S) = \phi'(x_0)^{-1}\widehat{T}_{\phi(x_0)}(\phi(U\cap S))$ satisfying

(i) $\|\phi'(x_0)v_0\| \le 1$,
(ii) $\langle -df(x_0), v_0\rangle > \frac{2}{3}|-df(x_0)|_{x_0}$.

According to the above lemma, $\exists\, \delta = \delta(x_0)$ such that $B(x_0,\delta) \subset U$, and $\forall\, x \in B(x_0,\delta)\cap S$ the vector field

$$v_0(x) = \phi'(x)^{-1}\left[\phi(x_0) - \phi(x) + \phi'(x_0)v_0\right] \in T_x(S)$$

satisfies

$$\|\phi'(x)v_0(x)\| < \frac{4}{3},$$

and

$$\langle -df(x), v_0(x)\rangle > \frac{2}{3}|-df(x)|_x.$$

We find a locally finite partition of unity $\{\beta_\ell \mid \ell \in \Lambda\}$ associated with a refinement covering $\mathcal{B}$ of the covering $\{B(x_0,\delta(x_0)) \mid x_0 \in S^*\}$. Set

$$W(x) = \sum \beta_\ell(x)V_\ell(x)$$

where $V_\ell(x)$ is the v.f. defined on $B(x_\ell,\delta(x_\ell))$, which includes the support of $\beta_\ell(x)$. Set

$$V(x) = \frac{3}{2}|-df(x)|_x W(x).$$

We obtain

(1) $V(x) \in T_x(S)$,
(2) $\|\phi'(x)V(x)\| < 2|-df(x)|_x$,
(3) $\langle -df(x), V(x)\rangle > |-df(x)|_x^2$,

$\forall\, x \in S^*$.

2. *ODE on closed sets.*

We define deformations by the semiflow for the p.g.v.f. on a closed set S. The question is under what conditions does the semiflow remain in the set S?

Definition 6.5. A subset $S \subset M$ is said to be a (locally) semi-invariant set with respect to a vector field V if the semiflow $x(t)$ remains in S for all (small) $t \ge 0$:

$$\begin{cases} x' = V(x(t)) \\ x(0) = x \in S. \end{cases}$$

The following local existence and uniqueness theorem for semiflow on a closed set was obtained by Brézis [Bre1] and Martin, cf. [Mar1].

Theorem 6.3 (Brézis-Martin). *Let A be an open subset of X, and let $B \subset A$ be closed in A. If $V : A \to X$ is a locally Lipschitzian mapping, then $\forall\, x \in B$, $\exists\, \delta > 0$ and $\tau(t)$ satisfying*

$$\begin{cases} \dot{\tau}(t) = V(\tau(t)) & \forall\, t \in [0;\delta) \\ \tau(0) = x \in B; \quad \tau(t) \in B \end{cases}$$

if and only if

$$\lim_{h \downarrow 0} h^{-1} d(x + hV(x), B) = 0$$

i.e., $V(x) \in T_x(B) \quad \forall\, x \in B$.

Lemma 6.5. *Let S be a closed subset of M. Then S is a semi-invariant set with respect to a locally Lipschitzian v.f. V on M, if and only if*

$$\lim_{h \downarrow 0} h^{-1} d\left(\phi(x) + h\phi'(x) \cdot V(x); \phi(U \cap S)\right) = 0 \tag{6.5}$$

$\forall\, x \in S$ $\forall$ chart (ϕ, U).

Proof. Assume that (6.5) holds. The following ODE in X

$$\begin{cases} \dot{y}(t) = \phi'(x)V(x)|_{x=\phi^{-1}(y(t))} \\ y(0) = \phi(x) \end{cases}$$

has a local solution y_ϕ. Since $\phi(U \cap S)$ is a closed set in $\phi(U)$ and $(\phi' \cdot V) \circ \phi^{-1}$ is tangent to $\phi(U \cap S)$, by the Brézis-Martin Theorem, $(U \cap S)$ is locally invariant with respect to $\phi' \circ V$. Setting $x_\phi(t) = \phi^{-1}(y_\phi(t))$, it follows that $x_\phi(t) \in U \cap S$.

It is not difficult to verify that x_ϕ does not depend on the special choice of ϕ in a neighborhood of $t = 0$, and it is easy to extend the solution to a maximal solution which remains in S via the method of continuation.

Conversely, if S is a semi-invariant set with respect to V, then for any chart (ϕ, U), $\phi(U \cap S)$ is semi-invariant with respect to $(\phi' \cdot V) \circ \phi^{-1}$. (6.5) follows directly from the Brézis-Martin Theorem.

3. *Deformation theorems and critical groups*

With the above preparatory work, the first and second deformation theorems do hold for any locally convex closed subset S of a C^2 paracompact Banach manifold.

The proofs are just the same as in Section 3 and hence are omitted. Similarly, the critical groups for isolated critical points with respect to S are well-defined.

4. *Morse relations for functions with isolated critical points*

Let S be a locally convex closed subset S of a C^2 paracompact Banach manifold. Let $f \in C^1(\Omega, \mathbb{R}^1)$, where Ω is a neighborhood of S. Assume that f satisfies the (PS) condition with respect to S and has only finitely many critical points $\{x_1, x_2, \dots, x_n\}$ with respect to S in $f^{-1}[a,b] \cap S$, with critical value $\{c_1, c_2, \dots, c_m\}$ satisfying $a < c_j < b$, $j = 1, 2, \dots, m$.

Relative to any coefficient field for homology, set

$$M_q = M_q(f; [a,b]) = \sum_{j=1}^{m} \text{ rank } C_q\,(f, x_j)\,,$$

and

$$R_q = R_q(f; [a,b]) = \text{ rank } H_q\,(f_r, f_a)\,, \quad q = 0, 1, 2, \dots .$$

We have

$$\sum_{q=0}^{\infty} M_q t^q = \sum_{q=0}^{\infty} R_q t^q + (1+t)Q(t),$$

where $Q(t)$ is a formal series with nonnegative coefficients.

Remark 6.2. Readers might be puzzled about the different conditions in the two subsections. The functions in the first subsection are under strong boundary conditions, but in the second subsection there are no restrictions. The underlying spaces in the first are manifolds with smooth boundaries, but in the second are locally convex closed sets. After all, we have the same conclusion, i.e., the Morse inequalities.

The point is that the meanings of the critical points in Sections 6.1 and 6.2 are different. In Section 6.1 they correspond to the variational equation, i.e., the Euler-Lagrange equation, while in Section 6.2 they correspond to variational inequality.

Remark 6.3. Morse theory on convex sets was initially studied by M. Struwe [Str1] and K. C. Chang, J. Eells [ChE1] in the Plateau problem. It was developed in K. C. Chang [Cha7] to study the variational inequality problems. The introduction of local convexity in critical point theory first appeared in T. Q. Wang [WaT1].

7. Equivariant Morse Theory

Let us assume a compact Lie group G and a smooth manifold on which the group G acts. Equivariant Morse theory studies the Morse relations and the Morse handle body theorem for G-invariant functions. Noticing that for a G-invariant function, if x is a critical point, then the points on the G orbit containing x are also critical points. It is hard to say regarding the isolatedness of critical points if G is continuous. In this section, we shall introduce the notions of isolatedness and nondegeneracy for critical orbits.

7.1. Preliminaries

Let G be a compact Lie group. A G space (or manifold) M is a topological space (or manifold resp.) with a continuous G action, i.e., $\psi : G \times M \to M$ with $\psi(g,x)$, written as $g \cdot x$, such that $e \cdot x = x$, and $(g_1 \cdot g_2) \cdot x = g_1 \cdot (g_2 \cdot x)\ \forall\, g_1, g_2 \in G$, $\forall\, x \in M$.

$\forall\, x \in M$, we call $\mathcal{O}(x) = \{g \cdot x \mid g \in G\}$ a G orbit. The set of all G orbits is called the orbit space. Endowed with the quotient topology, it is denoted by M/G or simply $\overline{M}$. The subgroup of G, defined by $G_x = \{g \in G \mid g \cdot x = x\}$, is called the isotropy subgroup at x.

If x has the isotropy subgroup H, then $g \cdot x$ has the isotropy subgroup $g \cdot H \cdot g^{-1}$. Thus a conjugacy class of isotropy subgroups is attached to each orbit.

G is called a free action if $G_x = e\ \forall\, x \in M$.

Let H be a subgroup of G, and M be a G space, we denote

$$M^H = \{x \in M \mid h \cdot x = x \quad \forall\, h \in H\},$$

i.e., the set of points fixed under H.

M^G is called the fixed point set of G, which is also denoted by Fix_G.

A set $A \subset M$ is called G-invariant, if $g \cdot x \in A$, $\forall\, x \in A$, $\forall\, g \in G$.

A G pair (X,Y) is a pair of G-invariant spaces (X,Y) with $Y \subset X$.

A function $f : M \to \mathbb{R}^1$ is called G-invariant, if $f(g \cdot x) = f(x); \forall\, x \in M, \forall\, g \in G$.

A map $F : (X,Y) \to (X',Y')$ between two G pairs is called G equivariant, if

$$F(g \cdot x) = g \cdot F(x) \quad \forall\, x \in X, \quad \forall\, g \in G.$$

Thus, a G equivariant map F induces a map

$$\overline{F} = F/G : (\overline{X},\overline{Y}) \to \left(\overline{X}',\overline{Y}'\right).$$

Let $\pi : E \to B$ be a fiber bundle. (E,π,B) is called a G bundle, if $\forall\, g \in G$, $g : E \to E$ is a differentiable bundle map, such that $gE_x = E_{g\cdot x}\ \forall\, x \in B$.

Thus, if M is a G-manifold, the tangent bundle TM is a G bundle, with $g \cdot X = d_x\psi(g,x) \cdot (X)$, $\forall\, X \in T_x(M)$, $\forall\, x \in M$.

(E,π,B) is called a Riemannian G vector bundle, if the G vector bundle possesses a Riemannian metric, and the G-action is isometric.

In the following, we always assume that M is a Hilbert Riemannian manifold, with a Riemann metric on TM. Let $\Sigma \subset M$ be a compact connected submanifold; then $T\Sigma$, the tangent bundle of Σ, is a subbundle of TM, and then the normal bundle $N\Sigma$, which is the orthogonal complement to $T\Sigma$, is also a subbundle of TM.

If in addition M is a G manifold and Σ in G invariant, then both $T\Sigma$ and $N\Sigma$ are all G-bundles.

Let $f \in C^1(M, \mathbb{R}^1)$ be a G invariant. It gives rise to a G-equivariant gradient vector field df

$$\langle df(g \cdot x), g \cdot X \rangle = \langle df(x), X \rangle$$

$\forall\ (g, x) \in G \times M$, $\forall\ X \in T_x(M)$, i.e.,

$$g^* \cdot df \cdot g = df.$$

Since the action g on $T_x(M)$ is unitary, $\mathrm{d}g^* = g^{-1}$, we obtain

$$df \cdot g = g \cdot df.$$

Analogously, the Hessian $d^2 f$ is also G-equivariant, if $f \in C^2$

It is obviously seen that the level sets f_c, $f^{-1}(c)$, and the critical sets K, $K_c = K \cap f^{-1}(c)$ are all G-invariant. And a critical orbit $\mathcal{O} = \mathcal{O}(x)$ is a G submanifold of M.

It follows that $T_x(\mathcal{O}) \subseteq \ker\ d^2 f(x)$ and that the induced bounded self-adjoint operator $d^2 f(x) : N_x(\mathcal{O}) \to N_x(\mathcal{O})$, satisfies

$$g^* \cdot d^2 f(g \cdot x) \cdot g = d^2 f(x).$$

7.2. Equivariant Deformation

For a G-invariant function, we shall improve the first and second deformation theorems to make the deformations equivariant. Namely, we shall prove

Theorem 7.1. (First G-equivariant deformation theorem) *Let M be a $C^2 G$-Hilbert Riemannian manifold. Suppose that $f \in C^1(M, \mathbb{R}^1)$ is G-invariant and satisfies the $(PS)_c$ condition. Assume that N is a G-invariant closed neighborhood of K_c for some $c \in \mathbb{R}^1$. Then there exist constants $\bar{\epsilon} > \epsilon > 0$ and a G-equivariant continuous map $\eta : [0,1] \times M \to M$ such that*

(1) $\eta(t, \cdot)\ |_{Cf^{-1}[c-\bar{\epsilon}, c+\bar{\epsilon}]} = \mathrm{id}\ |_{Cf^{-1}[c-\bar{\epsilon}, c+\bar{\epsilon}]}$
(2) $\eta(0, \cdot) = \mathrm{id}$,
(3) $\eta(1, f_{c+\epsilon} \backslash N) \subset f_{c-\epsilon}$
(4) $\forall\ t \in [0,1]$, $\eta(t, \cdot) : M \to M$ *is a G-homeomorphism,*
(5) $t \mapsto f \circ \eta(t, x)$ *is nonincreasing* $\forall\ t$.

Theorem 7.2. (Second G-equivariant deformation theorem). *Let M be a C^2G-Hilbert Riemannian manifold. Suppose that $f \in C^1(M, \mathbb{R}^1)$ is G invariant and satisfies the (PS)$_c$ condition, $\forall\, c \in [a,b]$. Assume that a is the only critical value of f in $[a,b)$, and that any connected component of K_a is always a part of a certain critical orbit. Then f_a is a G-equivariant strong deformation retract of $f_b \backslash K_b$.*

The proofs of these theorems are just modifications of those given in Section 3. A new ingredient is to construct a G-equivariant locally Lipschitzian p.g.v.f. We carry this out as follows:

Lemma 7.1. *Assume that $f \in C^1(M, \mathbb{R}^1)$ is G-invariant. Then there exists a $C^{1-0}G$-equivariant p.g.v.f.*

Proof. We have already a C^{1-0} p.g.v.f. $X(x)$ from Theorem 3.1. Let us define

$$\widehat{X}(x) = \int_G g^{-1} \cdot X(g \cdot x)\, d\mu,$$

where $d\mu$ is the right invariant Haar measure on G. Since

$$\begin{aligned}\widehat{X}\,(g' \cdot x) &= \int_G g^{-1} X\,(g \cdot g' \cdot x)\, d\mu \\ &= g' \int_G (g \cdot g')^{-1}\, X\,(g \cdot g' \cdot x)\, d\mu \\ &= g' \widehat{X}(x)\ \forall\, g' \in G,\end{aligned}$$

$\widehat{X}$ is G equivariant.

We shall prove $\widehat{X} \in C^{1-0}$. It is only required to prove that $\forall$ orbits $G(x_0)$, $\exists$ a neighborhood U, such that X is uniformly Lipschitzian on it. Actually, $\forall\, x \in G(x_0)\ \exists$ an open ball $B(x, \delta_x)$ and a constant $C_x > 0$ such that

$$\|X(y) - X(z)\| \le C_x d(y,z) \quad \forall\, y, z \in B(x, \delta_x)$$

where $\|\cdot\|$ is the Finsler structure on TM. Since $G(x_0)$ is compact we find a finite covering $U = \cup_{i=1}^n B(x_i, \delta_{x_i}) \supset G(x_0)$.

Let

$$\begin{aligned}\delta = &\ \text{Min}\ \left\{\delta_{x_i}, 1 \le i \le n\right\}, \quad \text{and} \\ C = &\ \text{Max}\ \left\{C_{x_i}, \frac{4}{\delta} \sup_{y \in U} \|X(y)\|, 1 \le i \le n\right\}.\end{aligned}$$

We conclude

$$\|X(y) - X(z)\| \le C d(y,z) \quad \forall\, y, z \in U.$$

In fact, if $d(y,z) \geq \delta/2$, then

$$\begin{aligned}\|X(y) - X(z)\| &\leq \|X(y)\| + \|X(z)\| \\ &\leq 2 \sup_{y \in U} \|X(y)\| \\ &\leq \frac{\delta}{2} \cdot C \leq Cd(y,z).\end{aligned}$$

Otherwise, $d(y,z) < \delta/2$, y and z must fall into a certain neighborhood $B(x_i, \delta_{x_i})$, and hence

$$\|X(y) - X(z)\| \leq C_{x_i} d(y,z) \leq Cd(y,z).$$

Finally, we prove that $\widehat{X}$ is again a p.g.v.f. Since g is isometric

$$\begin{aligned}\|\widehat{X}(x)\| &\leq \int_G \|X(g \cdot x)\| \, d\mu \\ &\leq 2 \int_G \|df(g \cdot x)\| \, d\mu \\ &= 2 \int_G \|g \cdot df(x)\| \, d\mu \\ &= 2\|df(x)\|,\end{aligned}$$

and

$$\begin{aligned}\langle df(x), \widehat{X}(x)\rangle &= \int_G \langle df(x), g^{-1} X(g \cdot x)\rangle \, d\mu \\ &= \int_G \langle df(g \cdot x), X(g \cdot x)\rangle \, d\mu \\ &\geq \int_G \|df(g \cdot x)\|^2 \, d\mu \\ &= \|df(x)\|^2.\end{aligned}$$

By virtue of Lemma 7.1, the flow defined by the G-equivariant p.g.v.f. $\widehat{X}$ must be G-equivariant. (The uniqueness of the solution of ODE). The rest of the proofs of Theorems 7.1 and 7.2 is exactly the same as in Section 3.

7.3. The Splitting Theorem and Handle Body Theorem for Critical Manifolds

Definition 7.1. A connected submanifold $\mathcal{O} \subset M$ is called a critical manifold, if $df(x) = \theta \; \forall \, x \in \mathcal{O}$ (may not have group action) and $f|_{\mathcal{O}} = \text{const.}$ A critical manifold $\mathcal{O} \subset M$ is called isolated if there is a neighborhood U of $\mathcal{O}$ such that $U \cap K = \mathcal{O}$.

In the following, we always assume that $\mathcal{O}$ is compact.

Now we shall study the local behavior of the function f near its isolated critical manifold. The restriction of the tangent bundle TM on $\mathcal{O}$ now is split into a direct sum $T\mathcal{O} \oplus N\mathcal{O}$, the tangent bundle and the normal bundle of the manifold $\mathcal{O}$. The Riemannian metric on TM reduces a metric on the normal bundle $N\mathcal{O}$.

The exponential map exp, regarding the geodesic sprays on M, yields a diffeomorphism from the normal disk bundle $N\mathcal{O}(\epsilon) = \{(x, v) \in N\mathcal{O} | \|\sigma\| \leq \epsilon\}$ to the tubular neighborhood $N(\epsilon) = \{x \in M | \operatorname{dist}(x, \mathcal{O}) \leq \epsilon\}$ of $\mathcal{O}$, for some $\epsilon > 0$ because $\mathcal{O}$ is compact. In this sense, we shall not distinguish $N\mathcal{O}(\epsilon)$ from $N(\epsilon)$.

In order to split the normal bundle $N\mathcal{O}$ into mutually orthogonal bundles, we need

Lemma 7.2. *Let H be a Hilbert space and Q, R be two orthogonal projections. If $\|Q - R\| < 1$, then $QR : RH \to QH$ is an isomorphism.*

Proof. QR is an injection. Otherwise, $\exists\ v \in RH \backslash \{\theta\}$ with $Qv = \theta$. Therefore $\|(Q - R)v\| = \|v\|$, which contradicts $\|Q - R\| < 1$. QR is also a surjection. Otherwise, $\exists\ v \in QH \cap (RH)^{\perp} \backslash \{\theta\}$, then $(v, QRw) = (Rv, w) = 0\ \forall\ w \in H$, which implies $Rv = \theta$. Again, we get $\|Q - R)v\| = \|v\|$; this is impossible.

Lemma 7.3. *Let $\xi = (E, \pi, N)$ be a Hilbert vector bundle on a connected manifold N, with a Riemannian metric $\langle\ ,\ \rangle$. Let P be an orthogonal bundle projection with respect to $\langle\ ,\ \rangle$. Let I be a section satisfying $I_x =$ id in $\mathcal{L}(E_x, E_x)\ \forall\ x \in N$. Assume $Q = I - P$. Then $P\xi = (PE, \pi|_{PE}, N)$ and $Q\xi = (QE, \pi|_{QE}, N)$ are Hilbert vector bundles with*

$$\xi = P\xi \oplus Q\xi.$$

Proof. It suffices to prove that $\forall\ x \in N\ \exists$ a bundle chart $\psi : U \times H \to \pi^{-1}(U)$, $x \in U$, and an orthogonal projection $\widetilde{P} \in \mathcal{L}(H, H)$ such that $\forall\ y \in U$, $\psi_y : H \to E_y$, $\psi_y(v) = \psi(y, v)$ maps $\widetilde{P}H$, $\widetilde{Q}H$ isometrically onto PE_y and QE_y resp. where H is a Hilbert space, and $\widetilde{Q} = I - \widetilde{P}$.

From the definition, we have a bundle chart $\phi : U \times H \to \pi^{-1}(U)$, $x \in U$, and then define

$$P_y = \phi_y^{-1} P \phi_y \quad \forall\ y \in U.$$

For small U, $\|P_y - P_x\| < 1\ \forall\ y \in U$. By Lemma 7.2

$$P_x P_y : P_y H \to P_x H$$

is an isomorphism. Similarly, set $Q_y = \phi_y^{-1} Q \phi_y$, we have

$$Q_x Q_y : Q_y H \to Q_x H \text{ is an isomorphism.}$$

$$\begin{array}{ccccc} E_x & \xleftarrow{\phi_x} & H & \xrightarrow{\phi_y} & E_y \\ P\downarrow & & P_x \downarrow P_y & & \downarrow P \\ E_x & \overset{\longleftarrow}{\underset{\phi_x^{-1}}{\dashrightarrow}} & H & \overset{\longrightarrow}{\underset{\phi_y^{-1}}{\dashleftarrow}} & E_y \end{array}$$

Thus the map

$$B_y = P_x P_y + Q_x Q_y$$

is invertible in $\mathcal{L}(H,H) \cap \mathcal{L}(P_y H, P_x H) \cap \mathcal{L}(Q_y H, Q_x H)$.

We define $A_y \in \mathcal{L}(H,H)$ by

$$\langle \phi_y v, \phi_y w \rangle_{E_y} = (A_y B_y v, B_y w)_H \quad \forall\, v, w \in H.$$

Then $y \mapsto A_y$ is C^1 and positive definite.

Finally, set $\psi_y = \phi_y \circ B_y^{-1} A_y^{-\frac{1}{2}}$, and $\widetilde{P} = P_x$. The maps $\psi(y,v) = \psi_y(v)$, and $\widetilde{P}$ are just what we need.

Now we extend the splitting theorem to isolated critical manifolds. First, we need

Assumption 7.1. $f \in C^2(M, \mathbb{R}^1)$, $\mathcal{O}$ is an isolated critical manifold, and there exists $\epsilon > 0$ such that

$$\sigma\left(d^2 f(x)\right) \cap ([-\epsilon, \epsilon] \backslash \{0\}) = \emptyset, \quad \forall\, x \in \mathcal{O}. \tag{7.1}$$

Furthermore, $\dim \ker(d^2 f(x)) = \text{const.}\ \forall\, x \in \mathcal{O}$.

Noticing that $d^2 f(\cdot)$ is a well-defined quadratic form on the normal bundle $N\mathcal{O}$, let χ_* $(* = +, -, 0)$ be the characteristic functions of the intervals $[\epsilon, +\infty)$, $(-\epsilon, \epsilon)$ and $(-\infty, -\epsilon]$ resp. Then the orthogonal bundle projections

$$P_* = \chi_*\left(d^2 f(\cdot)\right) \quad * = +, -, 0 \tag{7.2}$$

are well-defined on the normal bundle $N\mathcal{O}$.

Lemma 7.4. *Under Assumption (7.1),*

$$\xi = \xi^+ \oplus \xi^- \oplus \xi^0 \tag{7.3}$$

where $\xi = (N\mathcal{O}, \pi, \mathcal{O})$, $\quad \xi^* = \left(P_* N\mathcal{O}, \pi|_{P_* N\mathcal{O}}, \mathcal{O}\right)$, $* = +, -, 0$.

This is a consequence of Lemma 7.3.

Theorem 7.3. *Under Assumption* (7.1), *there exist* $\epsilon > 0$, *a local homeomorphism* $\Phi : N\mathcal{O}(\epsilon) \to N(\epsilon)$, *and a* C^1 *map* $h : \xi^0(\epsilon) \to \xi^+ \oplus \xi^-$ *such that*

$$f \circ \Phi(p) = \frac{1}{2}(\|P_+(x)v\|^2 - \|P_-(x)v\|^2) + f \circ \exp(P_0(x)v + h(P_0(x)v))$$

$\forall\, p = (x, v) \in N\mathcal{O}(\epsilon)$, $x = \pi p$, $v \in N\mathcal{O}_x(\epsilon)$, *where* $\xi^0(\epsilon) = \{(x, v) \in \xi^0 |\ x = \pi x,\ v \in \xi^0_x,$ *with* $\|v\| \le \epsilon\}$. P_* *and* ξ^* *are defined in* (7.2) *and* (7.3) *resp.*

Proof. $\forall\ x \in \mathcal{O}$, the function $f \circ \exp|_{\xi_x}$ possesses θ as an isolated critical point. The splitting theorem (5.1) is exploited to obtain a local homeomorphism $\phi_x : \xi_x(\epsilon) \to \xi_x(\epsilon)$, and a C^1 map $h_x : \xi^0_x(\epsilon) \to \xi^+_x \oplus \xi^-_x$, such that

$$f \circ \exp_x \phi_x(v) = \frac{1}{2}\left(\|P_+(x)v\|^2 - \|P_-(x)v\|^2\right) + f \circ \exp_x\left(P_0(x)v + h_x\left(P_0(x)v\right)\right)$$

$\forall\, v \in \xi_x(\epsilon)$.

By a standard method, we verify that Φ and h can be defined by $\Phi_x = \exp_x \phi_x$, and $h|_{\xi^0_x(\epsilon)} = h_x$ respectively.

Corollary 7.1. *Assume that* $f \in C^2(M, \mathbb{R}^1)$ *is* G*-invariant, and that* $\mathcal{O}$ *is an isolated critical orbit of* f. *Suppose that* $\exists\ x_0 \in \mathcal{O}$, *such that* $\mathcal{O}$ *is either an isolated point in* $\sigma(A)$ *or not in* $\sigma(A)$, *where* $A = d^2 f(x_0)|_{\xi_{x_0}}$. *Then the conclusion of Theorem 7.3 holds. In addition,* Φ *and* h *are all* G*-equivariant.*

Proof. Since

$$d^2 f(g \cdot x) = g^{-1} d^2 f(x) g \quad \forall\, x \in M,$$

the spectrum $\sigma(d^2 f(x)|_{\xi_x})$ is G-invariant $\forall\ x \in \mathcal{O}$, so Assumption (7.1) holds. It remains to verify G equivariance. Since $z = h_x(y)$ solves the equation uniquely:

$$d_z f \circ \exp_x(y + z) = \theta,$$

we conclude that $gh_x(y) = h_{g\cdot x}(gy)$, provided by the equivariance of df.

Similarly, the map η_x solves the equation uniquely:

$$\frac{1}{2}\left(A(x)\eta_x(u, y), \eta_x(u, y)\right) = f \circ \exp_x\left(u + h_x(y) + y\right) - f \circ \exp_x\left(h_x(y) + y\right),$$

where $y \in \xi^0_x(\epsilon)$, $u \in (\xi^+_x \oplus \xi^-_x)(\epsilon)$, $A(x) = d^2 f(x)$, and $\eta_x(u, y) \in (\xi^+_x \oplus \xi^-_x)$. The map ϕ_x is just the inverse of the map $(u, y) \mapsto (\eta_x(u, y), y)$. Since

$$A(g \cdot x) = g^{-1} A(x) g,$$

we obtain

$$\eta_{g\cdot x}(gu, gy) = g\eta_x(u, y).$$

Thus

$$\Phi_{g\cdot x}(gv) = g\Phi_x(v) \quad \forall\, x \in \mathcal{O}, \quad \forall\, v \in \xi_x(\epsilon).$$

Definition 7.2. Under Assumption (7.1), the dimension of the fiber of the vector bundle ξ^- is called the Morse index of f with respect to the isolated critical manifold $\mathcal{O}$, and is denoted by ind $(f, \mathcal{O})$.

$\mathcal{O}$ is called nondegenerate if the dimension of the fiber of the vector bundle ξ^0 is zero, i.e., $\xi^0 = \mathcal{O}$. f is called nondegenerate if it has only nondegenerate critical manifolds.

Definition 7.3. Let $\xi = (E, \pi, \mathcal{O})$ be a vector bundle over a compact connected manifold $\mathcal{O}$. The disk bundle

$$\xi(r) = \{(x, v) \in E \mid v \in E_x,\ x = \pi x \quad \text{with} \quad \|v\| \le r\}, r > 0,$$

is called a handle bundle.

Theorem 7.4. *Suppose that $f \in C^2(M, \mathbb{R}^1)$ satisfies the (PS) condition, and that c is a critical value with $K_c = \{\mathcal{O}_1, \mathcal{O}_2, \ldots, \mathcal{O}_\ell\}$ consisting of nondegenerate critical manifolds. Then there exist $\epsilon > 0$ and homeomorphisms $h_j : \xi_j^-(\epsilon) \to M$, $j = 1, 2, \ldots, \ell$, such that*

$$f_{c+\epsilon} \simeq f_{c-\epsilon} \cup \bigcup_{j=1}^{\ell} h_j\left(\xi_j^-(\epsilon)\right),$$

with

$$f_{c-\epsilon} \cap h_j\left(\xi_j^-(\epsilon)\right) = f^{-1}(c-\epsilon) \cap h_j\left(\xi_j^-(\epsilon)\right) = h_j\left(\partial\xi_j^-(\epsilon)\right),$$

where $\partial\xi_j^-(\epsilon)$ is the $m_j - 1$ spherical bundle, $m_j =$ ind $(f, \mathcal{O}_j)$, ξ_j^- is the associated negative subbundle of $N\mathcal{O}_j$, $j = 1, 2, \ldots, \ell$.

The proof is the same as Theorem 4.4, because we have already established the generalized Morse lemma (Theorem 7.3). Actually, the deformation is constructed along the fibers.

Corollary 7.2. *Suppose that the function f in Theorem 7.4 is G-invariant, and $\mathcal{O}_j$, $j = 1, 2, \ldots, \ell$, are nondegenerate critical orbits. Then the homeomorphisms $h_j : \xi_j^-(\epsilon) \to M$, $j = 1, 2, \ldots, \ell$, are G-equivariant, and the deformation is also G-equivariant.*

In the proof of the corollary, G-equivariance follows from the construction along fibers and Theorems 7.1 and 7.2.

The Gromoll-Meyer pair (W, W_-) for an isolated critical manifold $\mathcal{O}$ is naturally extended. (Only $W \cap K = \{p\}$ is changed to be $W \cap K = \mathcal{O}$.) The construction of such a pair (W, W_-) for isolated critical manifold is similar to the construction in (5.11) and (5.12). We only want to replace the ball B_δ by the tubular neighborhood $N(\delta)$ of $\mathcal{O}$, and to replace the function g (defined in Section 5) by

$$g(x) = \lambda f(x) + \text{ dist } (x, \mathcal{O})^2.$$

In particular, if f is G-invariant, $\mathcal{O}$ is an isolated critical orbit; then the Gromoll-Meyer pair (W, W_-) may be chosen G-invariant, and the flow in the definition is G-equivariant.

7.4. G Cohomology and G Critical Groups

The extension of nondegeneracy of critical manifolds has the following advantage:

Suppose that $E \xrightarrow{\pi} M$ is a fibring, and f is a nondegenerate function on M in the sense of Definition 7.2. Then it is easy to see that the pull back $\pi^* f$ on E is again nondegenerate. Further, the index of $\mathcal{O}$ as a nondegenerate critical manifold of M equals the index of $\pi^{-1}\mathcal{O}$ as a critical manifold of E, i.e., ind $(f, \mathcal{O}) =$ ind $(\pi^* f, \pi^{-1}\mathcal{O})$.

Let us consider a compact Lie group action G. If the G action is free, then the orbit space $\overline{M} = M/G$ is also a manifold, the projection

$$\pi : M \to M/G$$

is a smooth fibration, with fiber G, and there is not any difficulty in carrying out the Morse theory on the orbit space $\overline{M}$.

However, if the action is not free, then $\overline{M}$ possesses singularities, and one cannot do the same as above.

Consider any smooth principal G bundle E over a base manifold B, and the following diagram:

$$\begin{array}{ccccc} E & \xleftarrow{\pi_2} & E \times M & \xrightarrow{\pi_1} & M \\ \downarrow{\scriptstyle\pi} & & \downarrow{\scriptstyle\pi} & & \downarrow{\scriptstyle\pi} \\ E/G & \underset{q}{\longleftarrow} & E \times_G M & \underset{\eta}{\longrightarrow} & M/G \end{array}$$

of the G-actions on M and E, and G operates diagonally on $E \times M$, i.e., $g(\sigma, x) = (g\sigma, gx)\ \forall\ (\sigma, x) \in E \times M$, where $E \times_G M = (E \times M)/G$.

Since the action on E is free, this diagonal action is also free. On the other hand, a G-invariant f on M clearly lifts to a G-invariant f on $E \times M$, and hence to a smooth function f_E on $E \times_G M$. Now $E \times_G M$ is itself a fiber space over the base $B = E/G$ with fiber M and structure group G. Noticing that

$$\pi^{-1}(E \times_G \mathcal{O}) = \pi_1^{-1}(\mathcal{O})$$

for any critical manifold $\mathcal{O}$ of f, $E \times_G \mathcal{O}$ is the corresponding critical manifold of f_E. We have

$$\operatorname{ind}(f, \mathcal{O}) = \operatorname{ind}(f_E, E \times_G \mathcal{O}).$$

There are many principal G-bundles E we may choose; among them we single out a universal G-bundle that is unique up to homotopy which possesses the following important property: The total space E is contractible. Such bundles always exist cf. Housmoller [Hou1]. We are satisfied ourselves to give the following few examples:

$G =$	$E = E_G =$	$B = B_G =$		
$\mathbb{Z}_2$	S^∞	$\mathbb{R}P^\infty$	real	
S^1	S^∞	$\mathbb{C}P^\infty$	complex	projective
$SU(2)$	S^∞	$\mathbb{H}P^\infty$	quaternion	space

In the universal case, we shall write $M_G = E \times_G M$, $B_G = E/G$, $E_G = E$, and $f_G = f_E$. M_G is called the homotopy quotient of M by G.

The advantage of this choice is that the map

$$\eta : M_G \to M/G$$

is a homotopy equivalence, if G acts freely on M. In summary, we consider the function f_G on M_G, the homotopy quotient, which is G-free, instead of f on M/G. However, there is an (equivalently) alternative way to consider the problem, that is the concepts of G cohomology.

Definition 7.4. Given a G pair (X, Y) and a coefficient field $\mathbb{K}$, let

$$H_G^*(X, Y; \mathbb{K}) = H^*(X_G, Y_G, \mathbb{K}),$$

where X_G and Y_G are the homotopy quotients of X and Y by the group G. We call H_G^* the G cohomology.

It was proved by A. Borel that the G cohomology enjoys most of the properties of the cohomology. More precisely, the exactness, the homotopy, and the excision axioms hold, but not the dimension axiom:

$$H_G^*(pt) = H^*(B_G).$$

Furthermore if $F : (X, Y) \to (X', Y')$ is a G-equivariant map, then $F \times 1 : (X, Y) \times E_G \to (X', Y') \times E_G$ induces a G-equivariant map on the homotopy quotient:

$$F_G : (X_G, Y_G) \to (X'_G, Y'_G),$$

and hence a homomorphism:

$$F_G^* : H_G^*(X', Y') \to H_G^*(X, Y)$$

cf. T. Tom Dieck [Die1].

Now we are able to define the G-critical groups, and present some computations. At the end, the Morse inequalities are established.

Definition 7.5. Suppose that U is a G-invariant neighborhood of an isolated critical orbit $\mathcal{O}$, such that $K \cap (f_c \cap U) = \mathcal{O}$, where $c = f(\mathcal{O})$. Then for any coefficient ring $\mathbb{K}$,

$$C_G^*(f, \mathcal{O}) = H_G^*(f_c \cap U, (f_c, \backslash \mathcal{O}) \cap U, \mathbb{K})$$

is called the qth G critical group of $\mathcal{O}$, $q = 0, 1, 2, \ldots$.

And by the same proof as Theorem 5.2, we have

$$C_G^*(f, \mathcal{O}) = H_G^*(W, W_-; \mathbb{K}) \tag{7.4}$$

where (W, W_-) is a G-Gromoll-Meyer pair of $\mathcal{O}$.

Example 1. Suppose that $\mathcal{O}$ is an isolated critical orbit, corresponding to a local minimum of a G-invariant function f. Then

$$C_G^*(f, \mathcal{O}) = H_G^*(\mathcal{O}) = H^*(\mathcal{O}_G). \tag{7.5}$$

Example 2. Suppose that the normal bundle $\xi = N\mathcal{O}$ is a trivial bundle. Then we have the following formula:

$$C_G^q(f, \mathcal{O}) = \bigoplus_{j=0}^{q} \widetilde{C}_G^{q-j}(f, \mathcal{O}) \otimes H_G^j(\mathcal{O}) \tag{7.6}$$

where $\widetilde{C}_G^i(f, \mathcal{O}) = C_G^i(f \circ \exp) |_{\xi_x}, \theta)$, $i = 0, 1, \ldots$, $\forall\, x \in \mathcal{O}$.

Claim. We shall prove that $\forall\, x \in \mathcal{O}$, the function $f \circ \exp\ |_{\xi_x}$ possesses an isolated critical point θ. In fact, if $v \in \xi_x(\epsilon)$ is a critical point of $f \circ \exp\ |_{\xi_x}$, then the normal derivative at v must be zero. On the other hand, $f \circ \exp$ is G invariant, and the tangent derivative is also zero. It follows that $d(f \circ \exp)(v) = \theta$. However, we have assumed that $\mathcal{O}$ is isolated, therefore $v = \theta$.

Denote $(W, W_-)_x = (W, W_-) \cap \exp \xi_x(\epsilon)\ \forall\, x \in \mathcal{O}$, for a suitable $\epsilon > 0$ such that $\exp^{-1}(W, W_-) \subset \xi(\epsilon)$. Obviously, $\exp^{-1}(W, W_-)_x$ is a GM pair for the restriction$f \circ \exp\ |_{\xi_x}$. Noticing that

$$(W, W_-)_{g \cdot x} = g(W, W_-)_x$$

and g is a diffeomorphism, we obtain

$$C_G^* \left(f \circ \exp\ |_{\xi_x}, \theta\right) = C_G^* \left(f \circ \exp\ |_{\xi_{g \cdot x}}, \theta\right).$$

Thus the critical group $\widetilde{C}_G^*(f, \mathcal{O})$ is well-defined.

By the assumption that $\xi = N\mathcal{O}$ is a trivial bundle, we know that (W, W_-) possesses a product structure $(W, W_-)_x \times \mathcal{O}$. Using the Künneth formula and (7.4), we obtain

$$\begin{aligned} C_G^q(f, \mathcal{O}) &= H_G^q(W, W_-; \mathbb{K}) \\ &= \bigoplus_{j=0}^{q} H_G^{q-j}\left((W, W_-)_x; \mathbb{K}\right) \otimes H_G^j(\mathcal{O}) \\ &= \bigoplus_{j=0}^{q} C_G^{q-j}\left(f \circ \exp|_{\xi_x}, g\right) \otimes H_G^j(\mathcal{O}) \\ &= \bigoplus_{j=0}^{q} \widetilde{C}_G^{q-j}(f, \mathcal{O}) \otimes H_G^j(\mathcal{O}). \end{aligned}$$

Example 3. (Nondegenerate critical orbits) Let us recall the orientation of a sphere bundle and Thom's isomorphism theorem.

Consider a r sphere bundle $\xi = ((E, \dot{E}), \pi, B)$,

$$\begin{array}{ccc} (D^{r+1}, S^r) & \longrightarrow & \left(E, \dot{E}\right) \\ & & \downarrow \pi \\ & & B \end{array}$$

where B is connected, and D^{r+1} is the $r+1$ disk, $\partial D^{r+1} = S^r$. ξ is said to be orientable if there is an element $\theta \in H^{r+1}(E, \dot{E}; \mathbb{K})$ such that, $\forall\ x \in B$, the restriction of θ to $(\pi^{-1}(p), \pi^{-1}(p) \cap \dot{E})$ is a generator of $H^{r+1}(\pi^{-1}(p), \pi^{-1}(p) \cap \dot{E}; \mathbb{K})$. Such a cohomology class is called an orientation class over $\mathbb{K}$ of the sphere bundle ξ.

(Thom's Isomorphism Theorem). *Let $\xi = ((E, \dot{E}), \pi, B)$ be an r-sphere bundle. Then for any $\mathbb{K}$ module Λ, there are natural isomorphisms*

$$\Phi : H_*\left(E, \dot{E}; \Lambda\right) \to H_{*-r-1}(B; \theta \otimes \Lambda), \Phi(z) = \pi_*(\theta \cap z)$$

and

$$\Phi^* : H^*(B; \theta \otimes \Lambda) \to H^{*+r+1}(E, \dot{E}; \Lambda), \Phi^*(v) = \pi^* v \cup \theta$$

where θ is the orientation bundle of ξ, cf. Spanier [Spa1], p.259.

Theorem 7.5. *Assume that $f \in C^2(M, \mathbb{R}^1)$ is G-invariant, and that $\mathcal{O}$ is a nondegenerate critical orbit, with ind $(f, \mathcal{O}) = \lambda$, and $c = f(\mathcal{O})$. Then we have*

$$C^*_G(f, \mathcal{O}) = H^{*-\lambda}_G(\mathcal{O}, \theta^- \otimes \mathbb{K}),$$

where θ^- is the orientation bundle of ξ^-.

Proof. We apply Theorem 7.3. There is a G-homeomorphism $\Phi : N\mathcal{O}(\epsilon) \to N(\epsilon)$ such that

$$f \circ \Phi(p) = \frac{1}{2}\left(\|P_+(x)v\|^2 - \|P_-(x)v\|^2\right) + c$$

where $p = (x, v)$, $x = \pi p$ and $v \in \xi_x(\epsilon)$. Applying the deformation employed in Theorem 4.1, we obtain

$$\begin{aligned} C^*_G(f, \mathcal{O}) &= H^*_G\left(f_c \cap \xi(\epsilon), (f_c \backslash \mathcal{O}) \cap \xi(\epsilon); \mathbb{K}\right) \\ &= H^*_G\left(\xi^-(\epsilon), \xi^-(\epsilon) \backslash \mathcal{O}; \mathbb{K}\right). \end{aligned}$$

The pair $(\xi^-(\epsilon), \xi^-(\epsilon)\backslash\mathcal{O})$ is then a λ sphere bundle:

$$\begin{array}{ccc} (D^\lambda, S^{\lambda-1}) & \longrightarrow & (\xi^-(\epsilon), \xi^-(\epsilon)\backslash\mathcal{O}) \\ & & \downarrow \\ & & \mathcal{O} \end{array}$$

Thom's isomorphism theorem is applied to obtain

$$C^*_G(f, \mathcal{O}) = H^{*-\lambda}_G\left(\mathcal{O}, \theta^- \otimes \mathbb{K}\right).$$

Furthermore, if ξ^- is G-orientable (if G is connected, this simply means that ξ^- is orientable. Otherwise we assume that G preserves the orientation on ξ^- also), then we have

$$C^*_G(f, \mathcal{O}) = H^{*-\lambda}_G(\mathcal{O}, \mathbb{K}).$$

In all of these examples the computations finally reduce to the computation of the G cohomology of the G orbit $\mathcal{O}$, i.e., $H^*_G(\mathcal{O})$. Noticing that $\mathcal{O} = \mathcal{O}(x_0) = G/G_{x_0}$, where G_{x_0} is the isotropy subgroup at $x_0 \in M$, the map η defined in the commutative diagram in Section 7.4 is induced on an orbit space by projection. The mapping $E \to \mathcal{O}(x_0)$ given by $v \mapsto [v, x_0]$, where $[v, x_0] \in E \times_G M$ is the orbit containing (v, x_0), induces a homeomorphism $E/G_{x_0} \to \eta^{-1}(\mathcal{O}(x_0))$. Thus $\eta^{-1}(\mathcal{O}(x_0))$ has the homotopy type of the classifying space $B_{G_{x_0}}$. Therefore

$$H^*_G(\mathcal{O}) = H^*\left(\eta^{-1}\left(\mathcal{O}\left(x_0\right)\right)\right) = H^*\left(B_{G_{x_0}}\right).$$

In case G_{x_0} is trivial, we have $H^q_G(\mathcal{O}) = H^q(\{pt\}) = \delta_{q0}$.

For example, let $M = \mathbb{R}^2$, $G = \mathbb{Z}_2$, acting on $\mathbb{R}^2$, be reflecting about the origin θ. Then $E_G = S^\infty$, $B_G = \mathbb{R}P^\infty$,

$$\begin{array}{lll} \forall\, x_0 \neq \theta, & G_{x_0} = \{e\}, & H^q_G(\mathcal{O}(x_0), \mathbb{Z}_2) = \delta_{q0}\mathbb{Z}_2 \quad \forall\, q; \\ \text{but for } x_0 = \theta, & G_\theta = \mathbb{Z}_2, & H^q_G(\mathcal{O}(\theta), \mathbb{Z}_2) = \mathbb{Z}_2 \quad \forall\, q. \end{array}$$

The advantage in introducing the G homology lies in providing more information about the critical orbits.

The Morse inequalities are then apparently extended.

Theorem 7.6. *Suppose that $f \in C^1(M, G)$ is G-invariant and satisfies the (PS) condition in $f^{-1}[a,b]$, where a, b are regular values. Assume that f has only isolated critical orbits $\{\mathcal{O}_j \mid j = 1, 2, \ldots, \ell\}$ in $f^{-1}[a,b]$. Then*

$$\sum_{i=0}^{\infty}\sum_{j=1}^{\ell} \operatorname{rank} C^i_G(f, \mathcal{O}_j)\, t^i = \sum_{i=0}^{\infty} \operatorname{rank} H^i_G(f_b, f_a)\, t^i + (1+t)Q(t),$$

where $Q(t)$ is a formal series with nonnegative coefficients.

Proof. It follows from Theorems 7.1, 7.2 and the proofs of Theorems 4.2 and 4.3.

Theorem 7.7 (Bott). *Under the above assumptions, if, further, $f \in C^2$, and has only nondegenerate critical orbits $\mathcal{O}_j$, with* ind $(f, \mathcal{O}_j) = \lambda_j$, *$j = 1, 2, \ldots, \ell$, then we have*

$$\sum_{i=0}^{\infty}\sum_{j=1}^{\ell} \operatorname{rank} H^i_G\left(\mathcal{O}_j, \theta_j^- \otimes \mathbb{K}\right) t^{\lambda_j + i}$$

$$= \sum_{i=0}^{\infty} \operatorname{rank} H^i_G(f_b, f_a; \mathbb{K})\, t^i + (1+t)Q(t),$$

where θ_j^- is the orientation bundle of ξ_j^-, $\xi_j = N\mathcal{O}_j$, $j = 1, 2 \ldots, \ell$.

Proof. It follows from Theorems 7.5 and 7.6.

Remark 7.1. Equivariant Morse theory for nondegenerate critical manifolds with coefficient field $\mathbb{Z}_2$, was first studied by R. Bott [Bot1-2]. It was extended to complete G-Riemannian manifolds, and to Hilbert Riemannian manifolds by A. G. Wasserman [Was1] and W. Meyer [Mey1] respectively. Equivariant Morse theory for isolated critical orbits was done by Z. Q. Wang [WaZ1]. The use of G cohomology was started from M. Atiyah, R. Bott [AtB1], and followed by H. Hingston [Hin1], C. Viterbo [Vit1] and A. Floer [Flo1].

Finally, we shall give an equivariant version of Theorem 5.7.

Theorem 7.8. *Suppose that $f \in C^2(M, \mathbb{R}^1)$ is a G-invariant function, and that $d^2f(x)$ is a Fredholm operator $\forall\ x \in M$. Assume that c is a critical value, with $K_c = \{\mathcal{O}_1, \mathcal{O}_2, \dots, \mathcal{O}_\ell\}$. Then $\forall\ \epsilon > 0$ there exists a G-invariant function $g \in C^2(M, \mathbb{R}^1)$ satisfying the following:*

(1) $g(x) = f(x)$ *in* $M \backslash \cup_{j=1}^{\ell} N\mathcal{O}_j(\epsilon)$;
(2) $\|g - f\|_{C^2(M)} \le \epsilon$;
(3) g *has only nondegenerate critical orbits, concentrated in* $\cup_{j=1}^{\ell} N\mathcal{O}_j(\epsilon)$;
(4) *if* $m_j =$ ind $(f, \mathcal{O}_j)$ *and* $n_j = \dim\ker d^2f(\mathcal{O}_j)$*, then the Morse indices of those nondegenerate critical orbits of* g *in* $N\mathcal{O}_j(\epsilon)$ *are in* $[m_j, m_j + n_j]$.

The proof depends on a Meta theorem which is often advantageous in proving a theorem about compact Lie groups. It replaces the technique of doing a double induction on the dimension and number of components.

Meta Theorem. *Let P be a statement valued function defined for all compact Lie groups. If, whenever G is a compact Lie group, the truth of $P(H)$ for all compact sub-Lie groups $H \subsetneqq G$ implies the truth of $P(G)$ and if $P(\{e\})$ is true, then $P(G)$ is true for all compact Lie groups G.*

We need the following preparations.

Let $\pi : N\mathcal{O}(\epsilon) \to \mathcal{O}$ be the projection from the ϵ-tubular neighborhood $N\mathcal{O}(\epsilon)$ to the G-orbit $\mathcal{O} = Gx$ for some $x \in M$. (As before, we shall not distinguish the tubular neighborhood and the normal bundle.) Let $S = \pi^{-1}(x)$, and let G_x be the isotropy group at x.

Lemma 7.5.

(1) $\forall\ y \in S$, $G_y \subseteq G_x$.
(2) *S is a G_x-manifold.*
(3) *If f_0 is a G_x-invariant function defined on S, then the function*

$$f(z) = f_0(y) \qquad \text{if} \quad z = gy, \quad y \in S,$$

is well-defined on GS, and is G-invariant.

Proof. By definition, π is G-equivariant.

(1) $\forall\ y \in S$, $\forall\ h \in G_y$, i.e., $hy = y$, we have

$$x = \pi(y) = \pi(hy) = h\pi(y) = hx.$$

Therefore $h \in G_x$.

(2) We only want to verify that $G_x S \subseteq S$. Indeed, $\forall\, g \in G_x$, $\forall\, y \in S$,

$$\pi(gy) = g\pi(y) = gx = x.$$

It follows that $gy \in S$.

(3) It suffices to prove that f is well-defined. Indeed, if there exists $g \in G$ such that $gz = z$ for some $z \in S$, then

$$x = \pi(z) = \pi(gz) = g\pi(z) = gx.$$

It follows that $g \in G_x$. We have assumed that f_0 is G_x-invariant, so f is well-defined.

Proof of Theorem 7.8. By the partition of unity, one may assume that K_c is a single orbit $\mathcal{O}$.

Assume that the theorem is true for all compact sub-Lie groups H of G. We shall prove that it is true for G.

For $x \notin \text{Fix}_G$, let $\mathcal{O} = Gx$, $\pi : N\mathcal{O}(\epsilon) \to \mathcal{O}$, and $S = \pi^{-1}(x)$. According to Lemma 7.5, S is a G_x-manifold, where the isotropy group G_x is a sub-Lie group of G. By the assumption, we can modify the function $f|_S$ to a function $g|_S$ such that $g|_S$ satisfies (1) - (4) with $M = S$ and $G = G_x$. Then we apply Lemma 7.5, and the function $g|_S$ can be extended to $GS = N\mathcal{O}(\epsilon)$ and then to M as a G-invariant function g with only nondegenerate critical orbits, i.e., g satisfies (1)–(4).

If $x \in \text{Fix}_G$, then $\mathcal{O} = \{x\}$, $G_x = G$, and $N\mathcal{O}(\epsilon)$ is an ϵ-ball. We choose local coordinates at x, such that $x = \theta$, and $N\mathcal{O}(\epsilon)$ is the unit ball B in a Hilbert space E.

By Sard's Theorem in the nonequivariant case, one may assume that θ is a nondegenerate critical point of $f|_{\text{Fix}_G}$, and also θ is the unique critical point of f in B. Define

$$g(y) = f(y) + \frac{1}{2}\epsilon\lambda\left(\|y\|^2\right)\|Py\|^2$$

where $\lambda \in C^\infty(\mathbb{R}^1_+, \mathbb{R}^1)$ satisfies $\lambda(t) = 1$ as $t \le \frac{1}{4}$, and $\lambda(t) = 0$ as $t \ge \frac{1}{2}$, P is the orthogonal projection onto $\text{Fix}_G^\perp$, and $\epsilon > 0$ is small such that $d^2 g(y)$ is also a Fredholm operator.

Since $g = f$ on Fix_G, $g|_{\text{Fix}_G}$, has θ as the unique critical point, which is nondegenerate on Fix_G. We shall prove that θ is also nondegenerate on E. Indeed,

$$d^2 g(\theta) = d^2 f(\theta) + \epsilon P.$$

If $h \in \ker d^2 g(\theta)$, then

$$\left(d^2 f(\theta) + \epsilon P\right) h = \theta. \tag{7.7}$$

Since θ is nondegenerate for $f|_{\mathrm{Fix}_G}$, $h \in \mathrm{Fix}_G^{\perp} = (\ker P)^{\perp}$. And on $\mathrm{Fix}_G^{\perp}$, equation (7.7) is reduced to

$$P\left(d^2 f(\theta) + \epsilon I\right) Ph = \theta.$$

However, $d^2 f(\theta)$ is a Fredholm operator, 0 cannot be an accumulate point of $\sigma(d^2 f(\theta))$, which implies that for sufficiently small $\epsilon > 0$, $P(d^2 f(\theta) + \epsilon)P$ is invertible on $\mathcal{L}(\mathrm{Fix}_G^{\perp})$. Therefore $h = \theta$, i.e., θ is nondegenerate on E.

The case $G = \{e\}$ is just Theorem 5.7. The proof is completed by applying the Meta Theorem mentioned above.

Remark 7.2. Theorem 7.8 was proved by Wasseman [Was1] and Viterbo [Vit1].

CHAPTER II
Critical Point Theory

In the study of nonminimum critical points, a basic method is the so-called minimax principle. In this chapter we study the connections between Morse theory and a variety of concrete versions of the minimax principle. We point out that the minimax principle for relative homology classes is particularly suitable for Morse theory because certain critical groups for the critical points determined by this minimax principle can be proved being nontrivial; they then have contributions to Morse inequalities.

Section 1 is devoted to applying the homology minimax principle to the study of the linking method used initially by Rabinowitz, Benci-Rabinowitz and Ni. In Section 2, we estimate Morse indices for critical points, determined by various minimax principles in which the critical groups are not well-determined. In Section 3, we study the connections between Leray-Schauder degree theory with Morse theory, as well as the Ljusternik-Schnirelman category with the subordinate nontrivial homology classes. In Section 4 some theorems due to Ambrosetti-Rabinowitz, which were proved using genus, indices or pseudo-indices, are proved using homology methods. Section 5 consists of some abstract critical point theorems, which will be applied to concrete problems in the following chapters. Section 6 deals with perturbation theory and, although relatively independent, is useful in the applications.

1. Topological Link

A variety of critical point theorems, studied initially by Ambrosetti-Rabinowitz [AmR1], Benci-Rabinowitz [BeR1] and W. M. Ni [Ni1], is based on the following topological idea: There exist two linking sets which are separated by a level set. In this section we clarify the concept of linking from the points of view of the homotopy and the homology, and emphasize that the homology link is more interesting in Morse theory.

Throughout this section, M stands for a C^2-Finsler manifold, and $f \in C^1(M, \mathbb{R}^1)$.

Definition 1.1. Let D be a k-topological ball in M, and let S be a subset in M. We say that ∂D and S homotopically link, if $\partial D \cap S = \emptyset$ and $\varphi(D) \cap S \neq \emptyset$ for each $\varphi \in C(D, M)$ such that $\varphi|_{\partial D} = \mathrm{id}|_{\partial D}$.

The reason for the terminology comes from

Theorem 1.1. *Assume that ∂D and S homotopically link. If $f \in C(M, \mathbb{R}^1)$ satisfies*

$$f(x) > a \qquad \forall\, x \in S \tag{1.1}$$

$$f(x) \leq a \qquad \forall\, x \in \partial D \tag{1.2}$$

then $\pi_k(f_b, f_a) \neq 0$, where $b > \text{Max}\{f(x) \mid x \in \overline{D}\}$.

Proof. Since $D \subset F_b$, the class $[\partial D]$ which includes the map $\text{id}|_{\partial D}$ is trivial in $\pi_{k-1}(f_b)$, where $k = \dim D$. On the other hand, $\partial D \subset f_a$; but

$$\text{Max}_{x \in D} f \circ \varphi(x) > a \qquad \forall\, \varphi \in C(D, M)$$

provided that $\varphi(D) \cap S \neq \emptyset$. This means $\varphi(D)$ is not included in f_a; hence $[\partial D]$ is nontrivial in $\pi_{k-1}(f_a)$.

We observe the exact sequence

$$\cdots \longrightarrow \pi_k\,(f_b, f_a) \xrightarrow{\partial_*} \pi_{k-1}(f_a) \xrightarrow{i_*} \pi_{k-1}(f_b) \longrightarrow \cdots$$

where $i : f_a \to f_b$ is the inclusion, and ∂ is the boundary operator. It follows

$$[\partial D] \in \ker i_* = \text{Im}\ \partial_*.$$

Consequently $\pi_k(f_b, f_a)$ is nontrivial.

Similarly, we obtain

Definition 1.2. Let D be a k-topological ball in M, and let S be a subset in M. We say that ∂D and S homologically link, if $\partial D \cap S = \emptyset$ and $|\tau| \cap S \neq \emptyset$, for each singular k chain τ with $\partial \tau = \partial D$ where $|\tau|$ is the support of τ.

In the same way, we prove

Theorem 1.1′. *Assume that ∂D and S homologically link. If $f \in C(M, \mathbb{R}^1)$ satisfies (1.1) and (1.2), then $H_k(f_b, f_a) \neq 0$.*

There are many linking examples:

Example 1. Let Ω be an open neighborhood of a point z_0 in a Banach space. Set $z_1 \notin \overline{\Omega}$, $S = \partial D$, and $D = \overline{z_0 z_1}$, the segment joining z_0 and z_1. Then S and $\partial D = \{z_0, z_1\}$ homotopically link.

Claim. By connectedness, $S \cap \ell \neq \emptyset$, where ℓ is any path-connecting z_0 and z_1.

Example 2. Let X be a Banach space. $X = X_1 \oplus X_2$ is a direct sum decomposition where $\dim X_1 < +\infty$. Set

$$S = X_2, \quad \text{and} \quad D = B_1 \cap X_1$$

where B_1 is the unit ball centered at θ. Then ∂D and S link.

Claim. Obviously $S \cap \partial D = \emptyset$. We only want to show

$$\forall\, \varphi \in C(D, X) \text{ and } \varphi|_{\partial D} = \text{ id }|_{\partial D} \Rightarrow \varphi(D) \cap S = \emptyset. \tag{1.3}$$

Define a projection $P : X \to X_1$. It is equivalent to showing that $\exists\, x_0 \in D$ such that $P \circ \varphi(x_0) = \theta$.

Define a deformation

$$F(t, x) = tP \circ \varphi(x) + (1 - t)x \quad \forall\, (t, x) \in [0, 1] \times D.$$

Since $\theta \notin \partial D = F([0, 1] \times \partial D)$,

$$\deg(F(1, \cdot), D, \theta) = \deg(F(0, \cdot), D, \theta) = 1.$$

The equation $P \circ \varphi(x) = \theta$ is solvable, using Brouwer degree theory.

Example 3. Let $X = X_1 \oplus X_2$ be defined in Example 2. Let $e \in X_2$, $\|e\| = 1$, and let R_1, R_2, $\rho > 0$ with $\rho < R_1$. Set $S = X_2 \cap \partial B_\rho$, and $D = \{x + se \mid x \in X_1 \cap B_{R_2},\ s \in [0, R_1]\}$. Then ∂D and S homotopically link.

Claim. Obviously $S \cap \partial D = \emptyset$. It remains to prove (1.3).

Again we use the same projection P in Example 2 and define a new deformation:

$$F(t, x + se) = [(1 - t)x + tP \circ \varphi(x + se)] + \\ [(1 - t)s + t\|(I - P) \circ \varphi(x + se)\| - \rho]e.$$

It is easily seen that

$$F(0, x + se) = x + (s - \rho)e,$$
$$F(1, x + se) = P \circ \varphi(x + se) + [\|(I - P) \circ \varphi(x + se)\| - \rho]e,$$

and

$$F(t, x + se) = x + (s - \rho)e \neq \theta,\ \forall\, t \in [0, 1],\ \forall x + se \in \partial D.$$

Thus (1.3) is equivalent to finding $y_0 \in D$ satisfying

$$\begin{cases} P \circ \varphi(y_0) = \theta, \\ \|\varphi(y_0)\| = \rho; \end{cases}$$

i.e., $F(1, y_0) = \theta$. Since

$$\deg(F(1, \cdot), D, \theta) = \deg(F(0, \cdot), D, \theta) \\ = \deg\left(\text{id}_{X_1}, X_1 \cap B_{R_2}, \theta\right) \cdot \deg\,(\text{id } - \rho, (0, R_1), 0) = 1.$$

Consequently, (1.3) is solvable, i.e., S and ∂D homotopically link.

Actually the above three examples are of homological linking as well. We have

Theorem 1.2. *Suppose that the boundary ∂D of the k-topological ball D and S homotopically link. Assume that*

(1) *$S \cap D =$ single point,*
(2) *S is a path-connected orientable submanifold with codimension k,*
(3) *there exists a tubular neighborhood N of S such that $N \cap D$ is homeomorphic to D.*

Then ∂D and S homologically link.

Proof. $(N, \partial N)$ can be regarded as an orientable sphere bundle over the base space S with fibre (D^k, S^{k-1}), i.e.,

$$\begin{array}{ccc} (D^k, S^{k-1}) & \longrightarrow & (N, \partial N) \\ & & \swarrow \\ & & S \end{array}$$

According to Thom's isomorphism theorem (see Chapter 1, Section 7),

$$H_*(N, \partial N) \cong H_*(S) \otimes H_*\left(D^k, S^{k-1}\right).$$

Since S is path-connected, we have $H_0(S) \cong G$. It follows that

$$H_k(N, \partial N) \cong G.$$

Let $[\tau]$ be the generator of $H_k(N, \partial N)$. We choose a singular chain $\tau \in [\tau]$, such that $\partial\tau = \partial N \cap D$, according to (3).

Since N is a tubular neighborhood of S, we have

$$H_*(M, M \backslash S) \cong H_*(N, N \backslash S)$$

provided by the excision property. Let us apply a deformation retract along the exponential map. It follows that

$$H_*(N, N \backslash S) \cong H_*(N, \partial N).$$

From assumptions (1) and (3), we see that ∂D and $\partial\tau$ are homologous in $H_*(M, M \backslash S)$. Therefore $[D] = [\tau]$ is nontrivial in $H_k(M, M \backslash S)$, where $[D]$ is the relative singular homology class containing D. The nontriviality of $[D]$ implies that ∂D and S link homologically.

Thus the above three examples are also homological links.

Remark 1.1. In order to prove the homological links for the above three examples, only the Künneth formula is needed.

Theorem 1.3. (Minimax Principle). *Suppose that $\mathcal{F}$ is a family of subsets of M. Set*

$$c = \inf_{F\in\mathcal{F}} \sup_{x\in F} f(x).$$

Assume that (1) c is finite, (2) f satisfies the $(PS)_c$ condition, and (3) $\exists\ \varepsilon_0 > 0$, such that the family $\mathcal{F}$ is invariant under the family of maps $\Phi_{\varepsilon_0} = \{\varphi \in C(M,M) \mid \varphi \sim \mathrm{id},\ \varphi|_{f_{c-\varepsilon_0}} = \mathrm{id}_{f_{c-\varepsilon_0}}\}$. Then c is a critical value of f.

Proof. We prove it by contradiction. If c is not a critical value, then there exists $\varepsilon \in (0,\varepsilon_0)$ such that $K \cap f^{-1}[c-\varepsilon, c+\varepsilon] = \emptyset$. One finds $F_0 \in \mathcal{F}$ with $F_0 \subset f_{c+\varepsilon}$. There exists $\eta \in C(M,M)$ satisfying $\eta \sim \mathrm{id}$, $\eta|_{f_{c-\varepsilon_0}} = \mathrm{id}_{f_{c-\varepsilon_0}}$ and $\eta(f_{c+\varepsilon}) \subset f_{c-\varepsilon_0}$, according to the first deformation theorem. Set $F_1 = \eta(F_0)$; we have

$$F_1 \in \mathcal{F} \quad \text{and} \quad F_1 \subset f_{c-\varepsilon}.$$

This contradicts the definition of c.

Theorem 1.4. *Let $a < b$ be regular values of f. Set*

$$c = \inf_{Z\in\alpha} \sup_{x\in|Z|} f(x) \qquad \textit{with } \alpha \in \pi_k(f_b, f_a) \textit{ nontrivial}$$

(and

$$c^* = \inf_{\tau\in\alpha} \sup_{x\in|\tau|} f(x) \qquad \textit{with } \alpha \in H_k(f_b, f_a) \textit{ nontrivial}).$$

Assume that $c > a$ (and $c^* > a$ resp.), and that f satisfies the $(PS)_c$ (and $(PS)_{c^*}$) condition. Then c (and c^* resp.) is a critical value of f. Moreover, we have $c^* \le c$.

Proof. It suffices for us to verify that the families $\mathcal{F} = \{|Z| \mid Z \in \alpha\}$ (and $\mathcal{F}^* = \{|\tau| \mid \tau \in \alpha\}$) are invariant with respect to Φ_{ε_0}, where $\varepsilon_0 \in (0, c-a)$ (and $\varepsilon_0 \in (0, c^*-a)$ resp.). This is just the homotopy invariance of the homotopy (and homology) classes. The inequality $c^* \le c$ follows from $\mathcal{F} \subset \mathcal{F}^*$.

In particular, let α be the class in $\pi_k(f_b, f_a)$ (or $H_k(f_b, f_a)$) with $\partial_*\alpha = [\partial D]$ in Theorem 1.2. Then

$$c = \inf_{\varepsilon\in\Gamma} \sup_{x\in D} f\circ\varphi(x) \tag{1.4}$$

where $\Gamma = \{\varphi \in C(D,M) \mid \varphi|_{\partial D} = \mathrm{id}_{\partial D}\}$ (and
(1.5)

$$c^* = \inf\left\{ \sup_{x\in|\tau|} f(x) \mid \tau \text{ is a } k \text{ rel. singular chain in } (f_b, f_a) \text{ with } \partial\tau = \partial D \right\}).$$

The following assumption:

$$f(x) \geq d > a \qquad \forall\, x \in S, \tag{1.6}$$

guarantees that $c \geq d > a$ (and $c^* \geq d > a$), under which c and c^* are critical values.

Remark 1.2. The separation conditions (1.2) and (1.6) may be weakened to (1.2) and

$$f(x) \geq a \qquad \forall\, x \in S. \tag{1.7}$$

Claim. We shall prove that if $c = a$ then $K_c \neq \emptyset$. Otherwise we assume $K_c = \emptyset$. Since ∂D is compact and $\partial D \cap S = \emptyset$, $\exists$ two open neighborhoods $U_2 \supset \overline{U}_1 \supset U_1$ of ∂D such that $U_2 \cap S = \emptyset$. According to the first deformation theorem, there exist $\varepsilon > 0$ and a homeomorphism η such that

$$\begin{aligned} \eta &: (-f)_{-c+\varepsilon} \backslash U_2 \to (-f)_{-c-\varepsilon} \quad \text{and} \\ \eta\, |_{U_1} &= \mathrm{id}_{U_1}, \end{aligned}$$

i.e.,

$$\begin{aligned} \eta &: f^{c-\varepsilon} \backslash U_2 \to f^{c+\varepsilon} \quad \text{and} \\ \eta\, |_{U_1} &= \mathrm{id}_{U_1}, \end{aligned}$$

where $f^\alpha = \{x \in M \mid f(x) \geq \alpha\}$. Therefore

$$\eta(S) \subset f^{c+\varepsilon}.$$

Let $S_1 = \eta(S)$; then we have

$$\begin{aligned} &S_1 \cap \partial D = \emptyset \quad \text{and} \\ &S_1 \cap \varphi(D) = \eta\left(S \cap \eta^{-1}\varphi(D)\right) \neq \emptyset \quad \forall\, \varphi \in \Gamma, \end{aligned}$$

i.e., S_1 and ∂D link (homotopically).

Since $f(x) \geq c + \varepsilon \quad \forall\, x \in S_1$, we conclude that

$$c = \inf_{\varphi \in \Gamma} \sup_{x \in D} f \circ \varphi(x) \geq \inf_{x \in S_1} f(x) \geq c + \varepsilon.$$

This is a contradiction.

The above minimax principle (1.4) under condition (1.6) was obtained by Benci-Rabinowitz [BeR1], and Ni [Ni1]. The condition (1.6) was weakened to condition (1.1) by Chang [Cha8], and to condition (1.7) for $k = 1$ by G. J. Qi [Qi1]. Actually, under a more general linking definition (∂D may not be compact), the above conclusion remains true, cf. Y. H. Du [Du1] and N. Ghoussoub [Gho1].

Theorem 1.5. *Assume that $\alpha \in H_k(f_b, f_a)$ is nontrivial, and*

$$c^* = \inf_{Z\in\alpha} \sup_{x\in|Z|} f(x). \tag{1.8}$$

Suppose that f satisfies the $(PS)_{c^*}$ condition and that K_{c^*} is isolated. Then there exists $x_0 \in K_{c^*}$ such that $C_k(f, x_0) \neq 0$.

Proof. It suffices to prove that $H_k(f_{c^*}, f_{c^*}\backslash K_{c^*}) \neq 0$ using Theorem 4.2 in Chapter 1.

$\forall\, \varepsilon > 0\ \exists\, Z \in \alpha$ with support $|Z| \subset f_{c^*+\varepsilon}$. Let $y = \partial Z$ be the boundary of Z; then $|y| \subset f_a \subset f_{c^*}$. We shall prove that $[y] \in H_{k-1}(f_{c^*}\backslash K_{c^*})$ is nontrivial. Indeed, if $[y]$ is trivial, then there exists a singular k-chain τ such that $|\tau| \subset f_{c^*}\backslash K_{c^*}$ and $[\partial\tau] = [y]$. Due to the second deformation theorem, $\exists\ \eta \in C(M, M)$, satisfying $\eta \sim$ id and $\eta(f_{c^*}\backslash K_{c^*}) \subset f_{c^*-\varepsilon}$ for some $\varepsilon > 0$. Let $\tau_1 = \eta(\tau)$, we have $[\tau_1] = [\tau]$ and $|\tau_1| \subset f_{c^*-\varepsilon}$, which is in contradiction with the definition of c^*.

We observe the exact sequence

$$\longrightarrow H_k\left(f_{c^*+\varepsilon}, f_{c^*}\backslash K_{c^*}\right) \xrightarrow{\partial_*} H_{k-1}\left(f_{c^*}\backslash K_{c^*}\right) \xrightarrow{i_*} H_{k-1}\left(f_{c^*+\varepsilon}\right) \longrightarrow \cdots$$

where $i : f_{c^*}\backslash K_{c^*} \to f_{c^*+\varepsilon}$ is the inclusion. Since $i_*([y]) = 0$, i.e., $[y] \in \ker i_* = \operatorname{Im} \partial_*$, we conclude

$$H_k\left(f_{c^*}, f_{c^*}\backslash K_{c^*}\right) \cong H_k\left(f_{c^*+\varepsilon}, f_{c^*}\backslash K_{c^*}\right) \neq 0.$$

Corollary 1.1. *Under the assumptions of Theorem 1.5, if, in addition, all critical points in K_{c^*} are assumed to be nondegenerate, then there exists $x_0 \in K_{c^*}$ such that $ind(f, x_0) = k$.*

Theorem 1.5 can also be extended to G-invariant functions. Let G be a compact Lie group, and let M be a C^2-Hilbert Riemannian G-manifold. Suppose that $f \in C^1(M, \mathbb{R}^1)$ is G invariant, and that a, b are regular values of f. Assume that $[Z] \in H_G^k(f_b, f_a)$ is nontrivial. One defines

$$c = \inf_{Z\in[Z]} \sup_{x\in|Z|} f(x).$$

By the same proof, we have

Corollary 1.2. *Suppose that f satisfies $(PS)_c$ and that K_c consists of isolated critical orbits. Then there exists a critical orbit $O \subset K_c$ such that*

$$C_G^k(g, O) \neq 0.$$

As a consequence, we get an estimate of the Morse index for certain critical orbits in K_c.

Corollary 1.3. *Under the assumptions of Corollary 1.2, we assume further that $f \in C^2$, $d^2 f(x)$ is a Fredholm operator $\forall\, x \in M$, and that $K_c \cap \ \mathrm{Fix}_G = \emptyset$. If $H^*(B_{G_x}, \theta^-) = 0$ for $* \neq 0$, $\forall\, x \notin \mathrm{Fix}_G$, where θ^- is the orientation bundle of $\xi^-(Gx)$, then there is a critical orbit $O \subset K_c$ satisfying*

$$m \leq k \leq m + n$$

where $m = \ \mathrm{ind}(f, O)$ and $n = \dim \xi^0(O)$.

Proof. According to Theorem 7.8 in Chapter I, we consider the approximation function g, which has only nondegenerate critical orbits. Let $K_c = \{O_1, O_2, \ldots .O_\ell\}$, and let $m_i = \ \mathrm{ind}\,(f, O_i)$, $n_i = \dim \xi^0(O_i)$. Then g has critical orbits O_{is} with $\mathrm{ind}(g, O_{is}) = q_{is} \in [m_i, m_i + n_i]$, $i = 1, 2, \ldots, \ell$, $s = 1, 2, \ldots, j_i$.

One concludes that there exists at least a pair (i, s) such that $C_G^{q_{is}}(g, O_{is}) \neq 0$. Indeed

$$\bigoplus_{i=1}^{\ell} \bigoplus_{s=1}^{j_i} C_G^*\,(g, O_{is}) = H_G^*\,(g_{c+\varepsilon}, g_{c-\varepsilon}) = H_G^*\,(f_{c+\varepsilon}, f_{c-\varepsilon}),$$

so there exists a pair (i, s) such that $C_G^k(g, O_{is}) \neq 0$.

According to Theorem 7.5 in Chapter I, $C_G^k(g, O_{is}) = H_G^{k-q_{is}}(O_{is}, \theta^-(O_{is})) = H^{k-q_{is}}(B_{Gx_0}, \theta^-)$ where $x_0 \in O_{is}$. By the assumption that $H^*(B_{G_x}, \theta^-) = 0$ for $* \neq 0\ \forall\, x \notin \ \mathrm{Fix}_G$; it follows $k = q_{is}$. Therefore

$$m_i \leq k \leq m_i + n_i.$$

This is our conclusion.

Let us return to Example 1, in this case

$$c = c^* = \inf_{\ell \in \Gamma} \sup_{x \in \ell[0,1]} f(x),$$

where $\Gamma = \{\ell \in C([0,1], X) \mid \ell(i) = z_i,\ i = 0, 1\}$. According to Theorem 1.5, there exists a point $x_0 \in K_c$ such that

$$C_1\,(f, x_0) \neq 0. \tag{1.9}$$

Due to the geometric meaning, we call a point x_0 satisfying (1.9) a mountain pass point. The following theorem provides more precise information on mountain pass points.

Theorem 1.6. *Suppose that $f \in C^2(M, \mathbb{R}^1)$ and that x_0 is a mountain pass point. Assume that $d^2 f(x_0)$ is a Fredholm operator and that*

$$\dim \ker \left(d^2 f\left(x_0\right)\right)=1 \qquad \text{if} \quad 0 \in \sigma\left(d^2 f\left(x_o\right)\right) \tag{1.10}$$

then

$$\operatorname{rank} C_q\left(f, x_0\right)=\delta_{q 1}.$$

Proof. Let $k =$ ind (f, x_0).

1. If x_0 is nondegenerate, then $C_q(f, x_0) = \delta_{qk} G$. From (1.9), it follows $k = 1$. and then $C_q(f, x_0) = \delta_{q1} G$.

2. If x_0 is degenerate, then we have

$$C_q\left(f, x_0\right)=C_{q-k}\left(\tilde{f}, \theta\right) \qquad \forall\, q,$$

where $\tilde{f} = f|_N$, and N is the characteristic manifold of f, according to the shifting theorem. From (1.9), it follows that $k \leq 1$.

Case a. If $k = 1$, then $C_0(\tilde{f}, \theta) \neq 0$, which is equivalent to 0 being an isolated local minimum of $\tilde{f}$; then

$$C_q\left(\tilde{f}, \theta\right)=\delta_{q 0} G.$$

Thus

$$C_q\left(f, x_0\right)=\delta_{q 1} G.$$

Case b. If $k = 0$, then

$$C_q\left(f, x_0\right)=C_q\left(\tilde{f}, \theta\right), \qquad \forall\, q.$$

By assumption (1.10), the function $\tilde{f}$ is defined on a 1-manifold. And the only possibility for $C_1(\tilde{f}, \theta) \neq 0$ is the case in which x_0 is a local maximum of $\tilde{f}$. Thus

$$C_q\left(f, x_0\right)=C_q\left(\tilde{f}, \theta\right)=\delta_{q 1} G.$$

The proof is completed.

Remark 1.3. The condition (1.10) is essential. For example, the Monkey saddle is a mountain pass point, with critical groups $C_q = \delta_{q1} G \oplus G$, $\forall\, q \in \mathbb{N}$.

Remark 1.4. The idea of the homological link was introduced by J. Q. Liu [Liu1], from which he proved Theorem 1.5. C. Viterbo independently gave an estimate of the Morse indices for critical points defined by equivariant homology classes, i.e., Corollary 1.3, cf. [Vit1]. Theorem 1.2 is more or less related to V. Benci [Ben2] and X. F. Yang [Yan1].

2. Morse Indices of Minimax Critical Points

Minimax critical points, by definition, are independent of Morse theory. However, the Morse indices of these critical points play a role as a connection. We shall pay more attention to the estimate of the Morse indices of critical points as determined by some standard minimax methods.

All notation appearing in Section 1 will be used without change.

2.1. Link

Theorem 2.1. *Suppose that $f \in C^2(m, \mathbb{R}^1)$ satisfies the $(PS)_c$ condition, where $c > a$ is defined in (1.4). Assume that $K_c = \{x_1, x_2, \dots, x_\ell\}$ with Morse indices $\{k_1, k_2, \dots, k_\ell\}$, and that 0 is not an accumulate point of $\sigma(d^2 f(x_i))$ whenever $0 \in \sigma(d^2 f(x_i))$, $i = 1, 2, \dots, \ell$. Then*

$$\operatorname{Min}\{k_i \mid i = 1, 2, \dots, \ell\} \leq k \tag{2.1}$$

Proof. We prove (2.1) by contradiction.

Suppose that $k <$ Min $\{k_i \mid 1 \leq i \leq \ell\}$.

According to the splitting theorem, $\forall\, x_i$, there exist a neighborhood N_i, a homeomorphism $\Phi_i : B_r^\perp \times B_\delta^\circ \to N_i$, where $B_r^\perp$ and B_δ° are r-ball and δ-ball in $H_i^\perp$ and H_i where $H_i = \ker(d^2 f(x_i))$ respectively, and $r, \delta > 0$, such that in each N_i, the function f is written in the form

$$f(x) = c + \|y_+\|^2 - \|y_-\|^2 + \tilde{f}_i(y_0),$$

where $x = \Phi_i(y)$, $y = y_+ + y_- + y_0$, $y^\perp = (y_+, y_-) \in B_r^\perp$ and $y_0 \in B_\delta^\circ$, $i = 1.2, \dots, \ell$. One may assume that $\tilde{f}_i \mid_{B_\delta^\circ} \leq \frac{1}{2} r^2$.

$\forall\, \varepsilon > 0$, $\exists\, Z \in \alpha$, where $\alpha \in \pi_k(f_b, f_a)$, such that $|Z| \subset f_{c+\varepsilon}$. Using the deformation theorem, we have $Z_1 \in \alpha$ satisfying $|Z_1| \subset f_{c-\varepsilon_1} \cup \bigcup_{i=1}^{\ell} N_i$ for some $\varepsilon_1 > 0$. One may assume that $\varepsilon_1 < \frac{1}{2} r^2$. In order to simplify the notation, we omit the subscripts i, and the transformations Φ_i, and regard N_i as $B_r^\perp \times B_\delta^\circ$, if there is no confusion.

Let $g(y^\perp) = \|y_+\|^2 - \|y_-\|^2$; then $\forall\, \bar{y}_0 \in B_\delta^\circ$, the sections $f_c \mid_{y_0 = \bar{y}_0} = g_{-\tilde{f}(\bar{y}_0)}$, and $f_{c-\varepsilon_1} \mid_{y_0 = \bar{y}_0} = g_{-\varepsilon_1 - \tilde{f}(\bar{y}_0)}$. Let $B_r^- \subset B_r^\perp$ be the r-ball $\{y^\perp = (y_+, y_-) \in B_r^\perp \mid y_+ = \theta\}$. We shall prove that there exists $Z_2 \in \alpha$ satisfying

$$|Z_2| \subset \overline{\left(f_{c-\varepsilon_1} \backslash \bigcup_{i=1}^{\ell} N_i\right)} \cup \bigcup_{i=1}^{\ell} \Phi_i\,(Q_i)$$

where $Q_i = B_r^{k_i} \times \{\theta_+\} \times (B_\delta \cap H_i)$, and $B_r^{k_i}$ is the k_i-dimensional r-ball, $i = 1, 2, \dots, \ell$.

Indeed, let us denote

$$a = -\varepsilon_1 - \tilde{f}\,(y_0)\,, \quad b = \frac{1}{2}\left(\sqrt{\operatorname{Max}\{-a, 0\}} + \sqrt{\frac{r^2 - a}{2}}\right),$$

$$\varphi(s,b)=\begin{cases}0 & \text{if } 0\le s\le b\\ (s-b)ctg\,\alpha & \text{if } b<s\le\sqrt{\frac{r^2-a}{2}}\end{cases}$$

where $ctg\,\alpha=\sqrt{2(r^2+a)}/(\frac{\sqrt{r^2-a}}{2}-\sqrt{\text{Max}\{-a,0\}})$. We define deformations as follows:

$$\eta_1(t,y)=\begin{cases}y & \text{if } y\in f_{c-\varepsilon_1}\backslash G\\ y_-+y_0+\left[1+t\left(\frac{1}{|y_+|}\varphi(|y_-|,b)-1\right)\right]y_+ & \text{if } y\in G,\end{cases}$$

where $G=\{y\in B_r^\perp\times B_\delta^\circ \mid |y_+|\ge\varphi(|y_-|,b)\}$, and

$$\eta_2(t,y)=\begin{cases}y\\ \qquad \text{if } y\in f_{c-\varepsilon_1}\backslash\bigcup_{i=1}^{\ell}N_i\\ y_++y_0+\left(t\sqrt{r^2-|y_+|^2}+(1-t)|y_-|\right)\frac{y_-}{|y_-|}\\ \qquad \text{if } y\in\left(B_r^\perp\times B_\delta^\circ\right)\backslash G\\ y_++y_0+\frac{t\sqrt{r^2-|y_+|^2}+(1-t)\psi(|y_+|,b)}{\psi(|y_+|,b)}y_-\\ \qquad \text{if } y\in G,\text{ with } |y_+|\le\sqrt{\frac{r^2+a}{2}}\end{cases}$$

where $\psi(s,b)=tg\,\alpha\cdot s+b$, $t\in[0,1]$.

The composition of η_2 and η_1 yields $Z_2=(\eta_2\circ\eta_1)(1,Z_1)$, which satisfies

$$|Z_2|\subset E:=\overline{\left(f_{c-\varepsilon_1}\backslash\bigcup_{i=1}^{\ell}N_i\right)}\cup\bigcup_{i=1}^{\ell}\Phi_i\left(Q_i\right).$$

This means that there exists a continuous map $\sigma: D^k\to E$ such that $Z_2=\sigma(D^k)$. One may modify σ to be C^1 on $\sigma^{-1}\Phi_i(Q_i)$ by the partition of unity because Q_i is convex, $i=1,2,\dots,\ell$.

Since we have assumed that $k<\min k_i$, the projection $\pi_i|Z_2|$ of $|Z_2|$ onto $B_r^{k_i}$ cannot cover $B_r^{k_1}$. There must be at least one point $p_i\in\overset{\circ}{B}{}_r^{k_i}$ missing. Again, we project $\pi_i|Z_2|$ onto $\partial B_r^{k_i}$ from p_i, $1\le i\le\ell$. Then $|Z_2|$ is deformed into $f_{c-\varepsilon}\cup\bigcup_{i=1}^{\ell}\Phi_i(\partial B_r^{k_i}\times\{\theta_+\}\times(B_\delta\cap H_i))$, which gives $Z_3\in\alpha$ such that

$$\sup_{x\in|Z_3|}f(x)<c.$$

This is a contradiction.

By the same proof, we have

Theorem 2.1′. *Let M be a Hilbert-Riemannian manifold. Suppose that $f\in C^2(M,\mathbb{R}^1)$ satisfies the (PS) condition in $f^{-1}[a,b]$, and that*

$K \cap f^{-1}[a,b] = \{x_1, x_2, \dots, x_\ell\}$, *with* $k_i =$ ind (f, x_i) *and* $a < f(x_i) < b$, $1 \le i \le \ell$. *If 0 is not an accumulate point of* $\sigma(d^2 f(x_i))$ *whenever* $0 \in \sigma(d^2 f(x_i))$, $1 \le i \le \ell$. *Then*

$$\pi_k\,(f_b, f_a) = 0 \quad \textit{for } k < \text{ Min } \{k_i \mid i = 1, 2, \dots, \ell\}\,. \tag{2.2}$$

There is a dual way of defining critical values for Example 2.

Instead of the set family $\mathcal{F} = \{\varphi(D) \mid \varphi \in C(D,X) \mid \varphi\,|_{\partial D} = \text{ id}_{\partial D}\}$, we introduce a new family

$$\begin{aligned}\widehat{\mathcal{F}} = \{F \subset X \mid \text{ compact}, \forall\, \sigma \in C(F, X_1) \text{ with } \sigma\,|_{F\cap\partial D} = \text{ id}_{F\cap\partial D}, \\ \text{we have } \theta \in \sigma(F)\}\,.\end{aligned}$$

Since $\mathcal{F} \subset \widehat{\mathcal{F}}$ (using Brouwer degree theory), $\widehat{\mathcal{F}}$ is nonempty. Let us define

$$\hat{c} = \inf_{F \in \widehat{\mathcal{F}}} \sup_{x \in F} f(x) \tag{2.3}$$

then $\hat{c} \le c$. We conclude $\hat{c} \ge d$ if (1.6) holds. In fact, by taking σ equal to the projection onto X_1, $F \cap X_2 \ne \emptyset\; \forall\, F \in \widehat{\mathcal{F}}$, which implies $\sup_{x\in F} f(x) \ge \inf_{x \in X_2} f(x) \ge d$.

If, further, we assume (1.2) and that $f \in C^1(X, \mathbb{R}^1)$ satisfies the $(PS)_{\hat{c}}$ condition, then $\hat{c}$ is again a critical value.

Claim. By choosing $\varepsilon_0 \in (0, d-a)$, it is not difficult to verify that the family $\widehat{\mathcal{F}}$ is invariant under the class of maps Φ_{ε_0} (cf. Theorem 1.3). The conclusion follows from Theorem 1.3.

In order to obtain a counterpart of Theorem 2.1 for $\hat{c}$, we need

Lemma 2.1. *Let* $A \subset \mathbb{R}^n$ *be a compact set, and let* $\sigma \in C(A, \mathbb{R}^k)$ *with* $k > n$, *such that* $\theta \notin \sigma(A)$. *Then* σ *can be extended continuously to* $\tilde{\sigma} \in C(\mathbb{R}^n, \mathbb{R}^k \backslash \{\theta\})$.

Proof. Since A is compact and $\theta \notin \sigma(A)$, there exists $\varepsilon > 0$ such that $\sigma(A) \cap B_\varepsilon(\theta) = \emptyset$. According to Tietze's theorem, we have a continuous extension $\hat{\sigma}$ of σ, $\hat{\sigma}: \mathbb{R}^n \to \mathbb{R}^k$. Moreover, we may assume $\hat{\sigma} \in C^1(\mathbb{R}^n \backslash A, \mathbb{R}^k)$. By the assumption $k > n$, it follows from Sard's theorem that $\hat{\sigma}(\mathbb{R}^n \backslash A)$ is measure zero in $\mathbb{R}^k$. Therefore $B_\varepsilon(\theta)$ is not included in $\hat{\sigma}(\mathbb{R}^n)$, i.e., there exists $x_0 \in B_\varepsilon(\theta) \backslash \hat{\sigma}(\mathbb{R}^n)$. We project $\hat{\sigma}(\mathbb{R}^n) \cap B_\varepsilon(\theta)$ onto $\partial B_\varepsilon(\theta)$ from x_0. This gives the extension $\tilde{\sigma}$.

Theorem 2.2. *Let* X *be a Hilbert space with the direct sum decomposition* $X = X_1 \oplus X_2$, $\dim X_1 < +\infty$. *Assume that* $f \in C^2(X, \mathbb{R}^1)$ *satisfies the* $(PS)_{\hat{c}}$ *condition, where* $\hat{c}$ *is defined in* (2.3), *and that* (1.2) *and* (1.4) *hold. Suppose that* $K_{\hat{c}} = \{x_1, x_2, \dots, x_\ell\}$ *with Morse indices* $\{k_1, k_2, \dots, k_\ell\}$ *and* *nullities* $\{n_1, n_2, \dots, n_\ell\}$ *respectively, and that* $d^2 f(x_i)$, $i = 1, 2, \dots, \ell$, *are Fredholm operators. Then*

$$\text{Max}\ \{k_i + n_i \mid 1 \le i \le \ell\} > \dim X_1.$$

Proof. $\forall\ \varepsilon > 0$, $\exists\ F \in \widehat{\mathcal{F}}$ such that $F \subset f_{\hat{c}+\varepsilon}$. Applying the first deformation theorem, the splitting lemma, and the argument in Theorem 2.1, there is a deformation η satisfying

$$\eta\ |_{\partial D} = \ \text{id}_{\partial D}$$

and

$$\eta(F) \subset \overline{\left(f_{\hat{c}-\varepsilon} \backslash \bigcup_{i=1}^{\ell} N_i \right)} \cup \bigcup_{i=1}^{\ell} \Phi_i\,(Q_i)\,.$$

On the one hand, we have $\eta(F) \in \widehat{\mathcal{F}}$. On the other hand, let

$$A = \overline{\left(\eta(F) \backslash \bigcup_{i=1}^{\ell} N_i \right)} \cup \bigcup_{i=1}^{\ell} \Phi_i \left(\partial B_{r_i}^{k_i} \times \{\theta_+\} \times (B_\delta \cap H_i) \right).$$

We have $A \subset \overset{\circ}{f}_{\hat{c}}$, therefore $A \notin \widehat{\mathcal{F}}$. Consequently, $\exists\ \sigma : A \to X_1 \backslash \{\theta\}$ such that $\sigma\ |_{\partial D \cap A} = \ \text{id}_{\partial D \cap A}$.

Suppose that

$$\dim X_1 > \ \text{Max}\ \{k_i + n_i \mid 1 \le i \le \ell\}\,;$$

then the map $\sigma\ |_{\cup_{i=1}^{\ell}} \Phi_i(\partial B_{r_i}^{k_i} \times \{\theta_+\} \times (B_\delta \cap H_i))$ has a continuous extension

$$\tilde{\sigma} : \bigcup_{i=1}^{\ell} \Phi_i\,(Q_i) \to X_1 \backslash \{\theta\}$$

according to Lemma 2.1. Let us define

$$\hat{\sigma}(x) = \begin{cases} \sigma(x) & \forall\ x \in A \\ \tilde{\sigma}(x) & \forall\ x \in \eta(F) \backslash A. \end{cases}$$

Then $\hat{\sigma} : \eta(F) \to X_1 \backslash \{\theta\}$ is continuous. If we choose $r > 0$ so small such that

$$f\ |_{\eta(F) \backslash A} > a, \quad 1 \le i \le \ell,$$

then

$$\begin{aligned} \hat{\sigma}\ |_{\eta(F) \cap \partial D} &= \hat{\sigma}\ |_{A \cap \partial D} = \sigma\ |_{A \cap \partial D} = \ \text{id}_{A \cap \partial D} \\ &= \ \text{id}_{\eta(F) \cap \partial D}, \end{aligned}$$

which implies that $\eta(F) \notin \widehat{\mathcal{F}}$. This is a contradiction.

2.2. Genus and Cogenus

Let X be a Banach space. Let Σ be the family of compact symmetric subsets of X. Two integer valued functions $\gamma^+, \gamma^- : \Sigma \to \mathbb{N} \cup \{+\infty\}$ are defined as follows:

$$\gamma^+(A) = \sup\left\{n \in N \mid \exists\, \varphi : S^{n-1} \to A \quad \text{odd and continuous}\right\}$$

and

$$\gamma^-(A) = \inf\left\{n \in N \mid \exists\, \varphi : A \to S^{n-1} \quad \text{odd and continuous}\right\},$$

in which, if no such φ exists, then we define $\gamma^\pm(A) = +\infty$. $\gamma^+(A)$ is called the genus of A and $\gamma^-(A)$ is called the cogenus of A.

We have the following properties:

(1) $\forall\, \varphi : X \to X$ odd and continuous,

$$\gamma^\pm(A) \le \gamma^\pm(\varphi(A)). \tag{2.4}$$

(2) $\gamma^+(A) \le \gamma^-(A)$.

Claim. If not, $\exists\, A \in \Sigma$ such that $k = \gamma^+(A) > \gamma^-(A) = m$, then there exist continuous odd maps φ_1 and φ_2 satisfying

$$S^{k-1} \xrightarrow{\varphi_1} A \xrightarrow{\varphi_2} S^{m-1}.$$

According to the Borsuk-Ulam theorem, it is impossible.

Let us define two classes of families of subsets in Σ:

$$\mathcal{F}_k^\pm = \left\{A \in \Sigma \mid \gamma^\pm(A) \ge k\right\} \qquad \forall\, k \in \mathbb{N}.$$

Set

$$c_k^\pm = \inf_{A \in \mathcal{F}_k^\pm} \sup_{x \in A} f(x) \qquad \forall\, k \in \mathbb{N}. \tag{2.5}$$

If $c \in \{c_k^* \mid * = + \text{ or } -,\ k = 0, 1, 2, \dots\}$ is finite, and if $f \in C^1(X, \mathbb{R}^1)$ is odd and satisfies $(PS)_c$, then c is a critical value of f, according to Theorem 1.3.

The following relations are obvious:

$$c_k^- \le c_k^+, \qquad k = 0, 1, 2, \dots,$$

and

$$c_k^* \le c_{k+1}^* \qquad * = \pm, \quad k = 0, 1, 2, \ldots .$$

As a counterpart of the theorems in the previous subsection, we have

Theorem 2.3. *Let X be a Hilbert space, and let c be one of the c_k^*, $* = \pm$, $k = 0, 1, 2, \ldots$, defined in (2.5). Assume that $f \in C^2(X, \mathbb{R}^1)$ satisfies the $(PS)_c$ condition. Suppose that $K_c = \{\pm x_1, \pm x_2, \ldots, \pm x_\ell\}$ with Morse indices $\{k_1, k_2, \ldots, k_\ell\}$ and nullities $\{n_1, n_2, \ldots, n_\ell\}$ respectively, and that $\theta \notin K_c$. If $d^2 f(x_i)$, $i = 1, 2, \ldots, \ell$ are Fredholm operators, then*

$$\operatorname{Max}\{k_i + n_i \mid 1 \le i \le \ell\} \ge k - 1 \qquad \textit{as } * = -,$$

and

$$\operatorname{Min}\{k_i \mid 1 \le i \le \ell\} \le k - 1 \qquad \textit{as } * = +.$$

Proof. We prove them by contradiction. Assume that

$$k > \operatorname{Max}\{k_i + n_i \mid 1 \le i \le \ell\} + 1 \qquad \text{as } * = -,$$

and

$$k < \operatorname{Min}\{k_i \mid 1 \le i \le \ell\} + 1 \qquad \text{as } * = +.$$

$\forall\, \varepsilon > 0$, $\exists\, A_0 \in \mathcal{F}_k^*$ such that $A_0 \subset f_{c+\varepsilon}$. As in the proof of Theorem 2.2, we have a deformation η satisfying

$$\eta(A_0) \subset \overline{\left(f_{c-\varepsilon} \backslash \bigcup_{i=1}^{\ell} (N_i \cup (-N_i)) \right)} \cup \bigcup_{i=1}^{\ell} (\Phi_i (Q_i)) \cup (-\Phi_i (Q_i))$$

where $Q_i = B_r^{k_i} \times \{\theta_+\} \times (B_\delta \cap H_i)$ and $r > 0$ is small enough such that $N_i \cap (-N_i) = \emptyset$, $1 \le i \le \ell$. Thus $\eta(A_0) \subset \mathcal{F}_k^*$ and let

$$A = \overline{\left(\eta(A_0) \backslash \bigcup_{i=1}^{\ell} (N_i \cup (-N_i)) \right)}$$
$$\cup \bigcup_{i=1}^{\ell} \left(\pm \Phi_i \left(\partial B_{ri}^{k_i} \times \{\theta_+\} \times (B_\delta \cap H_i) \right) \right) \subset \mathring{f}_c.$$

For $* = -$. $\gamma^-(A) \le k - 1$, i.e., there exists an odd continuous map $\varphi : A \to S^{k-2}$. By the Dugundji Theorem, we may extend

$\varphi \mid_{\cup_{i=1}^{\ell} \pm\phi_i(\partial B_{r_i}^{k_i} \times \{\theta_+\}\times(B_\delta\cap H_i))}$ to $\varphi_1 : \bigcup_{i=1}^{\ell} \mathbb{R}^{k_i+n_i} \to B^{k-1}$ continuously, where B^{k-1} is the unit ball in $k-1$-dimensional Euclidean space. Let

$$\hat{\varphi}(x) = \frac{1}{2}\left(\varphi_1(x) - \varphi_1(-x)\right).$$

Then $\hat{\varphi}$ is an odd extension of $\varphi \mid_{\cup_{i=1}^{\ell}} \left(\pm\Phi_i\left(\partial B_{r_i}^{k_i} \times \{\theta_+\} \times (B_\delta \cap H_i)\right)\right)$ with $\hat{\varphi} : \bigcup_{i-1}^{\ell} \mathbb{R}^{k_i+n_i} \to B^{k-1}$. By the use of the same argument as in Lemma 2.1, we have a pair of points $\pm p_i \in \overset{\circ}{B}{}^{k-1}\backslash\hat{\varphi}(\mathbb{R}^{k_i+n_i})$.

Let $\tau_i^{\pm}$ be the projection onto ∂B^{k-1} from $\pm p_i$; then $\tau_i^{\pm} : B^{k-1}\backslash\{\pm p_i\} \to S^{k-2}$ is continuous, and $\tau_i^{-}(x) = -\tau_i^{+}(-x)$, $i = 1, 2, \ldots, \ell$. Define

$$\tilde{\varphi}(x) = \begin{cases} \varphi(x) & x \in A \\ \tau_i^{\pm}\hat{\varphi}(x) & x \in (\pm N_i) \cap \eta(A_0). \end{cases}$$

Then $\tilde{\varphi} : \eta(A_0) \to S^{k-2}$ is odd and continuous. Therefore $\gamma^-(\eta(A_0)) \le k-1$. This is impossible.

For $* = +$. We have $\gamma^+(A) \le k-1$ and $\gamma^+(\eta(A_0)) \ge k$. From the latter, there exists $\varphi_0 : S^{k-1} \to \eta(A_0)$ odd and continuous. Denoted by P_i^*, $* = +, -, 0$, the orthogonal projections onto the positive, negative and 0-subspaces of $d^2 f(x_i)$ respectively, we conclude that $P_i^- \varphi_0(S^{k-1})$ cannot cover $B_r^{k_i}$, because $k - 1 <$ Min $\{k_i \mid 1 \le i \le \ell\}$, $\ell = 1, 2, \ldots, \ell$. Thus, $\exists\, p_i \in \overset{\circ}{B}{}_r^{k_i}$ such that $\{\pm p_i\} \times (P_i^+ + P_i^0)X \cap \varphi_0(S^{k-1}) = \emptyset$, $i = 1, 2, \ldots, \ell$. Let $\tau_i^+ : B_r^{k_i}\backslash\{p_i\} \to \partial B_r^{k_i}$ be the projection onto the sphere from p_i, let $\tau_i^-(z) = -\tau_i^+(z)$, and let

$$\tau_i(x) = \begin{cases} \tau_i^{\pm}\left(P_i^- x\right) + \left(P_i^+ + P_i^0\right)x & \text{if } x \in \pm(B_r^{k_i}\backslash\{p_i\}) \times \left(\left(P_i^+ + P_i^0\right)X\right) \\ x & \text{if } x \notin \pm B_r^{k_i} \times \left(\left(P_i^+ + P_i^0\right)X\right), \end{cases}$$

for $i = 1, 2, \ldots, \ell$.

The functions $\tau_i : X\backslash\bigcup_{i=1}^{\ell}(\{\pm p_i\} \times (P_i^+ + P_i^0)X) \to X$, $i = 1, 2, \ldots, \ell$ are odd and continuous. Therefore $\tau_\ell \circ \cdots \circ \tau_1 \circ \varphi_0 : S^{k-1} \to A$ is odd and continuous which implies that $\gamma^+(A) \ge k$. Again it is impossible.

Remark 2.1. The same conclusion holds if $f \in C^1(S, \mathbb{R}^1)$, where S is the unit sphere of a Hilbert space, and if $\mathcal{F}_k^{\pm}$ are modified to be families of subsets on S.

Remark 2.2. Theorems 2.1 and 2.2 were obtained by Lazer-Solimini [LaS1], and Theorem 2.1′ improves a result due to J. T. Schwartz [ScJ1]. Lemma 2.1 can be found in L. Nirenberg [Nir1]. Theorem 2.3, under the assumption that f satisfies the Palais-Smale condition (not only $(PS)_c$), was given by A. Bahri [Bah1] for genus, and A. Bahri, P. L. Lions [BaL1] for cogenus. For some extensions, see for instance, Coffman [Cof1,2], S. Solimini [Sol2] and N. Ghoussoub [Gho1].

3. Connections with Other Theories

3.1. Degree Theory

Let M and N be oriented n-dimensional manifolds without boundary and let

$$T : M \to N$$

be a smooth map. If M is compact and N is connected, then the Brouwer degree $\deg(T)$ is well-defined [Nir 1].

First consider an open set $U \subset \mathbb{R}^n$, and a smooth vector field

$$V : U \to \mathbb{R}^n$$

with isolated zero at the point $p \in U$. The function

$$\overline{V}(x) = \frac{V(x)}{\|V(x)\|}$$

maps a small sphere centered at p into the unit sphere S^{n-1}. The degree of $\overline{V}$ is called the index of V at the zero p, denoted by the index (V, p).

For a smooth vector field V on an arbitrary manifold M (finite dimensional) with isolated zero at the point $p \in M$, the index of V at the zero p is defined as equal to the index of the corresponding vector field $dg^{-1} \circ V \circ g$ at zero $g^{-1}(p)$, where $g : U \to M$ is a parametrization of a neighborhood of p in M.

The celebrated Poincaré-Hopf Theorem studies the relationship between the indices of zeros of a smooth vector field on M and the Euler characteristic of the manifold M. Namely,

Theorem 3.1. (Poincaré-Hopf). *If M^m is a compact manifold with boundary ∂M, and if V is a smooth vector field on M which points outward at all boundary points and has only isolated zeros, then the sum of the indices at the zeros p_i, $i = 1, \dots, n$, of such a vector field is equal to the Euler characteristic of M, i.e.,*

$$\sum_{i=1}^{n} \text{index } (V, p_i) = \chi(M) \triangleq \sum_{q=1}^{m} (-1)^q \text{ rank } H_q(M).$$

We shall extend this theorem to bounded domains on an infinite dimensional Hilbert space for a special case of vector fields, the gradient vector fields. Let H be a real Hilbert space, and let D be a bounded open set in H. For a vector field $V : \overline{D} \to H$, with $\theta \notin V(\partial D)$ and $\varphi = \text{ id } - V$ compact, the Leray-Schauder degree

$$\deg(V, D, \theta)$$

is well-defined, cf. [Llo1]. Suppose that p is an isolated zero of V; then there exists a ball $B_\varepsilon(p_0)$ with radius $\varepsilon > 0$ such that V has no zeros in $B_\varepsilon(p)$ other than p. Hence it is possible to define

$$\text{index}\,(V,p) = \deg\,(V, B_\varepsilon(p), \theta)$$

which is independent of ε for $\varepsilon > 0$ sufficiently small in view of the excision property of the Leray-Schauder degree. The number index(V,p) is called the index of V at the point p. In case $\dim H < +\infty$, the definition of index(V,p) coincides with the previous one.

We start with a local result in which the connection between the index of the gradient vector field of a function f at its isolated critical point and the critical groups of that point is studied; namely:

Theorem 3.2. *Let H be a real Hilbert space, and let $f \in C^2(H, \mathbb{R}^1)$ be a function satisfying the (PS) condition. Assume that*

$$df(x) = x - T(x),$$

where T is a compact mapping, and that p_0 is an isolated critical point of f. Then we have

$$\text{ind}\,(df, p_0) = \sum_{q=0}^{\infty} (-1)^q \,\text{rank}\, C_q\,(f, p_0)\,. \tag{3.1}$$

Proof. 1. First, we assume that p_0 is nondegenerate. Since T is compact, we see that the Hessian

$$d^2 f(p_0) = \text{ id } - dT(p_0)$$

has only finite index j. By definition, and by Leray's formula

$$\text{index}\,(df, p_0) = (-1)^j.$$

In view of Theorem 4.1 in Chapter I, we have

$$C_q\,(f, p_0) = \delta_{q,j} G.$$

Thus,

$$\text{index}\,(df, p_0) = \sum_{q=0}^{\infty} (-1)^q \,\text{rank}\, C_q\,(f, p_0)$$

is proved in this special case.

2. For an isolated degenerate p_0 we may assume for simplicity that $p_0 = \theta$ and $f(p_0) = 0$. Let (W, W_-) be a Gromoll-Meyer pair constructed

in Chapter I, (5.11) and (5.12), and let $\delta > 0$ be sufficiently small such that $B_\delta \subset \overset{\circ}{W} \cap f^{-1}[-\frac{\gamma}{3}, \frac{\gamma}{3}]$, where γ is the real number appearing in Chapter I (5.11).

We shall define a function $\tilde{f}$ satisfying the (PS) condition such that

(1) $|f(x) - \tilde{f}(x)| < \frac{\gamma}{3}\ \forall\, x \in H$.
(2) $df(x) = d\tilde{f}(x)$ for x in a neighborhood of ∂W.
(3) In W, $\tilde{f}$ has only nondegenerate critical points $\{p_j\}_1^m$, finite in number, contained in B_δ.

Once the function $\tilde{f}$ is constructed, we obtain immediately

$$W_- = f_{-\gamma} \cap W \subset \tilde{f}_{-\frac{2}{3}\gamma} \cap W \subset f_{-\frac{\gamma}{3}} \cap W \subset f_{\frac{\gamma}{3}} \cap W \subset \tilde{f}_{\frac{2}{3}\gamma} \cap W \subset f_\gamma \cap W = W.$$

However, there are strong deformation retracts:

$$f_\gamma \cap W \to f_{\frac{\gamma}{3}} \cap W \quad \text{and} \quad f_{-\frac{\gamma}{3}} \cap W \to f_{-\gamma} \cap W$$

provided by the Gromoll-Meyer property. We have

$$H_*(W, W_-) = H_*\left(\tilde{f}_{\frac{2}{3}\gamma} \cap W, \tilde{f}_{-\frac{2}{3}\gamma} \cap W\right) \tag{3.2}$$

due to the exactness of the homological group sequence. Thus,

$$\begin{aligned} \text{index}\,(df, \theta) &= \deg(df, W, \theta) \\ &= \deg(d\tilde{f}, W, \theta) \qquad \text{(by (2))} \\ &= \sum_{j=1}^m \text{index}\left(d\tilde{f}, p_j\right) \qquad \text{(by (3))} \\ &= \sum_{j=1}^m \sum_{q=0}^\infty (-1)^q \operatorname{rank} C_q\left(\tilde{f}, p_j\right) \qquad \text{(by 1).} \end{aligned}$$

Applying Theorem 4.3 and Remark 4.1 of Chapter I to the function $\tilde{f}$ on W, we have

$$\begin{aligned} \sum_{j=1}^m \sum_{q=0}^\infty (-1)^q \operatorname{rank} C_q\left(\tilde{f}, p_j\right) &= \sum_{q=0}^\infty (-1)^q \operatorname{rank} H_q\left(\tilde{f}_{\frac{2}{3}\gamma} \cap W, \tilde{f}_{-\frac{2}{3}\gamma} \cap W\right) \\ &= \sum_{q=0}^\infty (-1)^q \operatorname{rank} H_q(W, W_-) \\ &= \sum_{q=0}^\infty (-1)^q \operatorname{rank} C_q(f, \theta). \end{aligned}$$

This is due to the fact that the negative gradient flow of $\tilde{f}$ directs inward on $\partial W\backslash W_-$, and hence also on $\partial W\backslash(W\cap\tilde{f}^{-1}(-\frac{2}{3}\gamma))$.

3. Finally, we shall construct a function $\tilde{f}$, satisfying the (PS) condition as well as conditions (1)–(3). We define

$$\tilde{f}(x) = f(x) + p(\|x\|)\,(x_0, x)\,,$$

where $p \in C^2(\mathbb{R}^1_+, \mathbb{R}^1)$ is a function satisfying

$$p(t) = \begin{cases} 1 & 0 \le t \le \delta/2 \\ 0 & t > \delta, \end{cases}$$

with $0 \le p \le 1$ and $|p'(t)| \le \frac{4}{\delta}$, $x_0 \in H$ is determined later. Let

$$\beta = \inf\{\|df\,(x)\|\ \ |x \in B_\delta\backslash B_{\delta/2}\};$$

then $\beta > 0$. We choose $x_0 \in H$ such that

$$0 < \|x_0\| < \min\left\{\frac{\beta}{6}, \frac{\gamma}{3}\right\}.$$

Then we have

$$\begin{aligned} &\left|f(x) - \tilde{f}(x)\right| < \frac{\gamma}{3} \quad \forall\, x \in H \\ &\left\|d\tilde{f}(x)\right\| \ge \frac{\beta}{6} \quad \forall\, x \in B_\delta\backslash B_{\delta/2} \\ &d\tilde{f}(x) = df(x) \quad \forall\, x \notin B_\delta. \end{aligned}$$

For smaller $\|x_0\|$, $d\tilde{f}$ is a k-set contraction mapping vector field with $k < 1$ (cf. Lloyd [Llo1]). Therefore $\deg(d\tilde{f}, W, \theta)$ is well-defined. The (PS) condition for the function $\tilde{f}$ is verified directly. On account of the Sard-Smale theorem, a suitable x_0 can be chosen such that $\tilde{f}$ is nondegenerate. The proof is complete.

Corollary 3.1. *Under the assumptions of Theorem 1.6, if p_0 is a mountain pass point, and if the smallest eigenvalue λ_1 of $d^2f(p_0)$ is simple whenever $\lambda_1 = 0$, then $\lambda_1 \le 0$, and*

$$\textit{index}\ (df, p_0) = -1.$$

Proof. If $\lambda_1 > 0$, then p_0 must be a local minimum, which implies $C_1(f, p_0) = 0$. This contradicts the assumption that p_0 is a mountain pass point. Therefore $\lambda_1 \le 0$.

In the case where $\lambda_1 = 0$, we have assumed $\dim \ker d^2 f(x_0) = 1$. It follows from Theorem 1.6 that $C_q(f, p_0) = \delta_{q1} G$.

In the case where $\lambda_1 < 0$, the Morse index $\operatorname{ind}(f, p_0) = k > 0$. From the mountain pass point assumption, it must be $k \leq 1$. Consequently, $k = 1$. Again, we apply Theorem 1.6, $C_q(f, p_0) = \delta_{q1} G$.

In both cases our conclusion follows from Theorem 3.2.

We generalize Theorem 3.2 as follows:

Theorem 3.3. *Under the assumption in Theorem 3.2, suppose that W is a bounded domain in H on which f is bounded. Assume that*

(1) $W_- \triangleq \{x \in \partial W \mid \eta(t, x) \notin W \ \forall\, t > 0\} = W \cap f^{-1}(a)$ *for some a, where $\eta(t, x)$ is the negative gradient flow of f emanating from x;*
(2) $-df|_{\partial W \setminus W_-}$ *directs inward.*

Then we have

$$\deg(df, W, \theta) = \chi(W, W_-). \tag{3.3}$$

Proof. Due to assumptions (1) and (2), $\theta \notin df(\partial W)$, the Leray-Schauder degree $\deg(df, W, \theta)$ is well-defined.

If f is nondegenerate on W, then the critical set K consists of finitely many isolated point $\{p_j\}_1^m$ since f is bounded on W, and assumption (2), as well as the (PS) condition holds. According to Theorem 3.2, we have

$$\begin{aligned} \deg(df, W, \theta) &= \sum_{j=1}^{m} \text{ index } (df, p_j) \\ &= \sum_{j=1}^{m} \sum_{q=o}^{\infty} (-1)^q \text{ rank } C_q\,(f, p_j) \\ &= \sum_{q=0}^{\infty} (-1)^q \text{ rank } H_q(W, W_-). \end{aligned}$$

The last equality follows from assumption (2), Theorem 4.3 of Chapter I, and the fact that $f(p_j) > a$, which is a consequence of assumption (1).

If f is degenerate, we perturb it as in Theorem 3.2. Since the critical set K is compact in $\overset{\circ}{W}$, we construct a C^2- function $p(x)$, satisfying:

$$p(x) = \begin{cases} 0 & x \in \Omega_\delta, \\ 1 & x \notin \Omega_{2\delta}, \end{cases}$$

where $\delta > 0$ such that $\operatorname{dist}(K, \partial W) \geq 2\delta$, K is the critical set of f in W, and

$$\Omega_\delta = \{x \in W \mid \operatorname{dist}(x, \partial W) < \delta\}.$$

We may assume that

$$|p(x)| \le 1 \quad \text{and} \quad |p'(x)| \le M_2 < +\infty.$$

Let

$$M_1 = \sup\{\|x\| \mid x \in W\}, \quad b = \inf\{f(x) \mid x \in W \backslash \Omega_\delta\}$$

and

$$\beta = \min\left\{\inf_{x \in \Omega_{2\delta}} \|df(x)\|, 1\right\}.$$

Then, by our assumptions, $b > a$ and $\beta > 0$. One defines

$$\tilde{f}(x) = f(x) + p(x)\,(x_0, x)$$

for suitable $x_0 \in H$, with

$$0 < \|x_0\| < \min\left\{\frac{b-a}{M_1}, \frac{\beta}{3}, \frac{\beta}{3M_1M_2}\right\}.$$

By the Sard-Smale Theorem, x_0 can be chosen such that $\tilde{f}$ is nondegenerate. Thus,

$$\begin{aligned}\left\|d\tilde{f}(x)\right\| &\ge \|df(x)\| - \|x_0\| - \|p'(x)\|\,\|x\| \cdot \|x_0\| \\ &\ge \beta - \frac{1}{3}\beta - M_1M_2\,\|x_0\| \\ &> \frac{1}{3}\beta > 0 \qquad \forall\, x \in \Omega_{2\delta}.\end{aligned}$$

Now the function $\tilde{f}$ satisfies the (PS) condition and assumptions (1) and (2). Indeed, the (PS) condition is easily verified by the above estimate, assumption (2) is trivial, and assumption (1) is verified by the following inequality:

$$\tilde{f}(x) \ge f(x) - \|x_0\|\,\|x\| \ge f(x) - \frac{b-a}{M_1} \cdot M_1 > a \quad \forall\, x \in W \backslash \Omega_\delta.$$

Since

$$\deg(df, W, \theta) = \deg\left(d\tilde{f}, W, \theta\right)$$

(the RHS is the generalized Leray-Schauder degree for a k-set contraction mapping vector field with $k < 1$), the proof is completed.

Remark 3.1. Theorem 3.2 was given by E. Rothe [Rot1] under stronger assumptions, and Corollary 3.1 was obtained by Hofer [Hof4] and Tian [Tia1] by a combination of the Poincaré-Hopf Theorem with the splitting theorem.

3.2. Ljusternik-Schnirelman Theory

Let X be a topological space. The category of a closed subset A of X is defined by $\mathrm{Cat}_X(A) = \inf\{m \in \mathbb{N} \cup \{+\infty\} \mid \exists$ contractible closed subsets $F_1, \dots, F_m$ in X, such that $A \subset \cup_{i=1}^m F_i\}$.

In the following let M be a C^1-Finsler manifold. We write simply $\mathrm{Cat}\,(M) = \mathrm{Cat}_M(M)$.

Ljusternik-Schnirelman Theorem. *Suppose that $f \in C^1(M, \mathbb{R}^1)$ is a function bounded from below, satisfying the (PS) condition. Then f has a least Cat(M) critical points.*

The topological invariant Cat(M) can be estimated by other topological invariants, for instance, the cup length of M. Namely, we have

$$\mathrm{Cat}(M) \geq CL(M) + 1.$$

For a proof and a more general result, see Corollary 3.5 below.

Now we extend this relation to estimate the number of critical points.

It is known that for a nontrivial class $[\tau] \in H_*(f_b, f_a)$, $a < b$, one can determine a critical value

$$c = \inf_{\tau \in [\tau]} \sup_{x \in |\tau|} f(x) \tag{3.4}$$

if $c \in (a, b)$ and if f satisfies the $(\mathrm{PS})_c$ condition.

It is natural to ask, if we have two homology classes $[\tau_1], [\tau_2] \in H_*(f_b, f_a)$, both nontrivial, and if c_1, c_2 are defined in the same way as in (3.4), are there two distinct critical points? Generally speaking, no, but we have

Lemma 3.1. *Suppose that $f \in C^1(M, \mathbb{R}^1)$, has regular values $a < b$. Let $[\tau_1], [\tau_2] \in H_*(f_b, f_a)$ be nontrivial classes. Suppose that $\exists\, \omega \in H^*(f_b)$, with $\dim \omega > 0$, such that*

$$[\tau_1] = [\tau_2] \cap \omega.$$

Set

$$c_i = \inf_{\tau_i \in [\tau_i]} \sup_{x \in |\tau_i|} f(x), \quad i = 1, 2. \tag{3.5}$$

Assume that f satisfies $(PS)_{c_i}$, $i = 1, 2$, and that $\exists$ a neighborhood N of K_{c_2} and a singular cochain $\hat{\omega} \in \omega$ such that supp $\hat{\omega} \cap N = \emptyset$. Then $c_1 < c_2$ are two distinct critical values.

Proof. By definition $c_1 \leq c_2$. It remains to prove $c_1 < c_2$.

$\forall \varepsilon > 0$, $\exists$ a singular relative closed chain $\tau_2 \in [\tau_2]$ such that $|\tau_2| \subset f_{c_2+\varepsilon}$. We choose a neighborhood N' of K_{c_2}, such that $N' \subset \overline{N}' \subset N$, and subdivide τ_2 into $\tau_2' + \tau_2''$, such that $|\tau_2'| \subset N$ and $|\tau_2''| \subset f_{c_2+\varepsilon}\backslash N'$. According to the assumption, supp $\hat{\omega} \cap N = \emptyset$, and we have

$$\tau_1 = \tau_2 \cap \widehat{\omega} = \tau_2'' \cap \hat{\omega},$$

which implies $|\tau_1| \subset f_{c_2+\varepsilon}\backslash N'$.

Since a and b are assumed to be regular, then $a < c_1 \le c_2 < b$.

According to the first deformation lemma, there exist constants $0 < \varepsilon < \bar{\varepsilon} <$ Min $\{b - c_2, c_1 - a\}$ and $\eta \in C([0,1] \times M, M)$ satisfying

$$\eta\left(1, f_{c_2+\varepsilon}\backslash N'\right) \subset f_{c_2-\varepsilon},$$

and

$$\eta(1,\cdot) \sim \text{id} \quad \text{in} \quad (f_b, f_a)\,.$$

Therefore

$$\eta\left(1, |\tau_1|\right) \subset f_{c_2-\varepsilon}.$$

However, $\eta(1, \tau_1) \in [\tau_1]$, and it follows

$$c_1 \le c_2 - \varepsilon < c_2.$$

We emphasize that the positive lower bound of the difference $c_2 - c_1$ depends on the behavior of df near K_{c_2}.

Corollary 3.2. *Suppose that there are positive constants d, $\bar{\varepsilon}$, and δ satisfying*

$$\|df(x)\| \ge d \quad \forall\, x \in f_{c_2+\bar{\varepsilon}}\backslash \left(f_{c_2-\bar{\varepsilon}} \cup \left(K_{c_2}\right)_\delta\right) \tag{3.6}$$

and

$$\bar{\varepsilon} < \mathit{Min}\ \left\{2\delta d^2, \delta d\right\}. \tag{3.7}$$

where $(K_c)_\delta = \{x \in M \mid \mathit{dist}\,(x, K_c) \le \delta\}$. Then

$$c_1 < c_2 - \bar{\varepsilon}/3.$$

Proof. The proof is the same as above. The only thing changed is that the first deformation theorem is replaced by Theorem 3.4 from Chapter I.

Let us recall the notion $[\tau_1] < [\tau_2]$, cf. Definition 1.1, Chapter I.

Corollary 3.3. *Suppose that $f \in C^1(M, \mathbb{R}^1)$ and that $a < b$ are regular values. Assume that $[\tau_1] < [\tau_2]$ are nontrivial classes in $H_*(f_b, f_a)$, and* *that $(PS)_{c_i}$ holds, where c_i is defined in* (3.5), $i = 1, 2$. *If* $\#K_{c_2} < \infty$, *then* $c_1 < c_2$ *are two distinct critical values.*

Proof. We follow the proof of Lemma 3.1. Since $\#K_{c_2} < \infty$, there exist two contractible neighborhoods of K_{c_2}, $N' \subset \overline{N}' \subset N$. We write $\tau_2 = \tau_2' + \tau_2''$ where $|\tau_2'| \subset N$ and $|\tau_2''| \subset f_{c_2+\varepsilon} \backslash N'$. Let $\omega \in H^*(f_b)$ be the cohomology class satisfying

$$\dim \omega > 0 \quad \text{and} \quad [\tau_1] = [\tau_2] \cap \omega.$$

The cap product does not change if we shrink N to a point because there exists $\hat{\omega} \in [\omega]$ which applies to any chain having support in N gives 0. Therefore $[\tau_2'] \cap \omega = 0$.

The remainder of the proof is the same as Lemma 3.1.

Theorem 3.4. *Suppose that $f \in C^1(M, \mathbb{R}^1)$ and that $a < b$ are regular values. Assume that f satisfies the (PS) condition, and that f has only isolated critical points in $f^{-1}[a, b]$. If there are m-nontrivial homology classes $[\tau_1] < [\tau_2] < \ldots < [\tau_m]$ in $H_*(f_b, f_a)$, then f has at least m-distinct critical values.*

Proof. It follows directly from Corollary 3.3.

Corollary 3.4. *Suppose that $f \in C^1(M, \mathbb{R}^1)$ is bounded from below and that f satisfies the (PS) condition. Then f has at least $CL(M) + 1$ critical points.*

Proof. According to Theorem 1.1 from Chapter I, $L(X, Y) = CL(X, Y) + 1$ for any topological pair (X, Y). We have $m = CL(M) + 1$ nontrivial homology classes $[\tau_1] < [\tau_2], \ldots, [\tau_m]$, by taking $Y = \emptyset$, and $X = M$. ($f_a = \emptyset$ if $a < \inf f$). The conclusion follows from Theorem 3.4.

Theorem 3.5. *Suppose that $f \in C^1(M, \mathbb{R}^1)$ and that $a < b$ are regular values. Assume that $[\tau_1] < [\tau_2] < \ldots < [\tau_m]$ are nontrivial homology classes in $H_*(f_b, f_a)$. Let*

$$c_i = \inf_{\tau_i \in [\tau_i]} \sup_{x \in |\tau_i|} f(x), \quad i = 1, 2, \ldots, m. \tag{3.8}$$

If $c = c_1 = c_2 = \ldots = c_m$, and if f satisfies the $(PS)_c$ condition, then we have

$$\operatorname{Cat}_M (K_c) \geq m.$$

Proof. Since K_c is compact, we may choose neighborhoods $N' \subset \overline{N}' \subset N$ of K_c satisfying the following conditions:

(1) $\operatorname{Cat}_M(N) = \operatorname{Cat}_M(K_c)$;

(2) there exist constants $0 < \varepsilon < \bar{\varepsilon} <$ Min $\{b - c, c - a\}$ and $\eta \in C(M, M)$ such that

$$\eta|_{f_{c-\bar{\varepsilon}}} = \mathrm{id}_{f_{c-\bar{\varepsilon}}},$$
$$\eta \sim \mathrm{id},$$

and

$$\eta\,(f_{c+\varepsilon}\backslash N') \subset f_{c-\varepsilon},$$

according to the first deformation theorem.

We prove our theorem by contradiction. If $\mathrm{Cat}_M(K_c) \leq m-1$, then there are $(m-1)$ contractible closed sets $\{B_j\}_1^{m-1}$ cover N, i.e., $\cup_{j=1}^{m-1} B_j \supset N$. Since $[\tau_i] < [\tau_2] < \ldots < [\tau_m]$, there exist $\omega_2, \omega_3, \ldots, \omega_m \in H^*(f_b)$, $\dim \omega_j > 0$ such that

$$[\tau_{j-1}] = [\tau_j] \cap \omega_j, \quad j = 2, 3, \ldots, m.$$

Since $\dim \omega_j > 0$ and B_j is contractible, we always may choose $\hat{\omega}_j \in \omega_j$, with $|\hat{\omega}_j| \cap B_j = \emptyset, \quad j = 2, 3, \ldots, m.$

We choose $\tau \in [\tau_m]$ which has support $|\tau| \subset f_{c+\varepsilon}$. Subdividing τ into $\tau = \sigma_1 + \sigma_2 + \ldots + \sigma_m$ such that

$$|\sigma_1| \subset f_{c+\varepsilon}\backslash N', \quad \text{and} \quad |\sigma_j| \subset B_j, \quad j = 2, 3, \ldots, m,$$

one has

$$\begin{aligned} \tau' &= \tau \cap (\hat{\omega}_2 \cup \ldots \cup \hat{\omega}_m) \\ &= \sigma_1 \cap (\hat{\omega}_2 \cup \ldots \cup \hat{\omega}_m). \end{aligned}$$

Hence $|\tau'| \subset f_{c+\varepsilon}\backslash N'$ and $\eta(|\tau'|) \subset f_{c-\varepsilon}$. However $\tau' \in [\tau_1]$; therefore $\eta(\tau') \in [\tau_1]$, which implies

$$c_1 \leq c - \varepsilon.$$

This is a contradiction.

Corollary 3.5. *Suppose that $f \in C^1(M, \mathbb{R}^1)$ and that $a < b$ are regular values. Assume that $[\tau_1] < [\tau_2] < \ldots < [\tau_m]$ are nontrivial homology classes in $H_*(f_b, f_a)$. Let c_i be defined as in (3.8), $i = 1, 2, \ldots, m$. Suppose that there are two open neighborhoods $N' \subset N$ of $\cup_{i=1}^m K_{c_i}$ satisfying dist $(N', \partial N) \geq \frac{7}{8}\delta$, $\delta > 0$ and that there are constants $\bar{b}, \bar{\varepsilon} > \varepsilon > 0$ and c such that*

(1) $\|df(x)\| \geq \bar{b} \quad \forall\, x \in f_{c+\bar{\varepsilon}}\backslash(f_{c-\bar{\varepsilon}} \cup N')$,
(2) $0 < \bar{\varepsilon} < \mathit{Min}\,\{\frac{1}{4}\delta\bar{b}^2, \frac{1}{8}\delta b\}$,
(3) $c - \varepsilon < c_1 \leq c_2 \leq \ldots \leq c_m < c + \varepsilon$.

Then $\mathrm{Cat}_M \overline{(N)} \geq m$.

Proof. This is a combination of the proofs of Theorem 3.5 and Corollary 3.2.

3.3. Relative Category

Definition 3.1. Let (X, Y) be a topological pair, and let Y be closed. A closed subset A of X is said to be of m category relative to Y, and we write $\mathrm{Cat}_{X,Y}(A) = m$, if

$$m = \inf\{n \in \mathbb{N} \cup \{+\infty\} \mid \exists \text{ closed subsets } F_0, F_1, \ldots, F_n, \\ \text{and } \exists\, h_i \in C([0,1] \times F_i; X),$$

such that

(1) $h_i(0,x) = x, \quad \forall\, x \in F_i, \quad 0 \le i \le m,$
(2) $h_0(1,x) \in Y, \quad \forall\, x \in F_0,$ and $h_0(t,x) = x\ \forall (t,x) \in [0,1] \times (Y \cap F_i),$
(3) $h_i(1,x) = x_i, \quad \forall\, x \in F_i,$ for some $x_i \in X, \quad 1 \le i \le m\}.$

We defined the cuplength of a topological pair (X, Y) in Section 1 Chapter I and related it to a notion concerning the length of the chain of subordinate nontrivial homology classes $L(X,Y)$, $L(X,Y) \ge CL(X,Y) + 1$. What is the relationship between $\mathrm{Cat}_{X,Y}(X)$ and $CL(X,Y)$?

Theorem 3.6. *$Cat_{X,Y}(X) \ge L(X,Y)$.*

Proof. Let $m = \mathrm{Cat}_{X,Y}(X)$. Next, assume that there are $[\tau_j] \in H_*(X,Y)$, $0 \le j \le m$, with $[\tau_0] < [\tau_1] < \ldots < [\tau_m]$. We have to prove $[\tau_0] = 0$. Assume, by contradiction, that $[\tau_0] \ne 0$. By definition we have $[\tau_{j-1}] = [\tau_j] \cap \omega_j$, $j = 1, 2, \ldots, m$, with $\omega_j \in H^*(X)$, $\dim \omega_j > 0$. Moreover there is a class $\omega_0 \in H^*(X,Y)$ such that

$$\begin{aligned} 0 \ne \langle [\tau_0], \omega_0 \rangle &= \langle [\tau_1] \cap \omega_1, \omega_0 \rangle \\ &= \ldots =< [\tau_m], \omega_m \cup \omega_{m-1} \cup \ldots \cup \omega_0 > . \end{aligned}$$

Consequently,

$$\omega_m \cup \omega_{m-1} \cup \ldots \cup \omega_0 \ne 0. \tag{3.9}$$

Consider the exact sequence

$$\to H^*(X, F_i) \xrightarrow{\eta_i} H^*(X) \xrightarrow{j_i^*} H^*(F_i) \to \ldots$$

for $i = 1, 2, \ldots, m$, and

$$\to H^*(X, F_0 \cup Y) \xrightarrow{\eta_0^*} H^*(X,Y) \xrightarrow{j_0^*} H^*(F_0 \cup Y, Y) \to \ldots .$$

We claim that η_i^* is surjective for $* \ge 1$, $i \ge 1$ and that η_0^* is surjective for $* \ge 0$. Indeed, provided by assumption (3), the injection j_i: $F_i \to X$ is homotopic to the constant map $F_i \mapsto x_i$, and we conclude that j_i^* :

$H^*(X) \to H^*(F_i)$ is trivial for $* \geq 1$, and then η_i^* is surjective for $* \geq 1$ and $i \geq 1$. Next we define

$$g(t,x) = \begin{cases} h_0(t,x) & \text{if } (t,x) \in [0,1] \times F_0 \\ x & \text{if } (t,x) \in [0,1] \times Y\,; \end{cases}$$

because of assumption (2), g is well-defined and continuous. Then the injection $j_0 \cdot (F_0 \cup Y, Y) \to (X,Y)$ is homotopic to the map $g(1,\cdot)$, which maps $F_0 \cup Y$ into Y. Again, by exactness, $\eta_0^* : H^*(X, F_0 \cup Y) \to H^*(X,Y)$ is surjective for $* \geq 0$. This proves the claim. Combining with the commutative diagram

$$\begin{array}{ccccccccc} H^*(X, F_0 \cup Y) \otimes & H^*(X,F_1) \otimes & \cdots \otimes & H^*(X,F_m) & \xrightarrow{\cup} & H^*\left(X, Y \cup \bigcup_{j=0}^{m} F_j\right) \equiv \{\theta\} \\ \downarrow \eta_0^* & \downarrow \eta_1^* & & \downarrow \eta_m^* & & \downarrow \\ H^*(X,Y) \otimes & H^*(X) \otimes & \cdots \otimes & H^*(X) & \xrightarrow{\cup} & H^*(X,Y) \end{array}$$

We find that $\omega_m \cup \omega_{m-1} \cup \ldots , \cup\, \omega_0 = 0$, which contradicts (3.9). This finishes the proof.

Corollary 3.6. *$Cat(X) \geq CL(X) + 1$.*

Claim. By definitions, $\text{Cat}_{X,\emptyset}(X) = \text{Cat}\,(X)$, and $CL(X,\emptyset) = CL(X)$. It is obvious that the following relative properties hold.

(1) If $Y = \emptyset$, then $F_0 = \emptyset$, and $\text{Cat}_{X,\emptyset}(A) = \text{Cat}_X(A)$.

(2) Let A, B be closed subsets of X, then

$$\text{Cat}_{X,Y}(A \cup B) \leq \text{Cat}_{X,Y}(A) + \text{Cat}_X(B).$$

(3) $\text{Cat}_{X,Y}(A) = 0 \Leftrightarrow \exists\, h : [0,1] \times A \to X$ such that
   (i) $h(0, \cdot\,) = \text{id}_A$,
   (ii) $h(1,A) \subset Y$,
   (iii) $h(t, \cdot\,)|_{A \cap Y} = \text{id}_{A \cap Y}$.

(4) If $\exists\, h \in C([0,1] \times A, X)$ such that $h(t, \cdot\,)|_{A \cap Y} = \text{id}_{A \cap Y}$, then $\text{Cat}_{X,Y}(A) \leq \text{Cat}_{X,Y}(B)$, where $B = h(1,A)$.

Remark 3.2. The relative category was initially studied by E. Fadell [Fad1]. The notion was consequently used by G. Fournier and M. Willem [FoW1]. Theorem 3.5 was independently proved by Chang, Long, Zehnder [CLZ1] and Fournier-Willem [FoW2]. A different definition was given by A. Szulkin [Szu1]. Similarly, but with a different idea, see J. Q. Liu [Liu2].

4. Invariant Functions

For a G manifold M, generally speaking, invariant functions possess more critical points than functions without symmetry. That is, if the action is free. As a matter of fact, an invariant function is regarded as a function defined on the quotient manifold M/G, and, in many cases, the quotient manifold M/G has a richer topology than the manifold M, e.g, $M = S^n$, $G = \mathbb{Z}_2$, which contains the antipodal map as an action, $M/G = P^n$, the real projective space. If the action is not free, then things are not so simple, but the fact that there are more critical points remains true. In this respect there are many interesting theories, among them, Ljusternik-Schnirelman category theory, genus and cogenus theory, cohomology index theory and pseudo index theory.

The purpose of this section is to present a different approach which is consistent with our Morse Theory. That is, in replacing index theory, we use the subordinate relative homology classes to obtain multiple critical points. Furthermore, a Galerkin approximation method is employed to avoid the pseudo index theory. For the sake of simplicity, we are only concerned with the two groups $G_1 = \mathbb{Z}_2$ and $G_2 = S^1$. It is well-known that $E = S^\infty$ is the universal total G-space for $G = G_i$, $i = 1, 2$, and that the classifying spaces B are P^∞ and $\mathbb{C}P^\infty$ respectively.

The cohomological rings $H^*(P^\infty, \mathbb{Z}_2)$ and $H^*(\mathbb{C}P^\infty, \mathbb{Q})$ read as follows:

$$H^*\left(P^\infty, \mathbb{Z}_2\right) = \mathbb{Z}_2(\omega), \quad \dim \omega = 1,$$

and

$$H^*\left(\mathbb{C}P^\infty, \mathbb{Q}\right) = \mathbb{Q}(\omega), \quad \dim \omega = 2,$$

where $R(\omega)$ is a polynomial ring generated by ω with the coefficient ring R, and $\mathbb{Q}$ is the rational field.

Let X be a Hilbert space, and let $G = G_1$ and G_2 act on X orthogonally (in the case where G_1, X is real) or unitarily (in the case where G_2, X is complex).

For $G = \mathbb{Z}_2$, one can find orthogonal subspaces X_1 and $X_2 = X_1^\perp$ such that $X_1 = \mathrm{Fix}_G$ and that the G action has the representation $x \mapsto -x \;\forall\, x \in X_2$. Thus G acts freely on the invariant subspace X_2.

For $G = S^1$, according to the Stone Theorem, $\forall\, g = e^{i\theta}$, $\theta \in [0, 2\pi]$, the unitary representation has the following spectral decomposition:

$$T(g) = \int_{-\infty}^{\infty} e^{i\lambda\theta} dE_\lambda,$$

where E_λ is the spectral family with spectrum $\sigma \subset \mathbb{Z}$. Let $X_1 = (E_{+0} - E_{-0})X$, which is the fixed point set Fix_G, and let $X_2 = X_1^\perp$.

It is worth noting that $\forall\, x \in X_2 \backslash \{\theta\}$, the isotropy group G_x is a finite group. Indeed, let $V \subset X_2$ be a finite dimensional invariant subspace.

Then V is isomorphic to the unitary space $\mathbb{C}^k$, and the group action is represented by the diagonal matrices

$$T(g) = \text{ diag } \left\{e^{i\lambda_1\theta}, e^{i\lambda_2\theta}, \dots, e^{i\lambda_k\theta}\right\}$$

where $\lambda_1, \dots, \lambda_k$ are nonzero integers. Let m be the greatest common factor of $\lambda_1, \dots, \lambda_k$; then $G_x = \mathbb{Z}_m$, $\forall\, x \in V\backslash\{\theta\}$.

Theorem 4.1. *Assume that*

(f$_1$) $f \in C^1(X, \mathbb{R}^1)$ *is G-invariant, $G = G_1$ or G_2.*
(f$_2$) $\exists$ *two regular values $a < b$, such that $(PS)_c$ hold for all $c \in [a, b]$.*
(f$_3$) *There exist G_d-invariant subspaces X_+ and X_- with*

$$j = \frac{1}{d} \text{ codim } X_+ < m = \frac{1}{d} \dim X_- < +\infty, \quad d = 1 \text{ or } 2,$$

where j and m are integers, such that
(1) $\text{Fix}_G \subset X_+$, $\text{Fix}_G \cap X_- = \{\theta\}$,
(2) $f(x) > a \quad \forall\, x \in X_+$,
(3) $f(x) < b \quad \forall\, x \in X_- \cap S_\rho$ *for some* $\rho > 0$,
(4) $\text{Fix}_G \cap f^{-1}[a, b] = \emptyset$.

Then f has at least $m - j$ distinct critical orbits.

The proof is separated into two cases: $G = \mathbb{Z}_2$ and $G = S^1$.

Case I. $G = \mathbb{Z}_2$. From (f$_3$) (2), (3) and (1), it follows that $f_a\backslash X_1 \subset X\backslash X_+$ and $X_- \cap S_\rho \subset f_b\backslash X_1$. We have the injections i, j and k as follows:

$$\begin{array}{ccc} (X_- \cap S_\rho, \emptyset)/G & \xrightarrow{\ i\ } & (f_b\backslash X_1, f_a\backslash X_1)/G \\ & \searrow^{k} & \downarrow{\scriptstyle j} \\ & & (X\backslash X_1, X\backslash X_+)/G\,. \end{array}$$

From (f$_3$) (1) and (4), the three pairs $(X_- \cap S_\rho, \emptyset)$, $(f_b\backslash X_1, f_a\backslash X_1)$ and $(X\backslash X_1, X\backslash X_+)$ are G-free. We shall figure out the homology groups of these pairs. We have

$$H_* \left((X_- \cap S_\rho)/G, \emptyset\right) = H_* \left(P^{m-1}\right)$$

and

$$H_* \left((f_b\backslash X_1)/G, (f_a\backslash X_1)/G\right) = H_* \left(f_b/G, f_a/G\right))$$

provided by the excision property.

$\forall\, x \in X$, we have the orthogonal decomposition $x = y + z$, where $y \in X_+^\perp$, $z \in X_+$. Let

$$\eta(t, x) = y + tz \quad \forall\, t \in [0, 1].$$

Then $\eta : (X\backslash X_+) \to X_+^\perp\backslash\{\theta\}$, which is G-equivariant. We obtain $H_*((X\backslash X_1)/G,\ (X\backslash X_+)/G) = H_*((X\backslash X_1)/G,\ (X_+^\perp\backslash\{\theta\})/G)$. Similarly,

$$H_*\left((X\backslash X_1)/G,\ \left(X_+^\perp\backslash\{\theta\}\right)/G\right) = H_*\left((X_2\backslash\{\theta\})/G,\ \left(X_+^\perp\backslash\{\theta\}\right)/G\right).$$

In summary, the following commutative diagram holds:

$$\begin{array}{ccc} H_*\left(P^{m-1},\emptyset\right) & \xrightarrow{\ i_*\ } & H_*\left(f_b/G, f_a/G\right) \\ & \searrow^{k_*} & \downarrow j_* \\ & & H_*\left(P^{n-1}, P^{j-1}\right) \end{array}$$

where $n = \dim X_2 \geq m$. As to the cohomology rings, we have

$$\begin{array}{ccc} H^*\left(P^{m-1}\right) & \xleftarrow{\ i^*\ } & H^*\left(f_b/G, X_1/G\right) \\ & \nwarrow^{k^*} & \uparrow j^* \\ & & H^*\left(P^{n-1}\right) \end{array}$$

It is easily seen that

$$k^*\omega^2 = (k^*\omega)^2,$$

where ω is the generator of $H^*(P^{n-1}) = \sum_{j=0}^{n-1}\omega^j$, and that $\dim k^*\omega \neq 0$.

Thus, we have nontrivial classes $[z_\ell] \in H_\ell(P^{m-1})$ for $\ell = 0, 1, \ldots, m-1$, satisfying $[z_{\ell-1}] = [z_\ell] \cap k^*\omega$, $\ell = 1, 2, \ldots, m-1$.

We shall prove that $k_*[z_\ell] \neq 0$ for $\ell = j, j+1, \ldots, m-1$. Indeed, it suffices to prove that $|z_\ell| \cap X_+ \neq \emptyset$, $\forall\ z_\ell \in [z_\ell]$, $\ell = j, j+1, \ldots, m-1$. If this is not true, i.e., there exists $\ell_0 \in [j, m-1]$ such that $|z_{\ell_0}| \cap X_+ = \emptyset$, then $|z_{\ell_0}|$ is deformed in $X_+^\perp \cap S_1$. It is impossible.

Since $k^* = i^* \cdot j^*$, we have

$$\begin{aligned} i_*\left[z_{\ell-1}\right] &= i_*\left([z_\ell] \cap k^*\omega\right) \\ &= i_*\left[z_\ell\right] \cap j^*\omega, \end{aligned}$$

and $\dim j^*\omega \neq 0$. Then we obtain $m-j$ subordinate classes

$$i_*\left[z_j\right] < i_*\left[z_{j+1}\right] < \ldots < i_*\left[z_{m-1}\right].$$

Theorem 3.4 is applied to provide $m-j$ distinct critical orbits of f with values in $[a,b]$, because $f^{-1}[a,b]/G$ is a manifold. The proof is complete.

Case II. $G = S^1$.

If we follow the proof given above, then the argument is stuck by the fact that the pairs $(X_- \cap S_\rho, \emptyset)$, $(f_b\backslash X_1, f_a\backslash X_1)$ and $(X\backslash X_1, X\backslash X_+)$ are no longer G-free.

Although $f^{-1}[a,b]/G$ is not a manifold, the minimax principle is still valid, due to the G-deformation theorem, cf. Chapter I, Theorem 7.1.

More algebraic topology is needed to derive the chain of subordinate classes. In the following, $\check{H}^*$ denotes the Čech-Alexander-Spanier cohomology functor. It is known (cf. Spanier [Spa1] pp. 340 and 420) that $\check{H}^*(Y) = H^*(Y)$, the singular cohomology if Y is a CW complex. We shall figure out $H^*((X\backslash X_1)/G)$ and $H^*((X_- \cap S_\rho)/G)$ via their counterparts in $\check{H}^*$.

Let us recall the following ([Spa1] p. 344).

Theorem 4.2. (Vietoris Begle). *Let $f : X' \to X$ be a closed continuous surjective map between paracompact Hausdorff spaces. Assume that there is $n \geq 0$ such that $\check{H}^{\#_q}(f^{-1}(x)) = 0 \;\forall\, x \in X$, and for $q < n$. Then*

$$f^* : \check{H}^q(X) \to \check{H}^q(X')$$

is an isomorphism for $q < n$ and a monomorphism for $q = n$.

From this we relate $\check{H}^*(Y/G, \mathbb{Q})$ with the G-cohomology $\check{H}^*_G(Y, \mathbb{Q})$ where $\mathbb{Q}$ is the rational field and $Y = X\backslash X_1$ or $X_- \cap S_\rho$.

Lemma 4.1. *Suppose that Y is a paracompact Hausdorff G space, and that $\forall\, x \in Y$, the isotropy group G_x is finite. Then the map $\tilde{j} : E \times_G Y \to Y/G$ induced by the projective $j : E \times Y \to Y$ induces isomorphisms*

$$\tilde{j}^* : \check{H}^q(Y/G, \mathbb{Q}) \to \check{H}^q_G(Y, \mathbb{Q})$$

in all dimensions q, where E is the universal total G space.

Proof. We make use of the filtration

$$E^1 \subset E^2 \subset \ldots \subset E^m \subset E^{m+1} \subset \ldots$$

of the universal total space E, and consider the diagram for each m:

$$\begin{array}{ccc} E^m \times Y & \xrightarrow{j^m} & Y \\ \downarrow & & \downarrow \\ E^m \times_G Y & \xrightarrow{\tilde{j}^m} & Y/G \end{array}$$

Note that $\tilde{j}^m$ is closed because E^m is compact and that $(\tilde{j}^m)^{-1}x = E^m/G_x$ $\forall\, x \in Y$, where G_x is the isotropy group. Since $\check{H}^q(E^m/G_x, \mathbb{Q}) = 0$ for $q < m$, applying the Vietoris Begle Theorem,

$$\left(\tilde{j}^m\right)^* : \check{H}^q(Y/G, \mathbb{Q}) \to \check{H}^q\left(E^m \times_G Y, \mathbb{Q}\right) \quad q < m$$

are isomorphisms. Then $\tilde{j}^*$ is just the composition of $(\tilde{j}^m)^*$ and the isomorphism $\check{H}^q(E^m \times_G Y, \mathbb{Q}) \cong \check{H}^q(E \times_G Y, \mathbb{Q})$, $q < m$.

Let $p : \xi \to B$ be an orientable r-dimensional vector bundle, and let $D\xi, S\xi$ be the associate disk and sphere bundles. According to Thom's isomorphism theorem, $\check{H}^r(D\xi, S\xi) \cong \mathbb{Z}$ possesses a generator $t(\xi)$, which induces the isomorphism

$$t(\xi)\cup : \check{H}^*(D\xi) \to \check{H}^{*+r}(D\xi, S\xi)$$

where $\cup$ is the cup product. Since the disk is contractible, $\check{H}^*(D\xi) \cong \check{H}^*(B)$, let $e(\xi) \in \check{H}^*(B)$ be the image of $t(\xi)$ under this isomorphism. The exactness of the cohomology sequence reads as follows:

$$\ldots \to \check{H}^*(D\xi, S\xi) \to \check{H}^*(D\xi) \to \check{H}^*(S\xi) \to \ldots .$$

It turns out to be the following Gysin sequence:

$$\ldots \to \check{H}^*(B) \underset{e(\xi)\cup}{\longrightarrow} \check{H}^{*+r}(B) \underset{p^*}{\longrightarrow} \check{H}^{*+r}(S\xi) \to \ldots ,$$

where p is the projection $S\xi \to B$.

Lemma 4.2. *If* $S^{2n-1} \subset X_2$, *then*

$$\check{H}^q_{S^1}\left(S^{2n-1}, \mathbb{Q}\right) = \begin{cases} \omega^j & q = 2j, \quad 0 \le j \le n-1 \\ 0 & \text{otherwise} \end{cases}$$

and

$$\check{H}^*_{S^1}\left(X_2 \backslash \{\theta\}, \mathbb{Q}\right) = \sum_{j=0}^{\dim X_2 - 1} \omega^j.$$

Proof. Consider the following diagram:

$$\begin{array}{ccc} E \times S^{2n-1} & \xrightarrow{\hat{p}} & E = S^\infty \\ \downarrow & & \downarrow \\ E \times_{S^1} S^{2n-1} & \underset{p}{\longrightarrow} & B = \mathbb{C}P^\infty, \end{array}$$

where p is the projection induced by the projection $\hat{p}$. We look at the vector bundle $\pi : \xi = E \times \mathbb{C}^n \to B = \mathbb{C}P^\infty$. The associate $2n-1$ sphere bundle $S\xi$ turns out to be $p : E \times_{S^1} S^{2n-1} \to B$. We have the Gysin sequence

$$\begin{array}{ccccc} \ldots \to \check{H}^q(D\xi, S\xi) & \to & \check{H}^q(B) & \to & \check{H}^q(S\xi) \to \ldots \\ \wr\| & \nearrow_{e(\xi)} & & & \\ \check{H}^{q-2n}(B) & & & & \end{array}$$

Thus, for $q < 2n-1$, we have the isomorphisms $\check{H}^q(B) \cong \check{H}^q(S\xi)$; and for $q = 2n-1$, from

$$\begin{array}{ccccccc} 0 \to \check{H}^{2n-1}(B) & \to & \check{H}^{2n-1}(S\xi) & \to & \check{H}^{2n}(D\xi, S\xi) & \to & \check{H}^{2n}(B) \\ \wr\| & & & & \wr\| & \nearrow_{e(\xi)} & \\ 0 & & & & \check{H}^0(B) & & \end{array}$$

it follows $\check{H}^{2n-1}(S\xi) \cong 0$.

For $q > 2n-1$, from Lemma 4.1,

$$\begin{aligned} \check{H}^q(S\xi, \mathbb{Q}) &= \check{H}^q\left(E \times_{S^1} S^{2n-1}, \mathbb{Q}\right) \\ &= \check{H}^q_{S^1}\left(S^{2n-1}, \mathbb{Q}\right) \\ &= \check{H}^q\left(S^{2n-1}/S^1, \mathbb{Q}\right) = 0. \end{aligned}$$

Therefore,

$$\check{H}^q_{S^1}\left(S^{2n-1}, \mathbb{Q}\right) = \begin{cases} \omega^j, & q = 2j, \quad 0 \le j \le n-1 \\ 0 & \text{otherwise;} \end{cases}$$

and then

$$\check{H}^q_{S^1}\left(S^{\infty}, \mathbb{Q}\right) = \begin{cases} \omega^j & q = 2j, \quad j = 0, 1, 2, \ldots \\ 0 & \text{otherwise.} \end{cases}$$

The conclusion follows.

After these preparations, the previous arguments can be modified to suit the case $G = S^1$. Indeed, since $(X_2 \backslash \{\theta\})/S^1$ and $(X_- \cap S_\rho)/S^1$ are CW complex, the singular cohomology and the Čech-Alexander-Spanier cohomology are isomorphic. We have $H^*(X_- \cap S_\rho)/S^1, \mathbb{Q}) \cong H^*(\mathbb{C}P^{m-1}, \mathbb{Q})$, and $H^*(X_2 \backslash \{\theta\})/S^1, \mathbb{Q}) \cong H^*(\mathbb{C}P^{n-1}, \mathbb{Q})$.

Again, from the commutative diagrams

$$\begin{array}{ccc} H^*\left((X_- \cap S_\rho)/S^1\right) \cong H^*\left(\mathbb{C}P^{m-1}\right) & \xleftarrow{i^*} & H^*\left((f_b \backslash X_1)/S^1\right) \\ & {}_{k^*}\nwarrow & \uparrow j^* \\ & & H^*\left(X \backslash X_1\right) \cong H^*\left(\mathbb{C}P^{n-1}\right) \end{array}$$

and

$$\begin{array}{ccc} H_*\left((X_- \cap S_\rho)/S^1, \emptyset\right) & \xrightarrow{i_*} & H_*\left(f_b/S^1, f_a/S^1\right) \\ & {}_{k_*}\searrow & \downarrow j_* \\ & & H_*\left((X \backslash X_1)/S^1, (X \backslash X_+)/S^1\right) \end{array}$$

we find nontrivial classes $[z_\ell] \in H_\ell((X_- \cap S_\rho/S^1)$, $\ell = 0, 1, \ldots, m-1$, satisfying $[z_{\ell-1}] = [z_\ell] \cap k^*\omega$, $\ell = 1, 2, \ldots, m-1$ and $k^*[z_\ell] \neq 0$ for

$\ell = j, j+1, \ldots, m-1$. Thus $i_*[z_j] < i_*[z_{j+1}] < \ldots < i_*[z_{m-1}]$ are the required $m-j$ subordinate classes.

The proof is complete.

Corollary 4.1. *Under the assumptions of Theorem 4.1, there are critical orbits* $\{O_\ell \mid \ell = j, j+1, \ldots, m-1\}$ *such that*

$$C_G^\ell(f, O_\ell) \neq 0, \quad \ell = j, j+1, \ldots, m-1.$$

A dual form of Theorem 4.1 can be formulated as follows. In replacing (f_2) and (f_3), we state

(f_2') There is a sequence of G_d invariant finite dimensional vector subspaces:

$$X^1 \subset X^2 \subset \ldots \subset X^n \subset \ldots, \qquad \dim X^n = nd,$$

with $\overline{\bigcup_{n=1}^\infty X^n} = X$, satisfying

(1) ($(\mathrm{PS})^n$ condition) $\exists\, n_0 \in \mathbb{N}$ such that $\forall\, n \geq n_0$, the restriction $f^n = f|_{X^n}$ satisfies the $(\mathrm{PS})_c$ condition $\forall\, c \in [a,b]$.

(2) ($(\mathrm{PS})^*$ condition) For any sequence $x_n \in X^n$, $n = 1, 2, \ldots$,

$$a \leq f(x_n) \leq b \quad \text{and} \quad df^n(x_n) \to \theta$$

imply a convergent subsequence.

(f_3') There exist G_d invariant subspaces X_+ and X_- with

$$j = \frac{1}{d} \operatorname{codim} X_+ < m = \frac{1}{d} \dim X_- < +\infty, \quad d = 1, 2,$$

where j and m are integers such that

(1) $\mathrm{Fix}_G \subset X_-$, $\mathrm{Fix}_G \cap X_+ = \{\theta\}$,

(2) $f(x) > a \quad \forall\, x \in X_+ \cap S_\rho$ for some $\rho > 0$

(3) $f(x) < b \quad \forall\, x \in X_-$,

(4) $\mathrm{Fix}_G \cap f^{-1}[a,b] = \emptyset$.

Theorem 4.3. *Under the assumptions* (f_1), (f_2'), *and* (f_3') *where* a, b *are regular values, the function* f *has at least* $m-j$ *distinct critical orbits.*

Proof. First, we change f to $-f$, $-a$ to b and $-b$ to a. Assumptions (2) and (3) of (f_3') are changed to be the following:

(2′) $f(x) < b \quad \forall\, x \in X_+ \cap S_\rho$ for some $\rho > 0$, and

(3′) $f(x) > a \quad \forall\, x \in X_-$.

According to (f_2'), we choose n large enough, such that both $X_+^\perp$ and X_- are included in X^n, and then restrict ourselves on the finite dimensional G_d-invariant subspace X^n. Thus

$$\begin{aligned} f^n(x) &< b \quad \forall\, x \in (X^n \cap X_+) \cap S_\rho, \text{ and} \\ f^n(x) &> a \quad \forall\, x \in X_-. \end{aligned}$$

Let $X_-^n = X_+ \cap X^n$ and $X_+^n = X_-$; then

$$\operatorname{codim} X_+^n = (n-m)d \quad \text{and} \quad \dim X_-^n = (n-j)d.$$

Also, $X^n \cap \operatorname{Fix}_G \subset X_- = X_+^n$, $(X^n \cap \operatorname{Fix}_G) \cap X_-^n = \operatorname{Fix}_G \cap X_+ = \{\theta\}$.

Now we apply Theorem 4.1 to f^n and conclude that there are $(m-j)$ subordinate homology classes

$$[z_j^n] < [z_{j+1}^n] < \ldots < [z_{m-1}^n]$$

which correspond to critical values

$$a \le c_j^n \le c_{j+1}^n \le \ldots \le c_{m-1}^n \le b \quad \text{for } n \text{ large.}$$

Let

$$c_\ell = \lim_{n\to\infty} c_\ell^n, \quad \ell = j, j+1, \ldots, m-1.$$

Then

$$a \le c_j \le c_{j+1} \le \ldots \le c_{m-1} \le b.$$

According to assumption (f_2'), (PS)*, c_ℓ, $j \le \ell \le m-1$, are all critical values. It remains to prove that if

$$c = c_k = c_{k+1} = \ldots = c_\ell, \quad j \le k \le \ell \le m-1,$$

then $\# K_c \ge \ell - k + 1$. We prove it by contradiction.

If $K_c = \{O_1, O_2, \ldots, O_s\}$, $s < \ell - k + 1$, then we choose neighborhoods U_α of O_α, $1 \le \alpha \le s$, such that

(1) $U_\alpha \cap U_\beta = \emptyset$ for $\alpha \ne \beta$,
(2) $\bigcup_{\alpha=1}^s U_\alpha \subset f^{-1}[a,b]$,
(3) Each U_α is in a G-tubular neighborhood of O_α in $f^{-1}[a,b]$.

We omit the subscripts α. Let $\pi : U \to O$ be the projection, and let $S = \pi^{-1}(p)\ \forall\, p \in O$. It is well known (cf. Chapter I, Lemma 7.15) that S is a G_p-manifold, where G_p is the isotropy group at p.

The orthogonal projection onto $(X^n)^\perp$ is denoted by $P_n^\perp$. Let $i : U/G_p \to X$ be the injection; then

$$X^n \cap S = \{x \in S \mid P_n^\perp i(x) = \theta\}$$

is a finite dimensional submanifold, provided n is large enough. Again, one may assume that the geodesic segment connecting p and the nearest point $p^n \in X^n \cap S$ is in S.

We choose suitable coordinates: $Y = Y_1 \oplus Y_2, \dim Y_1 = nd, \varphi : Y \cap B_1 \to S$, such that $\varphi(Y_1 \cap B_1) \subset S \cap X^n$, $\varphi(\theta) = p^n$, $p = \varphi(\bar{y}_2)$ and $\varphi(t\bar{y}_2) \in S \; \forall\, t \in [0,1]$, where B_1 is the unit ball and $\bar{y}_2 \in Y_2$. Then $\exists\, \bar{\delta} > 0$ such that $\varphi(V) \subset S$, where $V = \{(y_1, y_2) \in Y_1 \oplus Y_2 \mid \|y_1\| < \bar{\delta},\ \|y_2 - w\| < \bar{\delta},\ \exists t \in [0,1],\ w = t\bar{y}_2\}$.

Thus $N = \varphi(V)$ is an open neighborhood of p with the following property: $\exists$ a deformation retract

$$\eta : N/G_p \to X^n \cap N = N^n.$$

Indeed, $\eta = \varphi \circ \pi \circ \varphi^{-1}$, where π is the projection onto Y_1. Noticing that G_p is a finite rotation group with fixed point p, we may choose $\bar{\delta} > 0$ so small that $N^n = N^n/G_p$.

Let us introduce

$$\mathcal{N} = \bigcup_{\alpha=1}^{s} G \cdot N_\alpha \quad \text{and} \quad \mathcal{N}^n = \bigcup_{\alpha=1}^{s} G \cdot N_\alpha^n.$$

We choose $\delta > 0$ small enough such that there exists G-invariant open neighborhood $\mathcal{N}'$ of K_c with dist $(\partial\mathcal{N}, \mathcal{N}') \geq \frac{1}{8}\delta$. We conclude $\exists\, \bar{b} > 0,\ \bar{\varepsilon} > 0$ such that

$$\|df^n(x)\| \geq \bar{b} \quad \forall\, x \in (f^n)_{c+\bar{\varepsilon}} \setminus \left((f^n)_{c-\bar{\varepsilon}} \cup (\mathcal{N}^n)'\right) \text{ and}$$
$$0 < \bar{\varepsilon} < \text{Min}\left\{\frac{1}{4}\delta\bar{b}^2, \frac{1}{8}\delta\bar{b}\right\},$$

where $(\mathcal{N}^n)' = X^n \cap \mathcal{N}'$. Indeed, if not, $\exists\, x_n \in X^n \backslash (\mathcal{N}^n)'$ satisfying

$$f(x_n) \to c \quad \text{and} \quad \|df^n(x_n)\| \to 0,$$

then by (PS)*, $\exists\, x^* \notin \mathcal{N}'$ with $f(x^*) = c$, $df(x^*) = \theta$. This is impossible.

Therefore, $\forall\, \varepsilon \in (0, \bar{\varepsilon}/3)$, $\exists\, n_0$ large enough such that for $n \geq n_0$

$$c - \varepsilon < c_k^n \leq c_{k+1}^n \leq \ldots \leq c_\ell^n < c + \varepsilon$$

and

$$\bigcup_{t=k}^{\ell} K_{c_t^n} \subset (\mathcal{N}^n)'.$$

We exploit Corollary 3.5 and conclude that

$$\sum_{\alpha=1}^{s} \text{Cat}_{\bar{N}_\alpha^n}\left(\bar{N}_\alpha^n\right) \geq \ell - k + 1,$$

because $N^n_\alpha = N^n_\alpha/G_{p\alpha} = G \cdot N^n_\alpha/G$.

However, $\mathrm{Cat}_{\bar{N}^n_\alpha}(\bar{N}^n_\alpha) = 1$. Indeed $\overline{V}_\alpha$ is contractible, and so is $\bar{N}_\alpha$. Hence $\bar{N}^n_\alpha$ is contractible in $\bar{N}_\alpha/G_{p\alpha}$, i.e., $\exists\ \xi_\alpha : [0,1] \times \bar{N}^n_\alpha \to \bar{N}_\alpha/G_{p\alpha}$ satisfying $\xi_\alpha(0,\cdot) = \mathrm{id}_{\bar{N}^n_\alpha}$, $\xi_\alpha(1,\cdot) = \bar{p}_\alpha \in \bar{N}_\alpha/G_{p\alpha}$. Let us define $\tau_\alpha = \eta_\alpha \circ \xi_\alpha$. Then τ_α is a contraction in $\bar{N}^n_\alpha$.

Hence, we obtain a contradiction:

$$s = \sum_{\alpha=1}^{s} \mathrm{Cat}_{\bar{N}^n_\alpha}\left(\bar{N}^n_\alpha\right) \geq \ell - k + 1.$$

Corollary 4.2. *Suppose that $f \in C^1(X, \mathbb{R}^1)$ is even with $f(\theta) = 0$, and that*

(1) $\exists\ \rho, \alpha > 0$, and a finite dimensional linear subspace E such that $f|_{E^\perp \cap S_\rho} \geq \alpha$,

(2) $\exists$ a sequence of linear subspace $\widetilde{E}_m$, $\dim \widetilde{E}_m = m$, and $\exists\ R_m > 0$ such that

$$f(x) \leq 0 \quad \forall\, x \in \widetilde{E}_m \backslash B_{R_m}, \quad m = 1, 2, \dots.$$

If, further, f satisfies (PS) with respect to $\{\widetilde{E}_m \mid m = 1, 2, \dots\}$, then f possesses infinitely many distinct critical points corresponding to positive critical values.*

Proof. If not, f has at most ℓ critical points. Let $\dim E = j$; we choose $m - j > \ell$. Now, (f_1) is obviously true, with $\mathrm{Fix}_G = \{\theta\}$. Let $X^n = \widetilde{E}_n$; the $(\mathrm{PS})^n_c$ condition holds, using (2), so (f_2') is satisfied. Let $X_+ = E^\perp$, $X_- = \widetilde{E}_m$, $a = \alpha$, and $b = \mathrm{Max}_{x \in \tilde{E}_m} f(x) + 1$ then (f_3') holds. We apply Theorem 4.3. There are at least $m - j$ pairs of critical points. This is a contradiction.

Moreover, all critical values, obtained by the minimax principle, are greater than α, so they are positive.

Remark 4.1. Corollary 4.2 was given by Ambrosetti-Rabinowitz [AmR1], where the function f is assumed satisfying the (PS) condition rather than the (PS)* condition.

Remark 4.2. Theorems 4.1 and 4.3 were proved by many authors via index theory and pseudo index theory, and the index for $G = \mathbb{Z}_2$ is genus. The associated theorems were given by Clark [Cla1], Ambrosetti-Rabinowitz [AmR1], and V. Benci [Ben4]. S^1-index theory was first introduced by Fadell-Rabinowitz [FaR2]; see also V. Benci [Ben3], L. Nirenberg [Nir3], Costa-Willem [CoW1], Fadell, Husseini, Rabinowitz [FHR1]. Pseudo index theory was introduced by V. Benci [Ben4]; see also K. C. Chang [Cha13]. Other approaches can be found in Fadell [Fad1] (relative category) and Liu [Liu4] (pseudo category).

The new result is Corollary 4.1.

5. Some Abstract Critical Point Theorems

In this section, we shall give several abstract critical point theorems using Morse theory. Their applications will be studied in subsequent chapters.

Theorem 5.1. *Let M be a C^2 Hilbert-Riemannian manifold, and let $f \in C^2(M, \mathbb{R}^1)$ satisfy the (PS) condition. Suppose that $H_k(f_b, f_a) \neq 0$ for some $k \in \mathbb{N}$, where $b > a$ are regular values, and that $\{x_1, x_2, \ldots, x_\ell\} \subset K \cap f^{-1}[a, b]$ with Fredholm operators $d^2 f(x_i)$, $i = 1, 2, \ldots, \ell$. If either*

$$\text{(5.1)} \qquad \operatorname{ind}(f, x_i) > k \quad \textit{or} \quad \operatorname{ind}(f, x_i) + \dim \ker d^2 f(x_i) < k$$

for $i = 1, 2, \ldots, \ell$, then f has at least one more critical point x_0 with $C_k(f, x_0) \neq 0$.

Proof. If not, $K = \{x_1, x_2, \ldots, x_\ell\}$. From (5.1), it follows $C_k(f, x_i) = 0$, $i = 1, 2, \ldots, \ell$, provided by the shifting theorem. We apply the kth Morse inequality:

$$\begin{aligned} 0 = M_k(a, b, f) &= \sum_{i=1}^{\ell} \operatorname{rank} C_k(f, x_i) \geq \beta_k(a, b, f) \\ &= \operatorname{rank} H_k(f_b, f_a) > 0. \end{aligned}$$

This is impossible.

Corollary 5.1. *Suppose that the boundary ∂D of a k ball D and S homologically link in M, and that $\{x_1, x_2, \ldots, x_\ell\}$ are critical points of f satisfying (5.1). Then the conclusion of Theorem 5.1 holds.*

This is a combination of Theorems 1.2 and 5.1.

Remark 5.1. Corollary 5.1 includes a theorem due to Lazer Solimini [LaS1] as a special case, in which S and ∂D are as in Section 1, Example 2.

Let H be a real Hilbert space, and let A be a bounded self-adjoint operator defined on H. According to its spectral decomposition, $H = H_+ \oplus H_0 \oplus H_-$, where $H_\pm, H_0$ are invariant subspaces corresponding to the positive/negative, and zero spectrum of A respectively. Let $P_\pm, P_0$ be the projections (orthogonal) of these subspaces.

The following assumptions are given:

(H_1) $A_\pm := A\,|_{H_\pm}$ has a bounded inverse on $H_\pm$.
(H_2) $\gamma := \dim(H_- \oplus H_0) < \infty$.
(H_3) $g \in C^1(H, \mathbb{R}^1)$ has a bounded and compact differential $dg(x)$. In addition, if $\dim H_0 \neq 0$, we assume

$$g\,(P_0 x) \to -\infty \quad \text{as} \quad \|P_0 x\| \to \infty.$$

We shall study the number of critical points of the function

$$f(x) = \frac{1}{2}(Ax, x) + g(x),$$

or equivalently, the number of solutions of the operator equation

$$Ax + dg(x) = \theta.$$

Lemma 5.1. *Under the assumptions* (H_1), (H_2) *and* (H_3), *we have that*
(1) f *satisfies (PS) condition, and*
(2) $H_q(H, f_a) = \delta_{q\gamma} G$
for $-a$ *large enough, as* $f_a \cap K = \emptyset$.

Proof. 1. First, we verify that f satisfies (PS). For $\{x_n\}_1^\infty \subset H$, $df(x_n) \to \theta$, and $f(x_n)$ bounded, we shall find a convergent subsequence. In fact, from $df(x_n) \to \theta$, it follows $\forall\, \varepsilon > 0$, $\exists\, N = N(\varepsilon)$ such that for $n > N$

$$\left|\left(Ax_n, x_n^{\pm}\right) + \left(dg\,(x_n)\,, x_n^{\pm}\right)\right| \le \varepsilon \left\|x_n^{\pm}\right\|,$$

where $x_n^{\pm} = P_{\pm} x_n$. Hence $\|x_n^{\pm}\|$, and then $(Ax_n x_n)$ are bounded. Since

$$\begin{aligned} |g\,(P_0 x_n)| &\le |g\,(x_n) - g\,(P_0 x_n)| + |g\,(x_n)| \\ &\le m\left(\left\|x_n^{+}\right\| + \left|x_n^{-}\right|\right) + |g\,(x_n)| \end{aligned}$$

where $m = \sup\{\|dg(x)\| \mid x \in H\}$. If $f(x_n)$ is bounded, then $|g(x_n)|$, and therefore $|g(P_0 x_n)|$, is bounded. Thus $\|P_0 x_n\|$ is bounded. Since dg is compact there is a subsequence x_{n_i} such that $dg(x_{n_i})$ is convergent. By

$$df\left(x_{n_i}\right) = A_+ x_{n_i}^{+} + A_- x_{n_i}^{-} + dg\left(x_{n_i}\right) \to \theta$$

and by the boundedness of $A_{\pm}^{-1}$ we conclude that $x_{n_i}^{\pm}$ is convergent.

Since $\dim H_0$ is finite, there is a convergent subsequence $P_0 x_{n_i}$. The (PS) condition is verified.

2. Denote $\varepsilon_{\pm} = \inf\{(Ax_{\pm}, x_{\pm}) \mid \|x_{\pm}\| = 1\}$ which is positive, and $R_+ = \frac{m+1}{\varepsilon_+}$; let

$$\mathcal{M} = \left(H_+ \cap B_{R_+}\right) \times (H_0 \oplus H_-)\,.$$

From

$$(5.2) \qquad \begin{aligned} (df(x), x_+) &= (Ax_+, x_+) - (dg(x), x_+) \\ &\ge \varepsilon_+ \|x_+\|^2 - m\,\|x_+\| \end{aligned}$$

we know that f has no critical point outside $\mathcal{M}$, and that $-df(x)$ points inward to $\mathcal{M}$ on $\partial\mathcal{M}$.

Noticing that

$$\begin{aligned}
&-\frac{1}{2}\|A\|\,\|x_-\|^2 - m\left(\|x_-\| + R_+\right) + g\left(P_0 x\right) \\
&\qquad \le f(x) \le \frac{1}{2}\|A\|R_+^2 - \frac{1}{2}\varepsilon_-\|x_-\|^2 + m\left(\|x_-\| + R_+\right) + g\left(P_0 x\right),
\end{aligned}$$

we obtain

$$f(x) \to -\infty \Longleftrightarrow \|x_- + P_0 x\| \to \infty \quad \text{uniformly in} \quad x_+.$$

Thus, $\forall\, T > 0$, $\exists\, a_1 < a_2 < -T$, $R_1 > R_2 > 0$ such that

$$\begin{aligned}
\left(H_+ \cap B_{R_+}\right) \times \left((H_0 \oplus H_-)\backslash B_{R_1}\right) \subset f_{a_1} \cap \mathcal{M} \subset \left(H_+ \cap B_{R_+}\right) \\
\times \left((H_0 \oplus H_-)\backslash B_{R_2}\right) \subset f_{a_2} \cap \mathcal{M}
\end{aligned}$$

Also we find $T > 0$ such that $K \cap f_{-T} = \emptyset$.

The negative gradient flow of f defines a strong deformation retract

$$\tau_1 : \mathcal{M} \cap f_{a_2} \to \mathcal{M} \cap f_{a_1}.$$

Another strong deformation retract in $f_{a_2} \cap \mathcal{M}$

$$\tau_2 : \left(H_+ \cap B_{R_+}\right) \times \left((H_0 \oplus H_-)\backslash B_{R_2}\right) \to \left(H_+ \cap B_{R_+}\right) \times \left((H_0 \oplus H_-)\backslash B_{R_1}\right)$$

is defined by $\tau_2 = \xi(1, \cdot)$, where

$$\begin{aligned}
&\xi\left(t; x_+ + x_0 + x_-\right) \\
&\quad = \begin{cases} x_+ + x_0 + x_- & \text{if } \|x_0 + x_-\| \ge R_1 \\ x_+ + \frac{x_0 + x_-}{\|x_0 + x_-\|}\left(tR_1 + (1-t)\|x_0 + x_-\|\right), & \text{if } \|x_0 + x_-\| \le R_1 \end{cases}.
\end{aligned}$$

We compose these two strong deformation retracts, $\tau = \tau_2 \circ \tau_1$, and then obtain a strong deformation retract

$$\tau : \mathcal{M} \cap f_{a_2} \to \left(H_+ \cap B_{R_+}\right) \oplus \left((H_0 \oplus H_-)\backslash B_{R_1}\right)$$

and, again, the following deformation:

$$\begin{aligned}
&\eta\left(t, x_+ + x_- + x_0\right) \\
&\qquad = \begin{cases} x_+ + x_0 + x_- & \text{if } \|x_+\| \le R_+ \\ \frac{x_+}{\|x_+\|}\left(tR_+ + (1-t)\|x_+\|\right) + x_0 + x_- & \text{if } \|x_+\| > R_+ \end{cases}
\end{aligned}$$

is a strong deformation retract of the topological pair from

$$(H, f_{a_2}) \quad \text{to} \quad (\mathcal{M}, \mathcal{M} \cap f_{a_2})$$

provided by (5.2).

3. Finally, we have

$$\begin{aligned}
H_q\left(H, f_{a_2}\right) &\cong H_q\left(\mathcal{M}, \mathcal{M} \cap f_{a_2}\right) \\
&\cong H_q\Big(\left(H_+ \cap B_{R_+}\right) \times (H_0 \oplus H_-)\,, \\
&\qquad \left(H_+ \cap B_{R_+}\right) \times (H_0 \oplus H_-)\backslash B_{R_1}\Big) \\
&\cong H_q\left(H_0 \oplus H_+, (H_0 \oplus H_-)\backslash B_{R_1}\right) \\
&\cong H_q\Big((H_0 \oplus H_-) \cap B_{R_1}, \partial\left((H_0 \oplus H_-) \cap B_{R_+}\right)\Big) \\
&\cong \delta_{q\gamma} G.
\end{aligned}$$

Theorem 5.2. *Under the assumptions* (H_1), (H_2) *and* (H_3), *if f has critical points $\{p_i\}_{i=1}^n$ with*

$$\gamma \notin \bigcup_{i=1}^{n} \left[m_-(p_i), m_-(p_i) + m_0(p_i)\right]$$

where $m_-(p) = \mathrm{index}(f,p)$ and $m_0(p) = \dim\ker d^2 f(p)$, then f has a critical point p_0 different from $p_1, \dots, p_n$, with $C_\gamma(f, p_0) \neq 0$.

Proof. Directly follows from Lemma 5.1 and Theorem 5.1.

Remark 5.2. In Lemma 5.1 and Theorem 5.1, if $\dim H_0 = 0$, the boundedness of $dg(x)$ can be replaced by the following condition:

$$\|dg(x)\| = o(\|x\|) \quad \text{as} \quad \|x\| \to \infty. \tag{5.3}$$

Proof. Condition (5.3) implies that f has no critical point outside a big ball $B_R(\theta)$, $R > 0$.

Now we define a new function

$$\tilde{f}(x) = \frac{1}{2}(Ax, x) + \rho(\|x\|)g(x),$$

where

$$\rho(t) = \begin{cases} 1 & 0 \le t \le R_1 \\ 0 & t > R_2 \end{cases}, 0 \le \rho \le 1,$$

and satisfies the estimate

$$|\rho'(t)| \le \frac{3}{2}\frac{1}{R_2 - R_1}, \text{ where } R_2 > R_1 > R \text{ are suitably chosen.}$$

The new function $\tilde{f}$ possesses the same critical points as the function f, and satisfies the (PS) condition as well.

In fact, $f(x) = \tilde{f}(x)$ for $\|x\| \le R_1$, and $df(x) = Ax$ for $\|x\| > R_2$, we only want to verify that $\|d\tilde{f}(x)\| \neq \theta$ for $x \in B_{R_2}(\theta)\backslash B_{R_1}(\theta)$.

Let $\varepsilon = \frac{1}{5}\|A^{-1}\|^{-1}$ by assumption (5.3); $\exists\, R_0 > 0$ such that

$$\|dg(x)\| < \varepsilon\|x\| \quad \forall\, x \notin B_{R_0}.$$

The compactness of $dg(x)$ implies that $\exists\, M_\varepsilon > 0$ such that

$$\|dg(x)\| < \varepsilon\|x\| + M_\varepsilon \quad \forall\, x \in H.$$

Thus

$$|g(x)| < \varepsilon\|x\|^2 + M_\varepsilon\|x\| + |g(\theta)|.$$

Let

$$\begin{aligned} R_1 &> \max\left\{R, R_0, \frac{1}{\varepsilon}(4M_\varepsilon + 3)\right\}, \\ R_2 &= \max\{2, 1 + |g(\theta)|\}\, R_1\,; \end{aligned}$$

we have

$$\begin{aligned} \left\|d\tilde{f}(x)\right\| &= \left\|Ax + \rho'(\|x\|)g(x)\frac{x}{\|x\|} + \rho(\|x\|)dg(x)\right\| \\ &\ge \left\|A^{-1}\right\|^{-1}\|x\| - (\varepsilon\|x\| + M_\varepsilon) \\ &\quad - \frac{3}{2}\frac{1}{R_2 - R_1}\left(\varepsilon\|x\|^2 + M_\varepsilon\|x\| + |g(\theta)|\right) \\ &\ge 1 \quad \forall\, x \in B_{R_2}\backslash B_{R_1}. \end{aligned}$$

As to the (PS) condition, suppose that $d\tilde{f}(x_n) \to \theta$; then $\{x_n\} \subset B_{R_1}$ except for finitely many points, according to condition (5.3) and the invertibility of A. Since dg is compact, there exists a convergent subsequence $dg(x_n)$. Comparing this with the assumption $df(x_n) = d\tilde{f}(x_n) \to \theta$, and the boundedness of A^{-1}, we obtain a convergent subsequence.

Remark 5.3. In the case $H = \mathbb{R}^N$, $n = 1$, $p_1 = \theta$, and $\dim H_0 = 0$. This theorem is due to Amann Zehnder [AmZ1]. The above lemma, and the general statement with the condition $\gamma < m_-(p_i)$, $i = 1, \ldots, n$, is due to Chang [Cha1]. The above version is due to Z. Q. Wang [WaZ2].

Corollary 5.2. *Under the assumptions* (H_1), (H_2) *and* (H_3), *if f has a nondegenerate critical point p_0 with Morse index $m_-(p_0) \neq \gamma$, then f has a critical point $p_1 \neq p_0$.*

Moreover, if

$$m_0(p_1) \leq |m_-(p_0) - \gamma| \tag{5.4}$$

then f has one more critical point $p_2 \neq p_0, p_1$.

Proof. The first conclusion follows directly from Theorem 5.2 and then we have

$$C_\gamma(f, p_1) \neq 0.$$

However, by the shifting theorem

$$\gamma \in [m_-(p_1), m_-(p_1) + m_0(p_1)].$$

Condition (5.4) implies one of three possibilities:

(1) $m_-(p_0) \notin [m_-(p_1),\ m_-(p_1) + m_0(p_1)]$
and
$\gamma \in (m_-(p_1),\ m_-(p_1) + m_0(p_1))$,

(2) $\gamma = m_-(p_1)$,

(3) $\gamma = m_-(p_1) + m_0(p_1)$.

Using the splitting theorem and the critical group characterization of the local minimum and the local maximum, we see in both cases (2) and (3),

$$C_q(f, p_1) = \delta_{q\gamma} G.$$

The Morse inequalities in combination with the Betti numbers for the topological pair (H, f_a) (see Lemma 5.1), gives the existence of the critical point p_2.

In case (1), again by the splitting theorem and the critical group characterization of the local minimum and the local maximum, we obtain

$$C_{m_-(p_1)}(f, p_1) = C_{m_-(p_1)+m_0(p_1)}(f, p_1) = 0$$

and

$$\beta_q(H, f_a) = \delta_{q\gamma}$$

for $-a$ large enough, by Lemma 5.1.

However, case (1) implies either $m_-(p_0) < m_-(p_1)$ or $m_-(p_0) > m_-(p_1) + m_0(p_1)$.

If there were no other critical points, then in the first case, the $m_-(p_0) + 1^{\text{th}}$ Morse inequality would read as

$$-1 \geq 0.$$

This is a contradiction. And in the second case, both the $m_-(p_1) + m_0(p_1) - 1^{\text{th}}$ and the $m_-(p_1) + m_0(p_1)^{\text{th}}$ Morse inequalities would imply the equality

$$\sum_{q=m_-(p_1)}^{m_-(p_1)+m_0(p_1)} (-1)^q (\operatorname{rank} C_q(f, p_1) - \delta_{q\gamma}) = 0.$$

Again, the $m_-(p_0) + 1^{\text{th}}$ Morse inequality would read as

$$-1 \geq 0,$$

and this is also a contradiction.

To sum up, we have proved the existence of the third critical point.

Now we turn to a variant of Lemma 5.1 which provides more information on the number of critical points if the function f is defined on H attached by a compact manifold V.

Lemma 5.2. *Suppose that* (H_1) *and* (H_2) *hold. Let* V^n *be a* C^2 *compact manifold without boundary. Assume that* $g \in C^1(H \times V^n, \mathbb{R}^1)$ *is a function having a bounded (if* $\dim H_0 = 0$, $\|dg(x,v)\| = o(\|x\|)\ \forall\ v \in V$*) and compact* dg*, satisfying*

$$g(P_0 x, v) \to -\infty \quad \text{as} \quad \|P_0 x\| \to +\infty, \quad \text{if} \quad \dim H_0 \neq 0.$$

Let

$$f(x,v) = \frac{1}{2}(Ax, x) + g(x,v).$$

Then

(1) f *satisfies (PS) condition,*

(2) $H_q(H \times V^n, f_a) \cong H_{q-\gamma}(V^n)$

for $-a$ *large enough, with* $K \cap f_a = \emptyset$.

The proof is similar to the proof of the previous one. Now define

$$\mathcal{M} = \left(H_+ \cap B_{R_+}\right) \times (H_0 \oplus H_-) \times V^n.$$

By the same method, we eventually obtain

$$H_*(\mathcal{M}, f_a \cap \mathcal{M}) \cong H_*\left(B^\gamma, S^{\gamma-1}\right) \otimes H_*(V^n),$$

by the Künneth formula. Thus

$$\begin{aligned} H_q(H \times V^n, f_a) &\cong H_q(\mathcal{M}, f_a \cap \mathcal{M}) \\ &\cong H_{q-\gamma}(V^n). \end{aligned}$$

Theorem 5.3. *Under the assumptions of the above lemma, the function* f *has at least* $CL(V^n) + 1$ *critical points. If further,* $g \in C^2$, *and* f *is nondegenerate, then* f *has a least* $\sum_{i=0}^n \beta_i(V^n)$ *critical points, where* $\beta_i(V^n)$ *is the* i^{th} *Betti number of* V^n, $i = 0, 1, \ldots, n$.

Proof. Since

$$H_q(H \times V^n, f_a) \cong H_{q-\gamma}(V^n),$$

and
$$H^q(H \times V^n) \cong H^q(V^n),$$
we obtain $\ell+1$ nontrivial singular homology relative classes $[Z_{\ell+1}] < [Z_\ell] < \dots < [Z_1]$, with $Z_1, \dots, Z_{\ell+1} \in H_q(H \times V^n, f_a)$, where $\ell = CL(V^n)$.

The first conclusion follows from Theorem 3.4.

Moreover, if f is nondegenerate, by the same argument, we may assume $K \cap f_a = \emptyset$ for $-a$ large enough. The Morse inequalities now read as

$$m_q \geq \text{ rank } H_q(H \times V^n, f_a) = \text{ rank } H_{q-\gamma}(V^n) = \beta_{q-\gamma}(V^n), \qquad q = 0, 1, \dots .$$

Therefore there are at least $\sum_{j=0}^n \beta_j(V^n)$ critical points.

The same idea can be applied to study functions bounded from below. We have

Theorem 5.4. *Suppose that M is a C^2-Finsler manifold. Assume that $f \in C^1(M, \mathbb{R}^1)$, satisfying the (PS) condition, is bounded below. Suppose that there exists a critical point p_0, which is not the global minimum of f, with finite $\sum_{q=0}^\infty (-1)^q$ rank $C_q(f, p_0) \neq \chi(M) - 1$. Then f has at least three critical points.*

Proof. According to the (PS) condition and lower semi-boundedness, f has a global minimum p_1. Let $c_i = f(p_i)$, $i = 0, 1$. If f had no critical points other than p_0 and p_1 then for arbitrary $b > c_0$ there would be no critical point in $M \backslash f_b$, and the following identity would hold:

$$\chi(f_b) = \chi\left(f_b, f_{c_0 - \varepsilon}\right) + \chi\left(f_{c_0 - \varepsilon}\right)$$

where $0 < \varepsilon < c_0 - c_1$. Since there exists a strong deformation retract deforming M into f_b, and $f_{c_0-\varepsilon}$ into p_1, we would have

$$\chi(f_b) = \chi(M),$$

and
$$\chi\left(f_{c_0-\varepsilon}\right) = \chi(\{p_1\}) = 1.$$

But
$$\chi\left(f_b, f_{c_0-\varepsilon}\right) = \sum_{q=0}^{\infty} (-1)^q \text{ rank } C_q(f, p_0)$$

because p_0 is the unique critical point in $f^{-1}[c_0 - \varepsilon, b]$. This is a contradiction.

Corollary 5.3. *Let H be a Hilbert space, and $f \in C^1(H, \mathbb{R}^1)$ be bounded below with the (PS) condition. Suppose that $df(x) = x - T(x)$*

is a compact vector field, and p_0 is an isolated critical point but not the global minimum with index $(df, p_0) = \pm 1$. Then f has at least three critical points.

Remark 5.4. In the case $H = \mathbb{R}^n$, the corollary was proved by Krasnoselskii via degree theory, but was rediscovered by Castro Lazer in [CaL1] by a homology method. The above version was given by Chang [Cha1]. A little later, Amann presented a degree theoretic proof [Ama1] for Corollary 5.3. Theorems 5.3 and 5.4 are due to Chang [Cha1,5].

Now we turn to the study of bifurcation problems.

Let H be a Hilbert space and Ω be a neighborhood of θ in H. Suppose that L is a bounded self-adjoint operator on H, and that $G \in C(\Omega, H)$ with $G(u) = o(\|u\|)$ at $u = \theta$. We assume that G is a potential operator, i.e., $\exists\, g \in C^1(\Omega, \mathbb{R}^1)$, such that $dg = G$. Find solutions of the following equation with a parameter $\lambda \in \mathbb{R}^1$:

$$Lu + G(u) = \lambda u. \tag{5.5}$$

Obviously $u = \theta$, for all $\lambda \in \mathbb{R}^1$, is a solution of (5.5). We are concerned with the nontrivial solutions of (5.5) with small $\|u\|$.

Because (5.5) is the Euler equation of a function with parameter λ, the bifurcation phenomenon has its specific feature. We shall prove the following theorem due to Krasnoselskii [Kra1] and Rabinowitz [Rab2] via Morse theory, cf. [Cha6].

Theorem 5.5. *Suppose that $f \in C^2(\Omega, \mathbb{R}^1)$ with $df(u) = Lu + G(u)$, L being linear and $G(u) = o(\|u\|)$ at $u = \theta$. If μ is an isolated eigenvalue of L of finite multiplicity, then (μ, θ) is a bifurcation point for (5.5). Moreover, at least one of the following alternative occurs:*

1. *(μ, θ) is not an isolated solution of (5.5) in $\{\mu\} \times \Omega$.*
2. *There is a one-sided neighborhood Λ of μ such that for all $\lambda \in \Lambda\{\mu\}$, (5.5) possesses at least two distinct nontrivial solutions.*
3. *There is a neighborhood I of μ such that for all $\lambda \in I\backslash\{\mu\}$, (5.5) possesses at least one nontrivial solution.*

The proof depends upon the Lyapunov-Schmidt reduction. Let $X = \ker(L - \mu I)$, with $\dim X = n$; and let $P, P^\perp$ be the orthogonal projections onto X and $X^\perp$, respectively. Then (5.5) is equivalent to a pair of equations

$$\mu x + PG\left(x + x^\perp\right) = \lambda x \tag{5.6}$$

$$Lx^\perp + P^\perp G\left(x + x^\perp\right) = \lambda x^\perp \tag{5.7}$$

where $u = x + x^\perp$, $x \in X$, $x^\perp \in X^\perp$. Equation (5.7) is uniquely solvable in a small bounded neighborhood $\mathcal{O}$ of $(\mu, \theta) \in \mathbb{R}^1 \times X$, say $x^\perp = \varphi(\lambda, x)$

for $(\lambda, x) \in \mathcal{O}$, where $\varphi \in C^1(\mathcal{O}, X^\perp)$. Substitute $x^\perp = \varphi(\lambda, x)$ into (5.6), and

$$\mu x + PG(x + \varphi(\lambda, x)) = \lambda x, \tag{5.8}$$

which is again a variational problem on the finite dimensional space X. Let

$$\begin{aligned} J_\lambda(x) &= f(x + \varphi(\lambda, x)) - \frac{\lambda}{2}\left(\|x\|^2 + \|\varphi(\lambda, x)\|^2\right) \\ &= \frac{1}{2}(\mu - \lambda)\|x\|^2 + \frac{1}{2}(L\varphi, \varphi) - \frac{\lambda}{2}\|\varphi\|^2 + g(x + \varphi) \end{aligned} \tag{5.9}$$

where $dg = G$, with $g(\theta) = 0$. It is easy to verify that (5.8) is the Euler equation of J_λ, and that $\varphi(\lambda, x) = o(\|x\|)$ at $x = \theta$.

The problem is reduced to finding the critical points of J_λ near $x = \theta$ for fixed λ near μ, where $J_\lambda \in C^1(\Omega_1, \mathbb{R}^1)$, Ω_1 is a neighborhood of θ in X.

Proof of Theorem 5.5. Clearly $x = \theta$ is a critical point of J_λ $\forall$ λ such that $(\lambda, \theta) \in \mathcal{O}$. If θ is not an isolated critical point of J_μ, which corresponds to case (1) in the theorem, then there are only two possibilities:

(i) $x = \theta$ is either a local maximum or a local minimum of J_μ;
(ii) $x = \theta$ is neither a local maximum nor a local minimum of J_μ.

In case (i), suppose that θ is a local minimum of J_μ. For some $\varepsilon > 0$, $W = (J_\mu)_\varepsilon = \{x \in \Omega_1 \mid J_\mu(x) \leq \varepsilon\}$ is a neighborhood of θ, containing θ as the unique critical point. The negative gradient flow of J_μ preserves W, and therefore the negative gradient flow of J_λ preserves W for $|\lambda - \mu|$ small. Since W is contractible, $\chi(M) = 1$, $x = \theta$ is a local maximum of J_λ, for $\lambda > \mu$ and J_λ is bounded from below on W; we obtain two nontrivial critical points, according to Theorem 5.4, in particular, Corollary 5.3. Therefore, for each λ in a small right-hand side neighborhood of μ, there exist at least two distinct nontrivial solutions of (5.5).

Similarly, we prove that there exist at least two distinct nontrivial solutions of (5.5) for each λ in a small left-hand side neighborhood of μ, if θ is a local maximum of J_μ.

In case (ii), θ is neither a local maximum nor a local minimum of J_μ. We see that

$$C_0(J_\mu, \theta) = C_n(J_\mu, \theta) = 0, \tag{5.10}$$

according to Example 4 in Section 4 of Chapter I. Since

$$C_0(J_\lambda, \theta) = 1, \quad \text{for} \quad \lambda < \mu, \quad \text{and} \tag{5.11}$$

$$C_n(J_\lambda, \theta) = 1 \quad \text{for} \quad \lambda > \mu, \tag{5.12}$$

we conclude that there is a neighborhood I of μ such that for $\lambda \in I \backslash \{\mu\}$, J_λ possesses a nontrivial critical point. If not, $\exists$ $\lambda_m \to \mu$, say $\lambda_m > \mu$, such that J_{λ_m} has the unique critical point θ, then $C_n(J_{\lambda_m}, \theta) = 1$, $m = 1, 2, \ldots$, implies $C_n(J_\mu, \theta) = 1$ by Theorem 5.6 of Chapter I. This contradicts (5.10). Similarly for $\lambda_m < \mu$. This completes the proof.

Remark 5.5. A weaker result that (μ, θ) is a bifurcation point was proved by a simpler argument, cf. Berger [Ber1].

More information on the number of distinct solutions can be obtained if we assume, in addition, that the function f is G-invariant on some G manifold. For $G = \mathbb{Z}_2$ the reader is referred to E. Fadell and P. H. Rabinowitz [FaR1]; for $G = S^1$, see E. Fadell and P. H. Rabinowitz [FaR2], and A. Floer, E. Zehnder [FlZ1]. As to the general compact Lie group G, see T. Bartsch and M. Clapp [BaC1].

6. Perturbation Theory

We study two problems in this section:

(1) Given a C^2-function f, let Σ be a nondegenerate critical manifold of f. What becomes of Σ if we perturb f to $f + g$ where g is small?

In the first part of this section, we shall study this problem under the various metrics of $g : C^0, C^1$ and C^2.

(2) For a given function f, which does not have the (PS) condition, we perturb it to $f_\varepsilon = f + \varepsilon g$ such that for each $\varepsilon > 0$, f_ε possesses the (PS) condition. Under what conditions can one extend the critical point theory for the perturbed functions to the original one? This will be studied in the second part of this section.

6.1. Perturbation on Critical Manifolds

We start with the C^0-perturbation, i.e., g is assumed to be small in the C^0-norm. Because of the very flexibility of g, one cannot expect any strong conclusion.

Lemma 6.1. *Let $A \subset Y \subset B \subset A' \subset X \subset B'$ be topological spaces. Suppose that $H_*(B, A) \simeq H_*(B', A') \simeq 0$. Then $h_* : H_*(A', A) \to H_*(X, Y)$ is an injection.*

Proof. Observing the following diagrams:

$$\begin{array}{ccccccc} H_{q+1}(B', A') & \longrightarrow & H_q(A', A) & \xrightarrow{i_*} & H_q(B', A) & \longrightarrow & H_q(B', A') \\ & & \quad\searrow^{\alpha_*} & & \nearrow_{\beta_*}\quad & & \\ & & & H_q(X, A) & & & \end{array}$$

and

$$\begin{array}{ccccccc} H_q(B,A) & \longrightarrow & H_q(X,A) & \xrightarrow{(i_1)_*} & H_q(X,B) & \longrightarrow & H_{q-1}(B,A) \\ & & & (\alpha_1)_* \searrow \quad \nearrow (\beta_1)_* & & & \\ & & & H_q(X,Y) & & & \end{array}$$

where $i : (A', A) \to (B', A)$, i_1, α, β, α_1, β_1, are incursion maps. From the exactness of these sequences, and the assumptions $H_*(B, A) \simeq H_*(B', A') = 0$, i_* and $(i_1)_*$ are isomorphisms. However, $i_* = \beta_* \circ \alpha_*$ and $(i_1)_* = (\beta_1)_* \circ (\alpha_1)_*$. Therefore $(\alpha_1)_*$ and α_* are injections, so is $h_* = (\alpha_1)_* \circ \alpha_*$.

Theorem 6.1. *Suppose that $f \in C^1(M, \mathbb{R}^1)$ satisfies the (PS) condition, with an isolated critical value c. Assume that (a, b) is an interval containing c. Then there exists an $\varepsilon > 0$ such that for*

$$Sup\left\{|g(x) - f(x)| \mid x \in f^{-1}[a, b]\right\} < \varepsilon/3. \tag{6.1}$$

We have an injection $h_* : H_*(f_{c+\varepsilon}, f_{c-\varepsilon}) \to H_*(g_{c+\frac{\varepsilon}{2}}, g_{c-\frac{\varepsilon}{2}})$.

Proof. Choose $\varepsilon > 0$ such that c is the only critical value of f in $[c-\varepsilon, c+\varepsilon] \subset (a, b)$. (6.1) implies

$$f_{c-\varepsilon} \subset g_{c-\frac{\varepsilon}{2}} \subset f_{c-\frac{\varepsilon}{6}} \subset f_{c+\frac{\varepsilon}{6}} \subset g_{c+\frac{\varepsilon}{2}} \subset f_{c+\varepsilon}.$$

Applying Lemma 6.1, we obtain

$$h_* : H_*\left(f_{c+\frac{\varepsilon}{6}}, f_{c-\varepsilon}\right) \to H_*\left(g_{c+\frac{\varepsilon}{2}}, g_{c-\frac{\varepsilon}{2}}\right)$$

is an injection. Hence $H_*(f_{c+\varepsilon}, f_{c-\varepsilon}) \to H_*(g_{c+\frac{\varepsilon}{2}}, g_{c-\frac{\varepsilon}{2}})$ is an injection.

Theorem 6.2. *Suppose that $f \in C^1(M, \mathbb{R}^1)$, satisfying the (PS) condition, has only finitely many critical points in $f^{-1}[a, b]$, where a, b are regular values of f. Then there exists an $\varepsilon > 0$ such that*

$$M_q(f) \le M_q(g) \quad q = 0, 1, 2, \ldots$$

for all $g \in C^1(M, \mathbb{R}^1)$ satisfying (6.1) and the (PS) condition, where $M_q(\,\cdot\,)$, $\forall\, q$, are the Morse type numbers with respect to (a, b).

Proof. Straightforward.

Theorem 6.2 implies that the Morse type numbers are lower semi-continuous under C^0-perturbation. As a direct consequence, we have a result due to Ambrosetti-Coti Zelati-Ekeland [ACE1].

Corollary 6.1. *Assume that $f,\ g \in C^1(M, \mathbb{R}^1)$ satisfy the (PS) condition, with $c = \inf f > -\infty$, $d = \inf g > -\infty$, and that there exists $q > 0$ such that*

$$H_q\,(K_c(f)) \neq 0$$

where $K_c(f)$ is the critical set of f with critical value c.

Suppose that there exists $\varepsilon > 0$ such that

$$K(f) \cap f^{-1}(c, c+\varepsilon] = \emptyset \quad \text{and}$$
$$\operatorname{Sup}\{|g(x) - f(x)| \mid x \in f_{c+\varepsilon}\} < \frac{\varepsilon}{2}.$$

Then g has at least two critical points.

Proof. Obviously $K_d(g) \neq \emptyset$. We prove by contradiction that if $K(g) = K_d(g) =$ single point, then $M_q(g) = 0$. But

$$M_q(g) \geq M_q(f) = \text{ rank } H_q\,(f_{c+\varepsilon}) = \text{ rank } H_q\,(K_c(f)) > 0.$$

This is impossible.

Next, we turn to C^2-perturbation. It is equivalent to the C^1-perturbation of the variational equation $df(x) = \theta$. The inverse function theorem is applied.

Theorem 6.3. *Let M be a C^2-Hilbert-Riemannian manifold, and let $f : [-1,1] \times M \to \mathbb{R}^1$ be a C^2-function. Let Σ be a compact nondegenerate critical manifold for $f^0 = f(0, \cdot)$. Assume that $d^2 f^0(x)$ is a Fredholm operator $\forall\, x \in \Sigma$. Then there exist an $\bar{\varepsilon} > 0$ and a neighborhood U of Σ such that $\forall\, 0 < |\varepsilon| < \bar{\varepsilon}$. The function $f^\varepsilon = f(\varepsilon, \cdot)$ has at least $\operatorname{Cat}(\Sigma)$ critical points in U.*

Proof. We regard f^ε as a family of functions defined on the fibers of a normal disk bundle over Σ. The function f^0 has a nondegenerate critical point on each fiber. We shall prove by the inverse function theorem that f^ε has the same number of critical points.

1. We choose a tubular neighborhood W of Σ, which is diffeomorphic to a normal disk bundle $N\Sigma(r)$, $r > 0$, in the following sense: $\forall\, x \in W$, $\exists$ a unique decomposition, $z = Px \in \Sigma$ and $v = Qx \in N_z(\Sigma)(r)$ such that $x = \exp_z v$.

Suppose that M is modelled on a Hilbert space H, $\forall\, z \in \Sigma$. We consider the orthogonal projection

$$\pi_z : H \to \operatorname{Im} d^2 f^0(z).$$

Therefore $\forall\, x \in W$

$$df^\varepsilon(x) = \theta \iff \begin{cases} \pi_z df^\varepsilon(x) = \theta, & (6.2) \\ (I - \pi_z)\, df^\varepsilon(x) = \theta. & (6.3) \end{cases}$$

2. Let

$$\psi(\varepsilon, z, v) = \pi_z df^{\varepsilon}(x).$$

We have $\psi(0, z, 0) = \pi_z df^0(z) = \theta$, $\forall\ z \in \Sigma$, and

$$\psi'_v(0, z', 0) = \pi_{z'} d^2 f^0(z') = d^2 f^0(z'),$$

which is invertible from Im $d^2 f^0(z)$ into Im $d^2 f^0(z')$. By the implicit function theorem, one has $\bar{\varepsilon} > 0$, a neighborhood $U_0 \subset N\Sigma(r)_1$ and a C^1-map $\sigma : (-\varepsilon, \varepsilon) \times \Sigma \to U_0$ such that

$$\psi(\varepsilon, z, \sigma(\varepsilon, z)) = \theta$$

and $\sigma(o, z) = \theta$, which solves equation (6.2) uniquely in U_0. We shall prove that the section $z \mapsto \sigma(\varepsilon, z)$ of $N\Sigma(r)$ provides the critical points of f^{ε}.

3. Indeed, let

$$\Sigma_{\varepsilon} = \{x = \exp_z \sigma(\varepsilon, z) \mid z \in \Sigma\}.$$

Then Σ_{ε} is a compact connected manifold, with $\mathrm{Cat}(\Sigma_{\varepsilon}) = \mathrm{Cat}\,(\Sigma)$ since $\exp(\sigma(\varepsilon, \,\cdot\,))$ is a diffeomorphism between Σ and Σ_{ε}.

Note that

$$H \cong T_x M \cong T_z(\Sigma) \oplus N_z(\Sigma)$$

and

$$H \cong \ker d^2 f(z) \oplus\ \mathrm{Im}\ df^2(z),$$

where $\oplus$ denotes orthogonal direct sum, we have

$$\ker d^2 f(z) = T_z(\Sigma) \text{ and } \mathrm{Im}\ d^2 f(z) = N_z(\Sigma).$$

Let $U \subset W$ be the pull back of $U_0 \subset N\Sigma(r)$, then

$$\begin{aligned} U \cap K\,(f^{\varepsilon}) &= \{x \in \Sigma_{\varepsilon} \mid \text{in which (6.3) holds}\} \\ &= \{x \in \Sigma_{\varepsilon} \mid df^{\varepsilon}(x) \in T_z(\Sigma)\} \\ &= \{x \in \Sigma_{\varepsilon} \mid df^{\varepsilon}(x) \in T_x\,(\Sigma_{\varepsilon})\} \\ &= K\,(f^{\varepsilon}|_{\Sigma_{\varepsilon}})\,, \end{aligned}$$

where $K(f)$ denotes the critical set of f. This is due to the fact that as $|\varepsilon| > 0$ small, $T_z(\Sigma)$ is closed to $T_x(\Sigma_{\varepsilon})$.

4. $\forall\ 0 < |\varepsilon| < \bar{\varepsilon}$, the function $f^{\varepsilon}|_{\Sigma_{\varepsilon}}$ has at least $\mathrm{Cat}\,(\Sigma_{\varepsilon}) = \mathrm{Cat}\,(\Sigma)$ critical points, according to the Ljusternik-Schnirelman Theorem. The conclusion follows.

Finally, we study the C^1-perturbation.

Theorem 6.4. *Let M be a Hilbert-Riemannian manifold, and let $f,\ g \in C^1(M,\mathbb{R}^1)$. Suppose that both f and $f+g$ satisfy the (PS) condition, and that Σ is a connected nondegenerate critical manifold of f with finite index. If g is sufficiently small in the C^1-norm in a neighborhood U of Σ in M, then*

$$\#(K(f+g)\cap U) \geq CL(\Sigma)+1.$$

Furthermore, if $f,\ g \in C^2$, and $f+g$ is nondegenerate in U, then

$$\#K(f+g) \geq \sum_{i=0}^{\infty} \beta_i(\Sigma).$$

Proof. According to the (PS) condition, Σ is compact. With no loss of generality, we assume $K_c = \Sigma$. There exist $\delta > 0$ and two tubular neighborhoods $V,\ W$ of Σ satisfying

(1) $V \subset \overline{V} \subset W \subset \overline{W} \subset U$,

(2) $V \simeq N\Sigma(r_1),\ W \simeq N\Sigma(r_2),\ 0 < r_1 < r_2 < r_1+1$, where $N\Sigma(r)$ is the normal disk bundle of Σ,

(3) $W \subset \text{ int } (f_{c+\frac{\delta}{2}}\backslash f_{c-\frac{\delta}{2}})$ and $f^{-1}[c-\delta, c-\delta] \cap K = \Sigma$.

According to Theorem 7.4 of Chapter I,

$$(f_{c+\delta}, f_{c-\delta}) \sim \left(\xi^-(r), \dot{\xi}^-(r)\right),$$

where $(\xi^-(r), \dot{\xi}^-(r))$ is the negative sphere bundle over Σ.

Let us define $\varphi \in C^2(M,\mathbb{R}^1)$ satisfying $\varphi(x) = 1$ for $x \in V$, $\varphi(x) = 0$ for $x \notin W$, $0 \leq \varphi(x) \leq 1\ \forall\, x \in M$, and $\|d\varphi(x)\| < C$, a constant > 1.

According to the (PS) condition, there exists $\varepsilon_0 > 0$ such that

$$\|df(x)\| \geq \varepsilon_0 \quad \forall\, x \in W\backslash V.$$

Fixing $0 < \varepsilon < \text{ Min }\{\frac{\delta}{2}, \frac{\varepsilon_0}{2C}\}$, we obtain

$$\|d(f+\varphi g)(x)\| \geq \varepsilon_0 - \|dg(x)\| - \|d\varphi(x)\|\ |g(x)| > 0 \quad \forall\, x \in W\backslash V,$$

if $\|g\|_{C^1(U)} < \varepsilon$.

Thus the function $F(x) = (f+\varphi g)(x)$ satisfies the (PS) condition and possesses the same critical points as $f+g$ in W. Moreover, we have $(F_{c+\delta}, F_{c-\delta}) = (f_{c+\delta}, f_{c-\delta})$.

Using Thom's isomorphism theorem,

$$H_*\left(\xi^-(r), \dot{\xi}^-(r)\right) \simeq H_{*-k}(\Sigma),$$

where $k = \text{ ind}\,(f,\Sigma)$.

Since

$$\begin{aligned} H_*\left(F_{c+\delta}, F_{c-\delta}\right) &\simeq H_*\left(f_{c+\delta}, f_{c-\delta}\right) \\ &\simeq H_*\left(f_{c+\delta}\backslash \mathring{f}_{c-\delta}, f^{-1}(c-\delta)\right) \\ &\simeq H_*\left(\xi^-(r), \dot{\xi}^-(r)\right), \end{aligned}$$

and

$$H^*\left(f_{c+\delta}\backslash \mathring{f}_{c-\delta}\right) \simeq H^*\left(\xi^-(r)\right) \simeq H^*(\Sigma),$$

preserving the ring structure, there exist $\omega_1, \omega_2, \dots, \omega_\ell \in H^*(\Sigma)$, $\ell = CL(\Sigma)$, with $\dim \omega_i > 0$, $i = 1, 2, \dots, \ell$, such that $\omega_1 \cup \omega_2 \cup \dots \cup \omega_\ell \neq 0$, from which we obtain $[Z_0], [Z_1], \dots, [Z_\ell] \in H_*(F_{c+\delta}, F_{c-\delta})$ such that $[Z_0] < [Z_1] < \dots [Z_\ell]$.

Therefore there are at least $\ell + 1$ distinct critical points of F, and then, of $f+g$, in U, provided by Theorem 3.4.

Furthermore, if $f+g$ is nondegenerate, then $F = f + \varphi g$ is also. Since $H_*(F_{c+\delta}, F_{c-\delta}) \simeq H_{*-k}(\Sigma)$, we have at least $\sum_{i=0}^{\infty} \beta_i(\Sigma)$ critical points, provided by the Morse inequalities.

Remark 6.1. Theorems 6.1 and 6.2 were given by Marino Prodi [MaP1]. Theorem 6.3 was obtained by Reeken [Ree1]. A special case of Theorem 6.4, in which M is assumed to be finite dimensional, is a version of a theorem under the name of Conley Zehnder by A. Weinstein [Wei2], cf.[Cha15].

6.2. Uhlenbeck's Perturbation Method

Lemma 6.2. *Let M be a C^2-Finsler manifold, and let $f,\ g \in C^1(M, \mathbb{R}^1)$. Suppose that f is bounded below and that $g \geq 0$. For $\varepsilon \in (0,1]$, let $f^\varepsilon = f + \varepsilon g$. Assume that*

(1) $\|dg\|$ is bounded on sets on which g is bounded,

(2) f^ε satisfies the (PS)$_c$ condition, for some c.

Then the function $h = g(c-f)^{-1}$ satisfies the (PS)$_{\varepsilon^{-1}}$ condition and $K_{\varepsilon^{-1}}(h) = K_c(f^\varepsilon)$, where $K_b(f) =$ the critical set of f at level b.

Proof. By computation

$$\begin{aligned} dh &= (c-f)^{-1} dg + (c-f)^{-2} g\, df \\ &= (c-f)^{-2} g(df + h^{-1} dg). \end{aligned}$$

Thus

$$x_0 \in K_{\varepsilon^{-1}}(h) \Leftrightarrow \begin{cases} h(x_0) = \varepsilon^{-1} \\ dh(x_0) = \theta \end{cases}$$

$$\Leftrightarrow \begin{cases} f^\varepsilon(x_0) = f(x_0) + h(x_0)^{-1}g(x_0) = c \\ df^\varepsilon(x_0) = df(x_0) + h(x_0)^{-1}dg(x_0) = (c - f(x_0))^2 dh(x_0) = \theta \end{cases}$$

$$\Leftrightarrow x_0 \in K_c(f^\varepsilon).$$

Suppose that $\{x_j\}$ is a sequence along which

$$h(x_j) \to \varepsilon^{-1} \quad \text{and} \quad dh(x_j) \to \theta.$$

By the same algebraic computation, we have

$$f^\varepsilon(x_j) \to c \quad \text{and}$$

$$\begin{aligned} \|df^\varepsilon(x_j)\| &= \left\|(c - f(x_j))^2 g(x_j)^{-1} dh(x_j) + \left(\varepsilon - h(x_j)^{-1}\right) dg(x_j)\right\| \\ &\le \left|g(x_j) h^{-2}(x_j)\right| \, \|dh(x_j)\| + o(\|dg(x_j)\|). \end{aligned}$$

Since $0 \le g(x_j) \le (\frac{1}{\varepsilon} + 1)(c - f(x_j))$ is bounded, provided by the fact that f is bounded from below. Combining this with assumption (1), we obtain $df^\varepsilon(x_j) \to \theta$. Again, using assumption (2), it follows the $(\mathrm{PS})_{\varepsilon^{-1}}$ condition for the function h.

Corollary 6.2. *Under the conditions of the above theorem, if, further,* $K_c(f^\varepsilon) = \emptyset \ \forall \, \varepsilon \in (0, \varepsilon_0]$, *then* $(f^{\varepsilon_0})_c$ *is a strong deformation retract of* f_c.

Proof. Since $g \ge 0$ and $\varepsilon_0 > 0$, $(f^{\varepsilon_0})_c \subset f_c$. From the theorem,

$$K_c(f^\varepsilon) = \emptyset \Leftrightarrow K_{\varepsilon^{-1}}(h) = \emptyset \quad \forall \, \varepsilon \in (0, \varepsilon_0],$$

there is therefore no critical value of h between ε_0^{-1} and $+\infty$. The $(\mathrm{PS})_{\varepsilon^{-1}}$ condition for the function $h \ \forall \, \varepsilon \in (0, \varepsilon_0]$, implies that $(f^{\varepsilon_0})_c = h_{\varepsilon_0^{-1}}$ is a strong deformation of $f_c = h_\infty$.

Definition 6.1. Let M be a Finsler manifold, and let $f \in C^1(M, \mathbb{R}^1)$. We say that f satisfies the ε-deformation property, if

(1) $f(K(f))$ is closed;
(2) for any interval $[a, b] \subset \mathbb{R}^1$ on which f has no critical value, i.e., $K(f) \cap f^{-1}[a, b] = \emptyset$, there exists a family of functions $f^\varepsilon \in C^1(M, \mathbb{R}^1)$ with $f^\varepsilon \ge f \ \forall \, \varepsilon \in (0, 1]$ such that the level set $(f^\varepsilon)_a$ is a strong deformation retract of $f_b \ \forall \, \varepsilon \in (0, 1]$, i.e., $\exists \, \eta^\varepsilon : [0, 1] \times f_b \to f_b$, satisfying
   (i) $\eta^\varepsilon(0, \cdot\,) = \mathrm{id}$,
   (ii) $\eta^\varepsilon(1, f_b) \subset (f^\varepsilon)_a$, and
   (iii) $\eta^\varepsilon(t, \cdot\,)\,|_{(f^\varepsilon)_a} = \mathrm{id}_{(f^\varepsilon)_a}$.

Theorem 6.5. (ε-Minimax Principle) *Let $\mathcal{F}$ be a family of subsets of M, and let $f \in C^1(M, \mathbb{R}^1)$ satisfy the ε-deformation property. Set*

$$c = \inf_{F \in \mathcal{F}} \sup_{x \in F} f(x).$$

If (1) c is finite, (2) $\mathcal{F}$ is invariant with respect to η^ε $\forall\, \varepsilon \in (0,1]$ where η^ε is the strong deformation retract satisfying (i)–(iii) with any interval $[a,b]$ containing c as an interior point. Then c is a critical value of f.

Proof. Suppose that c is not a critical value provided by the closeness of $f(K(f))$, there exists $\delta > 0$ such that $K(f) \cap f^{-1}[c-\delta, c+\delta] = \emptyset$. Choosing $F_0 \in \mathcal{F}$ such that $F_0 \subset f_{c+\delta}$, we have $\eta^\varepsilon(1, F_0) \subset f^\varepsilon_{c-\delta}$, but

$$f(x) \le f^\varepsilon(x) \le c - \delta \quad \forall\, x \in \eta^\varepsilon(1, F_0) \in \mathcal{F}.$$

This is a contradiction.

Theorem 6.6. *Let M be a C^2-Finsler manifold modeled on a separable Banach space with differentiable norms. Let $f,\ g \in C^2(M, \mathbb{R}^1)$, satisfy the following assumptions:*

(1) *$f \ge -m > -\infty$, $g > 0$ and $f^\varepsilon = f + \varepsilon g$ satisfies the (PS) condition, $\forall \varepsilon > 0$.*

(2) *$\|dg(x)\|$ is bounded on sets on which g is bounded.*

(3) *$\overline{\bigcup_{0<\varepsilon\le\delta_0} K(f^\varepsilon) \cap (f^\varepsilon)^{-1}[a,b]}$ is compact for some $\delta_0 > 0$.*

(4) *$a, b \notin f(K(f))$. And $\forall\, x_0 \in K(f) \subset f^{-1}(a,b)$, $\exists$ a neighborhood U such that $K(f^\varepsilon) \cap U$ consists of a curve $x(\varepsilon)$, with $x(0) = x_0$, $\varepsilon \in [0, \delta]$, and $x(\varepsilon)$ are nondegenerate (in the sense of Section 4, Chapter I) and of the fixed index.*

Then f_a with handles, adjoined in a corresponding fashion to the critical points of f with values in (a,b), is a deformation retract of f_b. The dimensions of the handles correspond to the dimensions of the indices at these critical points.

Proof. First, assume $b < +\infty$. Then $\exists\ \delta > 0$ such that $a, b \notin f^\varepsilon(K(f^\varepsilon))\ \forall\, \varepsilon \in [0, \delta]$.

We claim that $(f^\varepsilon)^{-1}[a,b] \cap K(f^\varepsilon)$ consists of all these families of curves $x(\varepsilon)$.

In fact, if not, then $\exists\ \varepsilon_j \to 0$, $x_j \in K(f^{\varepsilon_j}) \cap (f^{\varepsilon_j})^{-1}[a,b]$; but x_j do not lie on any families $x(\varepsilon)$ defined in (4). However, by assumption (3), $x_j \to x^* \in K(f) \cap f^{-1}[a,b]$. This contradicts assumption (4).

Then Corollary 6.2 is applied: $(f^\delta)_a \simeq f_a$, $(f^\delta)_b \simeq f_b$. Since the Morse handle body theorem for f^δ is known (Chapter I, Section 4)

$$\left(f^\delta\right)_a \cup \bigcup_{j=1}^{s} h_j\left(D^{m_j}\right) \simeq \left(f^\delta\right)_b,$$

where $h_j(D^{m_j})$ denotes the attached handle, $j = 1, 2, \dots, s$. We obtain

$$f_a \cup \bigcup_{j=1}^{s} h_j\left(D^{m_j}\right) \simeq f_b.$$

Second, if $b = +\infty$, then $\exists\ b_j \to +\infty$ such that $b_j \notin f(K^f)$, and $\varepsilon_j \downarrow 0$ with the following properties.

(1) $\forall\ \varepsilon \in [o, \varepsilon_j]$, b_{j-1}, b_j are not critical values of f^ε,

(2) $\forall\ x_0 \in K(f^\varepsilon) \cap (f^\varepsilon)^{-1}[a, b_j]$, it lies on the unique one parameter family of nondegenerate critical points of f^ε near the critical points of f.

Then $(f^{\varepsilon_j})_{b_j}$ can be retracted to $(f^{\varepsilon_j})_{b_{j-1}}$ with handles corresponding to critical points of f^{ε_j} with values in (b_{j-1}, b_j). In addition, $(f^{\varepsilon_{j-1}})_{b_{j-1}}$ is a deformation retract of $(f^{\varepsilon_j})_{b_j}$. The sequence of retractions gives the desired result for $f_{+\infty}$.

Remark 6.2. The material of this subsection is taken from K. Uhlenbeck [Uhl2].

CHAPTER III

Applications to Semilinear Elliptic Boundary Value Problems

Semilinear elliptic boundary value problems have attracted great interest in the applications of critical point theory because they are good models to deal with multiple solutions problems with respect to both results and methods.

1. Preliminaries

Let us turn to some notation and basic facts in the theory of partial differential equations.

Let $\Omega \subset \mathbb{R}^n$ be a bounded open domain with smooth boundary $\partial\Omega$. For a nonnegative integer vector $\alpha = (\alpha_1, \dots, \alpha_n)$ we write

$$\partial^\alpha = \frac{\partial^{|\alpha|}}{\partial x_1^{\alpha_1} \dots \partial x_n^{\alpha_n}}$$

to denote the differential operator, with $|\alpha| = \alpha_1 + \dots + \alpha_n$.

Let $\mathcal{D}(\Omega)$ be the function space consisting of C^∞ functions with compact support in Ω, and let $\mathcal{D}'(\Omega)$ be the dual of $\mathcal{D}(\Omega)$, i.e., the Schwartz distribution space. For each integer $m \geq 0$, we denote

$$C^m(\overline{\Omega}) = \left\{u : \overline{\Omega} \to \mathbb{R}^1 \mid \partial^\alpha u \text{ is continuous on } \overline{\Omega},\ |\alpha| \leq m\right\},$$

with norm

$$\|u\|_m = \sum_{|\alpha| \leq m} \sup_{x \in \overline{\Omega}} |\partial^\alpha u(x)| .$$

For $p \geq 1$, and an integer $m \geq 0$, we denote

$$W_p^m(\Omega) = \left\{u \in L^p(\Omega) \mid \partial^\alpha u \in L^p(\Omega),\ |\alpha| \leq m\right\},$$

where L^p is the p-th power integrable Lebesgue space, and ∂^α is the differential operator in the distribution sense, with norm

$$\|u\|_{W_p^m} = \left\{ \sum_{|\alpha| \leq m} \|\partial^\alpha u\|_{L^p(\Omega)}^p \right\}^{\frac{1}{p}} .$$

$W_p^m(\Omega)$ is called the *Sobolev space*. In particular, if $p = 2$, $H^m(\Omega)$ stands for $W_2^m(\Omega)$. The closure of $\mathcal{D}(\Omega)$ in the space $W_p^m(\Omega)$ $(H^m(\Omega)$ and $C^m(\overline{\Omega})$ is denoted by $\overset{\circ}{W}{}_p^m(\Omega)$ $(H_0^m(\Omega), C_0^m(\overline{\Omega})$ respectively).

The dual space of $\overset{\circ}{W}{}_p^m(\Omega)$ (and $H_0^m(\Omega)$) is denoted by $W_{p'}^{-m}(\Omega)$ (and $H^{-m}(\Omega)$ resp.), where $\frac{1}{p} + \frac{1}{p'} = 1$.

The following inequalities are applied very frequently.

Poincaré inequality

$$\left(\int_\Omega |u|^p\,dx\right)^{\frac{1}{p}} \le C(\Omega)\left(\int_\Omega |\nabla u|^p\,dx\right)^{\frac{1}{p}} \quad \forall\, u \in \overset{\circ}{W}{}_p^1(\Omega)$$

where ∇u denotes the gradient of u, and $C(\Omega)$ is a constant independent of u.

Sobolev inequality. Suppose that for $1 \le p, r < \infty$ and integers $\ell \ge m \ge 0$, we have

(1) If $\frac{1}{p} \le \frac{1}{r} + \frac{\ell - m}{n}$, then the embedding $W_p^\ell(\Omega) \hookrightarrow W_r^m(\Omega)$ is continuous. If the inequality $\le$ is replaced by a strict inequality $<$, then the embedding is compact.

(2) If $\frac{1}{p} \le \frac{\ell - m}{n}$, then the embedding $W_p^\ell(\Omega) \hookrightarrow C^m(\overline{\Omega})$ is continuous. If the inequality $\le$ is replaced by a strict inequality $<$, then the embedding is compact.

For a function $f \in C(\overline{\Omega} \times \mathbb{R}^1, \mathbb{R}^1)$, suppose that there exist constants α, $C > 0$ such that

$$|f(x,t)| \le C\,(1 + |t|^\alpha)\,.$$

Then the following nonlinear operator:

$$u \mapsto f(x, u(x))$$

maps boundedly and continuously from $L^p(\Omega)$ to $L^q(\Omega)$, with $p = \alpha q$, for example, cf. [Ber1]. The operator is called the *Nemytcki operator*. Applying this result, in combination with the Sobolev inequality, we see that the functional

$$J(u) = \int_\Omega f(x, u(x))\,dx$$

is well-defined and continuous on the Sobolev space $H_0^1(\Omega)$ if

$$\alpha \le \frac{2n}{n-2}.$$

If we further assume that $f \in C^1(\overline{\Omega} \times \mathbb{R}^1, \mathbb{R}^1)$ satisfies the following growth condition:

$$|f_t'(x,t)| \le C\,(1 + |t|^\alpha) \text{ with } \alpha \le \frac{n+2}{n-2},$$

then the functional J is C^1 on the space $H_0^1(\Omega)$, with differential

$$(dJ(u), v) = \int_\Omega f'_t(x, u(x))v(x)\,dx, \quad \forall\, v \in H_0^1(\Omega).$$

Furthermore if $f \in C^2(\overline{\Omega} \times \mathbb{R}^1, \mathbb{R}^1)$, satisfies

$$|f''_{tt}(x,t)| \le C\,(1 + |t|^\alpha) \text{ with } \alpha \le \frac{4}{n-2},$$

then J is C^2 on $H_0^1(\Omega)$, with

$$d^2J(u)(v, w) = \int_\Omega f''_{tt}(x, u(x))v(x)w(x)\,dx \quad \forall\, v, w \in H_0^1(\Omega).$$

As for the Laplacian $-\Delta$ defined on $L^2(\Omega)$, with domain $D(-\Delta) = H^2(\Omega) \cap H_0^1(\Omega)$, it is a self-adjoint operator. The operator $-\Delta$ can be extended from $W_p^2 \cap \overset{\circ}{W}{}_p^1(\Omega)$ to $L^p(\Omega)$, $1 < p < \infty$, continuously.

It is known that $\ker(-\Delta) = \{\theta\}$, and that $K = (-\Delta)^{-1}$ maps $L^p(\Omega)$ into itself continuously, and is a compact operator. Also the operator K maps $L^2(\Omega)$ into $H_0^1(\Omega)$, such that

$$\int_\Omega u \cdot v\,dx = (Ku, v)_{H_0^1} \quad \forall\, v \in H_0^1(\Omega),\ u \in L^2(\Omega).$$

The following boundary value problem will be considered:

$$\text{(1.1)} \qquad \begin{cases} -\Delta u &= g(x, u(x)) \text{ in } \Omega \\ u|_{\partial\Omega} &= 0, \end{cases}$$

where the following growth condition on $g \in C(\overline{\Omega} \times \mathbb{R}^1, \mathbb{R}^1)$ is assumed:

$$\text{(1.2)} \qquad |g(x,t)| \le C\,(1 + |t|^\alpha) \quad \alpha \le \frac{n+2}{n-2}, \text{ if } n \ge 3$$

(and if $n \le 2$, α has no restriction) for constants C, $\alpha > 0$. We say that $u_0 \in H_0^1(\Omega)$ is a *weak solution* of (1.1) if

$$\int_\Omega [\nabla u_0 \cdot \nabla v - g\,(x, u_0(x))\,v(x)]\,dx = 0 \quad \forall\, v \in H_0^1(\Omega).$$

If we define the functional

$$\text{(1.3)} \qquad J(u) = \int_\Omega \left(\frac{1}{2}|\nabla u|^2 - G(x, u(x))\right) dx,$$

where

$$G(x,t) = \int_0^t g(x,\xi)\, d\xi, \tag{1.4}$$

then J is C^1 on $H_0^1(\Omega)$ with

$$(dJ(u),v) = \int_\Omega [\nabla u \cdot \nabla v - g(x,u(x)) \cdot v(x)]\, dx \quad \forall\, v \in H_0^1(\Omega).$$

This means that the differential equation (1.1) is just the *Euler equation* of the functional J and the weak solutions of (1.1) are critical points of J, and conversely.

Since it is well-known that weak solutions of (1.1) are classical solutions of (1.1), if the function g is smooth enough (cf. [GiT1]), it is enough to look for weak solutions of (1.1), i.e., the critical points of J.

The notion of sub- (or super-) solutions of the equation (1.1) is also important. We say $\underline{u} \in C^2(\Omega) \cap C(\overline{\Omega})$ (or $\bar{u}$) is a sub- (or super- resp.) solution if

$$\begin{cases} -\Delta \underline{u} \le g(x,\underline{u}(x)) \text{ in } \Omega \\ \underline{u}|_{\partial\Omega} \le 0 \end{cases} \qquad \begin{matrix} (\text{or } -\Delta \bar{u} \ge g(x,\bar{u}(x)) \text{ in } \Omega \\ \bar{u}|_{\partial\Omega} \ge 0). \end{matrix}$$

If furthermore $\underline{u} < \bar{u}$, then we say they are a pair of sub- or supersolutions. They are called strict if they are not solutions.

For second order elliptic operators, (especially for $-\Delta$) the maximum principle plays an important role in both qualitative and existence studies.

Theorem (Maximum Principle). *Suppose that $u \in C^2(\Omega)$ satisfies the equation*

$$-\Delta u \le 0 \quad \text{in } \Omega. \tag{1.5}$$

If u attains a maximum M at a point of Ω, then $u \equiv M$ in Ω.

Theorem. (Strong Maximum Principle) *Suppose that $u \in C^2(\Omega)$ satisfies (1.5). Suppose that $u \le M$ in Ω, and $u = M$ at a boundary point P. Assume that P lies on the boundary of a ball K_1 in D. If u is continuous in $\Omega \cup P$ and an outward directional derivative $\partial u/\partial n$ exists at P, then $\partial u/\partial n > 0$ at P unless $u \equiv M$.*

Corollary 1.1. *The operator $K = (-\Delta)^{-1}$ is positive, in the sense that it maps nonnegative functions to nonnegative functions. Particularly, $K : L^P(\Omega) \to C_0^1(\overline{\Omega})$ for $p > \frac{n}{2}$, maps nonnegative functions to the interior of the positive cone in $C_0^1(\overline{\Omega})$.*

For positive operators, we have the Krein-Rutman Theorem which asserts that the first eigenvalue $(-\Delta)$ is simple. More generally, we have

Theorem. (Kato-Hess [KaH1]) *Suppose that $m \in C(\overline{\Omega})$, and that there is a point $x_0 \in \Omega$ such that $m(x_0) > 0$. Then the equation*

$$\begin{cases} -\Delta u(x) = \lambda m(x) u(x) & x \in \Omega,\ \lambda \in \mathbb{R}^1 \\ u|_{\partial\Omega} = 0 \end{cases}$$

admits a principle eigenvalue $\lambda_1(m) > 0$, characterized by being the unique positive eigenvalue having a positive eigenfunction. Moreover, $\lambda_1(m)$ has the following properties:

(1) *if $\hat{\lambda} \in \mathbb{C}$ is an eigenvalue with $\mathrm{Re}\,\hat{\lambda} > 0$, then $\mathrm{Re}\,\hat{\lambda} \geq \lambda_1(m)$.*

(2) *$\mu_1(m) := 1/\lambda_1(m)$ is an eigenvalue of the operator $K\cdot(m\cdot) : L^2(\Omega) \to L^2(\Omega)$ with algebraic multiplicity 1.*

In the applications, sometimes we would consider the restriction $\widetilde{J}$ of J on a smaller Banach space $C_0^1(\overline{\Omega})$, where J is defined in (1.3). The functional $\widetilde{J}$ may lose the (PS) condition (on $C_0^1(\overline{\Omega})$, even if J has on $H_0^1(\Omega)$). However, by a bootstrap iteration, the following is proved in [Cha3].

Theorem 1.1. *Under assumption (1.2) with $\alpha < \frac{n+2}{n-2}$, if $n > 2$, suppose that $g \in C^1$, and that J satisfies the (PS) condition; then the functional $\widetilde{J}$ possesses the following properties:*

(1) *$\widetilde{J}(K)$ is a closed subset.*

(2) *For each pair $a < b$, $K \cap \widetilde{J}^{-1}(a,b) = \emptyset$ implies that $\widetilde{J}_a$ is a strong deformation retract of $\widetilde{J}_b \backslash K_b$, where K is the critical set of J (and also $\widetilde{J}$).*

Thus for any isolated $p_0 \in K$, we have

Corollary 1.2. *$C_*(\widetilde{J}, p_0) = C_*(J, p_0)$ with integral coefficients.*

Claim. For any open neighborhood U of p_0, let $V = \cup_{t\in\mathbb{R}^1}\eta(t,U)$, where η is the negative gradient flow of J. We have

$$\begin{aligned} C_*\left(J, p_0\right) &= H_*\left(J_0 \cap V, (J_0 \backslash \{p_0\}) \cap V; \mathbb{Z}\right) = H_*\left(\mathring{J}_{c+\varepsilon} \cap V, \mathring{J}_{c-\varepsilon} V; \mathbb{Z}\right) \\ &= H_*\left(\mathring{\widetilde{J}}_{c+\varepsilon} \cap V, \mathring{\widetilde{J}}_{c-\varepsilon} \cap V; \mathbb{Z}\right) = C_*\left(\widetilde{J}, p_0\right), \end{aligned}$$

using the Palais Theorem at the end of Chapter I, Section 1, where $c = f(p_0)$ and $\varepsilon > 0$ is suitably small.

2. Superlinear Problems

The classification of the semilinear elliptic BVPs into superlinear, asymptotically linear, and sublinear is very vague. Roughly speaking, it describes the growth of the function $g(x,u)$ with respective to u in (1.1). But sometimes $g(x,u)$ is superlinear in one direction, but sublinear in the other, so that it is not easy to classify them very clearly. Nevertheless, we follow the customary notation in the literature.

In the following, (1.2) is assumed ($\alpha < \frac{n+2}{n-2}$, subcritical, $\alpha = \frac{n+2}{n-2}$ is called critical).

Our first result in this section is the following.

Theorem 2.1. *Assume that the functional J defined in* (1.3) *satisfies the (PS) condition on the space $H_0^1(\Omega)$, and that J is unbounded below. Moreover, if there exists a pair of strict sub- and supersolutions of equation* (1.1), *then* (1.1) *possesses at least two distinct solutions.*

Before going into the proof, we recall a well-known result (cf. Amann [Ama1]) that if there is a pair of sub- and super- solutions $\underline{u} < \bar{u}$ of (1.1), then there is a solution $u_0 \in C$ of (1.1). One asks whether we can characterize the solution by the corresponding functional J?

Now we shall prove that J is bounded from below on $C_X = C \cap C_0^1(\overline{\Omega})$, where $C = \{u \in H_0^1(\Omega) \mid \underline{u}(x) \le u(x) \le \bar{u}(x) \text{ a.e.}\}$, and then attains its minimum, which is the variational characterization of u_0. Applying Example 1 from Chapter I, Section 4, we obtain the critical groups of u_0:

$$C_k(J, u_0) = \begin{cases} G & k = 0 \\ 0 & k \neq 0, \end{cases} \tag{2.1}$$

if it is isolated.

Lemma 2.1. *Suppose that $\underline{u} < \bar{u}$ is a pair of strict sub- and supersolutions of* (1.1). *Then there is a point $u_0 \in C_X$ which is a local minimum of the functional $\widetilde{J} = J|_{C_0^1(\overline{\Omega})}$. Moreover, if it is isolated, then*

$$C_k\left(\widetilde{J}, u_0\right) = \begin{cases} G & k = 0 \\ 0 & k \neq 0. \end{cases} \tag{2.2}$$

Proof. One may assume that $\underline{u}(x) < \bar{u}(x)$, without loss of generality. Define a new function

$$\hat{g}(x,\xi) = \begin{cases} g(x,\bar{u}(x)) \wedge (-\Delta\bar{u}(x)), & \xi > \underline{u}(x) \\ g(x,\xi), & \underline{u}(x) \le \xi \le \bar{u}(x) \\ g(x,\underline{u}(x)) \vee (-\Delta\underline{u}(x)), & \xi < \underline{u}(x) \end{cases}$$

where $a \vee b = \max\{a,b\}$, and $a \wedge b = \min\{a,b\}$. By definition, $\hat{g}(x,\xi) \in C(\overline{\Omega} \times \mathbb{R}^1)$ is bounded and satisfies:

$$g(x,\xi) = \hat{g}(x,\xi) \quad \text{for} \quad \underline{u}(x) \le \xi \le \bar{u}(x).$$

Let

$$\widehat{G}(x,\xi) = \int_0^\xi \hat{g}(x,t)\,dt.$$

Then $\widehat{G} \in C^1(\overline{\Omega} \times \mathbb{R}^1)$, and the functional

$$\widehat{J}(u) = \int_\Omega \left[\frac{|\nabla u|^2}{2} - \widehat{G}(x,u)\right] dx$$

defined on $H_0^1(\Omega)$ is bounded from below and satisfies the (PS) condition. Hence there is a minimum u_0 which satisfies

$$d\widehat{J}\,(u_0) = \theta,$$

i.e., u_0 satisfies the equation

$$\begin{cases} -\Delta u_0 = \hat{g}(x,u_0) \\ u_0|_{\partial\Omega} = 0. \end{cases}$$

According to the L^p regularity of solutions of elliptic BVP and the strong maximum principle, we see that $u_0 \in \overset{\circ}{C}_X$, the interior of C_X in the $C_0^1(\overline{\Omega})$ topology. (See Remark 2.1 below.) However, $\widetilde{J}|_{C_X} = J|_{C_X} = \widehat{J}|_{C_X}$; therefore u_0 is a local minimum of $\widetilde{J}$. (2.2) follows from Example 1, Chapter I, Section 4.

Under condition (1.2) J is well-defined on $H_0^1(\Omega)$. Since $C_0^1(\overline{\Omega})$ is dense in $H_0^1(\Omega)$, u_0 must be also a local minimum of J. In the case when it is isolated, (2.1) holds.

Remark 2.1. We verify $u_0 \in \overset{\circ}{C}_X$.

Claim. Since $\underline{u}$ is a strict sub-solution,

$$\begin{cases} -\Delta\,(u_0 - \underline{u})\,(x) \geq 0, \text{ but not identical to } 0, \text{ in } \Omega, \\ u_0 - \underline{u}|_{\partial\Omega} \geq 0. \end{cases}$$

It follows from the strong maximum principle, that $u_0 > \underline{u}$, and $\frac{\partial}{\partial n}(u_0 - \underline{u})|_{\partial\Omega} < 0$, where $\frac{\partial}{\partial n}$ is the outward normal derivative. Similarly, we have $u_0 < \bar{u}$, and $\frac{\partial}{\partial n}(\bar{u} - u_0)|_{\partial\Omega} < 0$. Therefore u_0 is an interior point of C_X in C_0^1 topology.

Proof of Theorem 2.1. We already have a local minimum so that it suffices to find another critical point. Since J is assumed to be unbounded below, $\exists\ u_1 \in H_0^1(\Omega)$ such that $J(u_1) < J(u_0)$. A weak version of a link (mountain pass) is easy to see.

$$J|_{\partial B(u_0,\delta)} \geq J(u_0) \qquad \text{for } \delta > 0 \text{ small.}$$
$$\max\{J\,(u_1)\,,J\,(u_0)\} \leq J\,(u_0)\,.$$

Exploiting Theorem 1.2 (or Remark 1.2) from Chapter II, there exists a different critical point.

We present an example for the application of Theorem 2.1.

Assume that

(g_1) (1.2) with $\alpha < \frac{n+2}{n-2}$;

(g_2) $\exists\, \theta > 2$ and $M > 0$ such that

$$\theta G(x,t) \le t\, g(x,t) \quad \forall\, x \in \Omega, \text{ for } |t| \ge M;$$

(g_3) $h \in L^{\frac{2n}{n+2}}(\Omega)$ is nonnegative, but not zero.

Theorem 2.2. *Under assumptions* (g_1), (g_2), *and* (g_3), *the equation*

$$\begin{cases} -\Delta u = g(x,u) - h & \text{in } \Omega \\ u|_{\partial\Omega} = 0 \end{cases} \tag{2.3}$$

possesses at least two solutions, if $g(x,t) \ge 0 \quad \forall\, (x,t) \in \Omega \times \mathbb{R}^1$, $\exists\, t_0 > 0$, $g(x,t_0) > 0$ *and* $g(x,0) = 0$.

The proof is just a verification of Theorem 2.1.

Lemma 2.2. *Under assumptions* (g_1) *and* (g_2), *for any* $h \in L^{\frac{2n}{n+2}}(\Omega)$, *the functional*

$$J(u) = \int_\Omega \left[\frac{1}{2}|\nabla u|^2 - G(x,u(x)) + h\,u(x)\right] dx \tag{2.4}$$

satisfies the (PS) condition on $H_0^1(\Omega)$.

Proof. Let $\{u_k\}$ be a sequence along which $|J(u_k)| \le C_1$ and $dJ(u_k) \to \theta$.

First, $\{u_k\}$ is bounded. In fact, $\exists\, C_2, C_3, C_4 > 0$ such that

$$\begin{aligned} C_1 &\ge \frac{1}{2}\|u_k\|^2 - \int_{|u_k(x)|\ge M} G(x,u_k(x))\,dx - |h|\cdot\|u_k\| - C_2 \\ &\ge \frac{1}{2}\|u_k\|^2 - \frac{1}{\theta}\int_{|u_k(x)|\ge M} u_k(x) g(x,u_k(x))\,dx - |h|\cdot\|u_k\| - C_2 \\ &\ge \left(\frac{1}{2}-\frac{1}{\theta}\right)\|u_k\|^2 + \frac{1}{\theta}\int_\Omega (\nabla u_k \nabla u_k - g(x,u_k)u_k)\,dx \\ &\quad - |h|\cdot\|u_k\| - C_3 \\ &\ge \left(\frac{1}{2}-\frac{1}{\theta}-\varepsilon\right)\|u_k\|^2 + \frac{1}{\theta}(dJ(u_k),u_k) - C_4, \end{aligned}$$

where $\|\cdot\|$, $|\cdot|$ and $(\ ,\ )$ stand for H_0^1 norm, $L^{\frac{2n}{n+2}}$ norm, and the $H_0^1(\Omega)$ inner product respectively. Since $dJ(u_k) \to \theta$, $|(dJ(u_k),u_k)| \le \varepsilon\|u_k\|$ if we choose $2\varepsilon < \frac{1}{2} - \frac{1}{\theta}$, then $\|u_k\|$ is bounded.

Let $p = \alpha + 1$, and consider the following maps:

$$\begin{array}{ccc} H_0^1(\Omega) & \xleftarrow{(-\Delta)^{-1}} & H^{-1}(\Omega) \\ {\scriptstyle i}\downarrow & & \uparrow{\scriptstyle i^*} \\ L^p(\Omega) & \xrightarrow[g(x,\cdot)]{} & L^{p'}(\Omega) \end{array}$$

where $\frac{1}{p} + \frac{1}{p'} = 1$. i is a compact embedding, as is i^*. Both $(-\Delta)^{-1}$ and $g(x, \cdot)$ are continuous.

The boundedness in $H_0^1(\Omega)$ of $\{u_k\}$ implies a convergent subsequence $(-\Delta)^{-1} \cdot i^* \cdot g(\cdot, u_{k'})$. Since

$$dJ\left(u_{k'}\right) = u_{k'} - (-\Delta)^{-1} \cdot i^* \cdot g\left(\cdot\, u_{k'}\right) \to \theta \quad \text{in } H_0^1,$$

finally, we obtain a convergent subsequence $\{u_{k'}\}$.

Proof of Theorem 2.2. It suffices to verify

(1) J is unbounded below.

(2) $\exists$ a pair of strict sub- and supersolutions for (2.3).

Claim (1). Since $g(x,t) \geq 0$ and $g(x,t_0) > 0$, so $G(x,t) > 0 \;\forall\, t \geq t_0$. For $t >$ Max $\{t_0, M\}$ we have

$$\frac{g(x,t)}{G(x,t)} \geq \frac{\theta}{t}. \tag{2.5}$$

Hence $G(x,t) \geq Ct^\theta$ for some constant $C > 0$. There exists a constant $C_1 > 0$ such that

$$J(u) \leq \int_\Omega \left\{ \frac{1}{2}|\nabla u|^2 - C\, u^\theta + h \cdot u \right\} dx + C_1$$

for any nonnegative $u \in H_0^1(\Omega)$.

Noticing $\theta > 2$, say, if we choose $u = t\varphi_1$, where $\varphi_1 > 0$ is the first eigenvector of $-\Delta$ with 0-Dirichlet boundary data, and $t > 0$,and let $t \to +\infty$, then

$$J(t\varphi_1) \to -\infty$$

Claim (2). The equation (2.3) has a strict supersolution 0, and a strict subsolution $\underline{u}$:

$$\begin{cases} -\Delta \underline{u} = -h & \text{in } \Omega \\ \underline{u}|_{\partial\Omega} = 0. \end{cases}$$

By the Maximum Principle $\underline{u} < 0$.

All conditions in Theorem 2.1 are fulfilled. The proof is complete.

Example 1. The equation

$$\begin{cases} -\Delta u = u^2 - h & \text{in } \Omega \\ u|_{\partial\Omega} = 0 \end{cases} \tag{2.6}$$

possesses at least two solutions, if (g_3) is satisfied.

Theorem 2.3. *Suppose* (g_1), (g_2) *with* $G(x,t) > 0$, $|t| \geq M$, *and*

(g_4) $g \in C^1(\overline{\Omega} \times \mathbb{R}^1)$ *with* $g(x,0) = g_t(x,0) = 0$.

Then equation (1.1) *possesses at least three nontrivial solutions.*

We need

Lemma 2.3. *Under the assumptions of Theorem 2.3, there exists a constant* $A > 0$, *such that*

$$J_a \simeq S^\infty, \text{ the unit sphere in } H_0^1(\Omega)$$

for $-a > A$ *where* J *is the functional* (1.3).

Proof. By the same deduction, but by assuming $G(x,t) > 0 \;\forall\, t,\ |t| \geq M$, we conclude

$$G(x,t) \geq C|t|^\theta \quad \forall\, t,\ |t| \geq M.$$

Thus $\forall\, u \in S^\infty$,

$$J(tu) \to -\infty \quad \text{as} \quad t \to +\infty.$$

We want to prove: $\exists\, A > 0$ such that $\forall\, a < -A$, if $J(tu) \leq a$, then $\frac{d}{dt}J(tu) < 0$.

In fact, set

$$A = 2M|\Omega| \max_{(x,t)\in\overline{\Omega}\times[-M,M]} |g(x,t)| + 1.$$

If $J(tu)(= \frac{t^2}{2} - \int_\Omega G(x,tu(x))\,dx) \leq a$, then

$$\begin{aligned}
\frac{d}{dt}J(tu) &= (dJ(tu),u) \\
&= t - \int_\Omega u(x)\cdot g(x,tu(x))\,dx \\
&\leq \frac{2}{t}\left\{\int_\Omega G(x,tu(x))\,dx - \frac{1}{2}\int_\Omega tu(x)\cdot g(x,tu(x))\,dx + a\right\} \\
&\leq \frac{2}{t}\left\{\left(\frac{1}{\theta}-\frac{1}{2}\right)\int_{|tu(x)|\geq M} tu(x)g(x,tu(x))\,dx + (A-1) + a\right\} \\
&\leq \frac{2}{t}\left\{\left(\frac{1}{\theta}-\frac{1}{2}\right)C\theta\int_{|tu(x)|\geq M} |t|^\theta|u(x)|^\theta\,dx - 1\right\} < 0.
\end{aligned}$$

The implicit function theorem is employed to obtain a unique $T(u) \in C(S^\infty, \mathbb{R}^1)$ such that

$$J(T(u)u) = a \quad \forall\, u \in S^\infty.$$

Next, we claim that $\|T(u)\|$ possesses a positive lower bound $\varepsilon > 0$. In fact, by (g_4), $g(x,0) = g'_t(x,0) = 0$, $J(t,u) = \frac{t^2}{2} - o(t^2)$ $\forall\, u \in S^\infty$. The conclusion follows.

Finally, let us define a deformation retract $\eta : [0,1] \times (H\backslash B_\varepsilon(\theta)) \to H\backslash B_\varepsilon(\theta)$, where $H = H_0^1(\Omega)$, and $B_\varepsilon(\theta)$ is the ε-ball with center θ, by

$$\eta(s,u) = (1-s)u + sT(u)u. \quad \forall\, u \in H\backslash B_\varepsilon(\theta).$$

This proves $H\backslash B_\varepsilon(\theta) \simeq J_a$, i.e. $J_a \simeq S^\infty$.

Proof of Theorem 2.5.

1. Provided by (g_4)

$$J(u) = \frac{1}{2}\|u\|^2 + o\left(\|u\|^2\right), \tag{2.7}$$

so, θ is a local minimum, and $C_q(J,\theta) = \delta_{q0} G$.

2. We find two nontrivial solutions. Let us define

$$g_+(x,t) = \begin{cases} g(x,t) & t \geq 0 \\ 0 & t \leq 0 \end{cases}$$

and

$$J_+(u) = \int_\Omega \left[\frac{1}{2}|\nabla u|^2 - G_+(x,u(x))\right] dx$$

where

$$G_+(x,t) = \int_0^t g_+(x,s)\, ds.$$

Again, $J_+ \in C^2(H_0^1(\Omega), \mathbb{R}^1)$ satisfies (PS), using Lemma 2.2. As in the proof of Theorem 2.2, we also have

$$J_+(t\varphi_1) \to -\infty \quad \text{as} \quad t \to +\infty$$

where $\varphi_1 > 0$ is the first eigenvector of $-\Delta$ with 0-Dirichlet data.

On the other hand $\exists\, \delta > 0$ such that

$$J_+|_{\partial B_\delta(\theta)} \geq \frac{1}{4}\delta^2$$

provided by (2.7). The mountain pass lemma (Theorem 1.4 from Chapter II) is applied to obtain a critical point $u_+ \in H_0^1(\Omega)$, with critical value $c_+ > 0$, which satisfies

$$\begin{cases} -\Delta u_+ = g_+(x,u_+) \\ u_+|_{\partial\Omega} = 0. \end{cases}$$

By using the Maximum Principle, $u_+ \geq 0$, so it is again a critical point of J.

Analogously, we define

$$g_-(x,t) = \begin{cases} g(x,t) & t \leq 0 \\ 0 & t > 0, \end{cases}$$

and obtain a critical point $u_- \leq 0$, with critical value $c_- > 0$. Chapter II (Theorem 1.7) and the Kato-Hess Theorem imply that $C_q(J_\pm, u_\pm) = \delta_{q1}G$. According to the Palais Theorem (cf. Section 1), we have

$$C_q\left(J_\pm, u_\pm\right) = C_q\left(\widetilde{J}_\pm, u_\pm\right) = C_q\left(\widetilde{J}, u_\pm\right),$$

where $\widetilde{J} = J|_{C_0^1}$; and again, $C_q(\widetilde{J}, u_\pm) = C_q(J, u_\pm)$. Therefore

$$C_q(J, u_\pm) = \delta_{q1}G.$$

3. Suppose that there were no more critical points of J. The Morse type numbers over the pair $(H_0^1(\Omega), J_a)$ would be

$$M_0 = 1,\ M_1 = 2,\ M_q = 0,\ q \geq 2,$$

but the Betti numbers

$$\beta_q = 0 \quad \forall\, q = 0, 1, 2, \ldots,$$

since $H_q(H_0^1(\Omega), J_a) \simeq H_q(H_0^1(\Omega), S^\infty) \simeq 0$.

This is a contradiction.

Example 2. The function

$$g(t) = \begin{cases} t^{\alpha_1} & t \geq 0 \\ -|t|^{\alpha_2} & t < 0 \end{cases}$$

satisfies all assumptions in Theorem 2.3, with $1 < \alpha_1, \alpha_2 < \frac{n+2}{n-2}$.

For odd g (in t), because of the symmetry of the functional J, one gets more solutions. Namely,

Theorem 2.4. *Under assumptions* (g_1), (g_2) *with* $G(x,t) > 0$ *for* $|t| \geq M$, *and*

(g_5) $g(x,t) = -g(x,-t) \quad \forall\, (x,t) \in \Omega \times \mathbb{R}^1$.

Equation (1.1) *possesses infinitely many pairs of solutions.*

Proof. The functional defined in (1.3), $J \in C^1(H_0^1(\Omega), \mathbb{R}^1)$, is even, and satisfies the (PS) condition.

According to (g_1), we have

$$J(u) \geq \frac{1}{2}\|u\|^2 - C\left(1 + \|u\|_{L^{\alpha+1}}^{\alpha+1}\right) \tag{2.8}$$

where $C > 0$ is a suitable constant. By using the Gagliardo-Nirenberg inequality,

$$\|u\|_{L^{\alpha+1}} \leq C_1 \|u\|_{H_0^1}^{\beta} \|u\|_{L^2}^{1-\beta} \tag{2.9}$$

where C_1 is a constant, and $0 < \beta < 1$ is defined by

$$\frac{1}{\alpha+1} = \beta\left(\frac{1}{2} - \frac{1}{n}\right) + (1-\beta)\cdot\frac{1}{2}.$$

Substituting (2.9) in (2.8), for $u \in \partial B_\rho(\theta)$, we have

$$J(u) \geq \frac{1}{2}\rho^2 - C_2 \rho^{(\alpha+1)\beta} \|u\|_{L^2}^{(1-\beta)(\alpha+1)} - C_3.$$

Let $\lambda_1 < \lambda_2 \leq \lambda_3 \leq \dots$, be the eigenvalues of $(-\Delta)$, associated with eigenvectors $\varphi_1, \varphi_2, \varphi_3, \dots$, and let $E_j = \operatorname{span}\{\varphi_1, \varphi_2, \dots, \varphi_j\}$, $j = 1, 2, \dots$. The variational characterization of the eigenvalues provides the estimates

$$\|u\|_{L^2} \leq \lambda_{j+1}^{-\frac{1}{2}} \|u\| \quad \forall\, u \in E_j^{\perp},\ j = 1, 2, \dots .$$

Hence

$$J(u) \geq \frac{1}{2}\left(1 - 2C_2 \rho^{(\alpha-1)} \lambda_{j+1}^{\delta}\right)\rho^2 - C_3 \quad \forall\, u \in \partial B_\rho(\theta) \cap E_j^{\perp}$$

where $\delta = -\frac{1}{2}(1-\beta)(1+\alpha) < 0$. Since $\lambda_j \to +\infty$ as $j \to \infty$, we choose ρ, j_0 such that

$$\begin{cases} 1 - 2C_2\rho^{\alpha-1}\lambda_{j_0+1}^{\delta} \geq \frac{1}{2}, \\ \rho^2 > 8C_3. \end{cases}$$

Thus

$$J(u) \geq \frac{1}{8}\rho^2 > 0 \quad \forall\, u \in \partial B_\rho(\theta) \cap E_{j_0}^{\perp}.$$

Since all norms on a finite dimensional space are equivalent, and since it was already known that

$$G(x,t) \geq C|t|^{\theta} \quad \forall\, t, |t| \geq M,$$

there exists $R_j > \rho$ such that

$$J(u) \leq 0 \quad \forall\, u \in E_j \backslash B_{R_j}(\theta) \quad j = 1, 2, \dots .$$

According to Corollary 4.2 in Chapter II, the proof will be finished if the (PS)* condition with respect to the subspaces $\{E_j | j = 1, 2, \dots\}$ is verified. However, the verification is similar to that of the (PS) condition given in Lemma 2.2.

Remark 2.2. Theorems 2.1 and 2.2 are taken from K. C. Chang [Cha3-4], and Theorem 2.3 from Z. Q. Wang [WaZ2].

Remark 2.3. There is a beautiful application of the Morse index estimates to the following perturbation result: The equation

$$\begin{cases} -\Delta u = |u|^{p-1}u - f(x) & \text{in } \Omega \\ u|_{\partial\Omega} = 0 \end{cases}$$

has infinitely many solutions, if $f \in L^2(\Omega)$, and $1 < p < \frac{n}{n-2}$. cf. A. Bahri, H. Berestycki [BaB1], A. Bahri, P-L. Lions [BaL1,2], Dong Li [DoL1], M. Struwe [Str1] and P. H. Rabinowitz [Rab4].

3. Asymptotically Linear Problems

3.1. Nonresonance and Resonance with the Landesman-Lazer Condition

First, we assume that the function g is of the form

$$g(x,t) = \hat{\lambda}t + \varphi(x,t), \tag{3.1}$$

where $\varphi \in C(\overline{\Omega} \times \mathbb{R}^1, \mathbb{R}^1)$, satisfying

$$\varphi(x,t) = o(|t|) \quad \text{as } |t| \to \infty \text{ uniformly in } x \in \Omega.$$

We study the BVP (1.1) via the abstract theorems of Section 5 in Chapter II. Set $H = H_0^1(\Omega)$, $A = \text{ id } - \hat{\lambda}(-\Delta)^{-1}$.

$$\Phi(x,t) = \int_0^t \varphi(x,s)\,ds$$

and

$$F(u) = -\int_\Omega \Phi(x, u(x))\,dx.$$

Problem (1.1) is equivalent to finding critical points of the functional

$$J(u) = \frac{1}{2}(Au, u) + F(u).$$

Theorem 3.1. *If $\hat{\lambda} \notin \sigma(-\Delta)$, the spectrum of $(-\Delta)$, then (1.1) possesses a solution. If we further assume that*

$$(3.2) \qquad \begin{cases} \varphi(x,0) = 0 & \varphi \in C^1\left(\overline{\Omega} \times \mathbb{R}^1, \mathbb{R}^1\right), \\ \varphi_t'(x,0) = \lambda_0 - \hat{\lambda} & |\varphi_t'(x,t)| \le C\left(1 + |t|^{\frac{4}{n-2}}\right) \end{cases}$$

and $\exists\, \bar{\lambda} \in \sigma(-\Delta)$ such that

$$(3.3) \qquad \text{either} \quad \hat{\lambda} < \bar{\lambda} < \lambda_0 \quad \text{or} \quad \lambda_0 < \bar{\lambda} < \hat{\lambda},$$

then (1.1) possesses a nontrivial solution.

Theorem 3.2. *If $\hat{\lambda} \in \sigma(-\Delta)$, and $\varphi(x,t)$ is bounded, and we assume the Landesman-Lazer condition that*

$$(3.4) \qquad \int_\Omega \Phi\left(x, \sum_{j=1}^{n_0} t_j \varphi_j(x)\right) dx \to +\infty \quad \text{as} \quad \sum_{j=1}^{n_0} t_j^2 \to \infty,$$

where $\operatorname{span}\{\varphi_1, \varphi_2, \dots, \varphi_{n_0}\} = \ker(-\Delta - \hat{\lambda} I)$. Then (1.1) possesses a solution. Furthermore, if (3.2) holds, and if $\exists\, \bar{\lambda} \in \sigma(-\Delta)$ such that $\hat{\lambda} < \bar{\lambda} < \lambda_0$, then (1.1) possesses a nontrivial solution.

Proof. Let $\lambda_1 < \lambda_2 \le \lambda_3 \le \dots$ be the eigenvalues associated with the eigenvectors $\varphi_1, \varphi_2, \varphi_3, \dots$ of $(-\Delta)$ with 0-Dirichlet data. Let

$$\begin{aligned} H_+ &= \operatorname{span}\left\{\varphi_j, \mid \lambda_j > \hat{\lambda}\right\}, \\ H_0 &= \ker\left(-\Delta - \hat{\lambda} I\right), \quad \text{and} \\ H_- &= \operatorname{span}\left\{\varphi_j \mid \lambda_j < \hat{\lambda}\right\}. \end{aligned}$$

Then (H_1) and (H_2) of Theorem 5.2 of Chapter II hold.

Since $dF(u) = (-\Delta)^{-1}\varphi(x, u(x))$, $F \in C^1(H, \mathbb{R}^1)$, and then $\|dF(u)\| = o(\|u\|)$ if (3.1), and $\|dF(u)\| \le C$, a constant, if $\varphi(x,t)$ is bounded. Note that

$$H_0^1 \xrightarrow{\quad i \quad} L^2 \xrightarrow{\varphi(x,\cdot)} L^2 \xrightarrow{(-\Delta)^{-1}} H_0^1.$$

So dF is compact. Therefore (H_3) also holds. We apply Theorem 5.2 of Chapter II (with $H_0 = \theta$) in the case $n = 0$, to obtain the existence of a solution in Theorem 3.1, and, in the case $n = 1$, to prove the existence of a nontrivial solution.

In fact, $\gamma = \dim(H_0 \oplus H_-) = \dim H_- = \operatorname{Max}\{j \mid \lambda_j < \hat{\lambda}\}$, and $m_-(\theta) = \operatorname{Max}\{j \mid \lambda_j < \lambda_0\}$, $m_0(\theta) = \dim\ker(-\Delta - \lambda_0 I)$. Condition (3.3) is equivalent to $\gamma \notin [m_-(\theta), m_-(\theta) + m_0(\theta)]$.

Similarly, Theorem 3.2 is proved by the fact that the Landesman-Lazer condition (3.4) implies

$$F(P_0u) \to -\infty \quad \text{as} \quad \|P_0u\| \to \infty,$$

where P_0 is the orthogonal projection onto H_0.

Remark 3.1. Problems in which $\hat{\lambda} \notin \sigma(-\Delta)$ are called nonresonance, and in which $\hat{\lambda} \in \sigma(-\Delta)$ are called resonance.

Theorems 3.1 and 3.2 are due to Amann-Zehnder [AmZ1] and Chang [Cha2] respectively. See also Liu [Liu3]. For a similar problem from the two point boundary value problem in ordinary differential systems, we can get a better estimate of the number of solutions.

Let us assume that $G \in C^2([0,\pi] \times \mathbb{R}^n, \mathbb{R}^1)$ satisfying

(1) $G(t,\theta) = 0, \quad G_u(t,\theta) = \theta,$

(2) $\exists$ integer k such that

$$k^2 I_n < A_\infty(t) < (k+1)^2 I_n,$$

where I_n is the $n \times n$ identity matrix and the limit $A_\infty(t) = \lim_{|u|\to\infty} d_u^2 G(t,u)$ exists.

(3) $\exists\, \tilde{k} \in \mathbb{Z}_+$ such that

$$\tilde{k}^2 I_n < A_0(t) = d_u^2 G(t,\theta) < \left(\tilde{k}+1\right)^2 I_n$$

where the notation $B < A$ means that the matrix $A - B$ is positive definite.

Theorem 3.3. [WaZ3] *Under assumptions* (1)–(3), *if* $k \neq \tilde{k}$, *then the equation*

$$\begin{cases} -\ddot{u} = G_u(t,u) \\ u(0) = u(\pi) = \theta \end{cases}$$

possesses at least two nontrivial solutions in $C^2([0,\pi], \mathbb{R}^n)$.

Proof. Again, we use the abstract framework, $H = H_0^2([0,\pi], \mathbb{R}^n)$, $A = \mathrm{id} - (-\frac{d^2}{dt^2})^{-1} A_\infty(t)\cdot$, $H_+ =$ span $\{e_i \sin jt \mid i = 1,2,\dots,n,\ j \geq k+1\}$ and $H_- =$ span $\{e_i \sin jt \mid i = 1,2,\dots,n,\ 0 \leq j \leq k\}$, where $\{e_i\}_1^n$ is the basis of $\mathbb{R}^n$.

Condition (2) implies that $G_u(t,u)$ is asymptotically linear. Indeed, for any $u_0 \in \mathbb{R}^n$ we have

$$\begin{aligned}
&\|G_u(t,u) - A_\infty(t)\cdot u\|_{\mathbb{R}^n} \\
&\quad \leq \|G_u(t,u_0) - A_\infty(t)u_0\|_{\mathbb{R}^n} \\
&\quad + \|G_u(t,u) - G_u(t,u_0) - A_\infty(t)(u-u_0)\|_{\mathbb{R}^n} \\
&\quad \leq \|G_u(t,u_0) - A_\infty(t)u_0\|_{\mathbb{R}^n} \\
&\quad + \mathrm{Sup} s \in [0,1]\, \|d^2G(t, su + (1-s)u_0) - A_\infty(t)\| \cdot \|u - u_0\|_{\mathbb{R}^n}.
\end{aligned}$$

Thus
$$\|G_u(t,u) - A_\infty(t)u\|_{\mathbb{R}^n} = o\left(\|u\|_{\mathbb{R}^n}\right) \quad \text{as} \quad \|u\|_{\mathbb{R}^n} \to \infty.$$
Let $F(u) = -\int_0^\pi G(t,u(t))\,dt + \frac{1}{2}\int_0^\pi \langle A_\infty(t)u(t), u(t)\rangle\,dt$; then
$$\|dF(u)\|_H = o(\|u\|_H)$$
and $dF(u)$ is compact. The following functional is considered:
$$J(u) = \frac{1}{2}(Au,u) + F(u).$$
By simple computations, we have
$$\left(d^2J(\theta)u,u\right) = \frac{1}{2}\int_0^\pi \left[|\dot{u}(t)|^2 - \langle A_0(t)u(t), u(t)\rangle\right]\,dt.$$

Condition (3) implies that θ is nondegenerate with $m_-(\theta) = (\tilde{k}+1)n$. But $\gamma = \dim H_- = (k+1)n \neq m_-(\theta)$. We conclude that there is a nontrivial solution $u_0(t)$.

Moreover, note that
$$\begin{aligned}\ker\left(d^2J(u_0)\right) = \{&u \in H_0^1\left([0,\pi],\mathbb{R}^n\right) \mid u(0) = u(\pi) = \theta \text{ and} \\ &-\ddot{u} = d_u^2G(t,u(t))u \text{ in } (0,\pi)\},\end{aligned}$$
which is of dimension $\le n$. The condition
$$|\gamma - m_-(\theta)| \ge m_0(u_0)$$
is fulfilled. Corollary 5.2 in Chapter II can be applied to obtain the second nontrivial solution.

3.2 Strong Resonance

Now we study a new class of resonance problem, called strong resonance, in which the Landesman-Lazer condition is dropped so that the (PS) condition does not hold. The function g is again written in the form (3.1):
$$g(x,t) = \hat{\lambda}t + \varphi(x,t) \quad \hat{\lambda} \in \sigma(-\Delta).$$
The Landesman-Lazer condition (3.4) placed on φ is replaced by the following:

(H) $\forall\, \xi_j \in \mathbb{R}^{m_0}$, $|\xi_j| \to \infty$, $\forall u_j \to u$ in $H_0^1(\Omega)$, and $\forall\, v \in H_0^1(\Omega)$;

we have

$$\lim_{j\to\infty}\int_\Omega \varphi\left(x, u_j(x) + \sum_{i=1}^{m_0} \xi_j^i e_i(x)\right) v(x)\,dx = 0, \tag{3.5}$$

and

$$\lim_{j\to\infty}\int_\Omega \Phi\left(x, u_j(x) + \sum_{i=1}^{m_0} \xi_j^i e_i(x)\right) v(x)\,dx = 0 \tag{3.6}$$

where $\{e_i(x)\}_1^{m_0}$ is an orthonormal basis of the eigenspace $\ker(-\Delta - \hat{\lambda} I)$, and $\xi_j = (\xi_j^1, \xi_j^2, \dots, \xi_j^{m_0})$.

Again we study the critical points of

$$J(u) = \frac{1}{2}(Au, u) + F(u)$$

as above. The only difference is that the assumptions $F(P_0 u) \to -\infty$ as $\|P_0 u\| \to \infty$, and the boundedness of $\|dF(u)\|$ are replaced by $F(u) \to 0$, and $dF(u) \to \theta$ as $\|P_0 u\| \to \infty$, where P_0 is the orthogonal projection onto $\ker(-\Delta - \hat{\lambda} I)$.

Lemma 3.1. *Under assumptions* (H_1) *and* (H_2) *of* Theorem 5.2 of Chapter II, *if we assume* (H_3')

(H_3') $F \in C^1(H, \mathbb{R}^1)$. dF is bounded and compact.

$$F(u_n) \to 0, \quad \text{and } dF(u_n) \to \theta \quad \text{as} \quad \|P_0 u_n\| \to \infty.$$

Then J satisfies the $(PS)_c$ condition $\forall\ c \neq 0$. Moreover, if $J(u_n) \to 0$, $dJ(u_n) \to \theta$ along a sequence u_n, then $\exists$ a subsequence (still denoted by u_n), with the property that either u_n converges, or

$$\|(I - P_0)\,u_n\| \to 0 \quad \text{as} \quad \|P_0 u_n\| \to \infty.$$

Proof. Suppose that

$$J(u_n) = \frac{1}{2}(Au_n, u_n) + F(u_n) \to c \tag{3.7}$$

$$dJ(u_n) = Au_n + dF(u_n) \to \theta. \tag{3.8}$$

Decompose u_n into $u_n^+ + u_n^- + u_n^0$, where $u_n^\pm$ is the orthogonal projection of u_n onto $H^\pm$, and $u_n^0 = P_0 u_n$. Then

$$\begin{aligned} \left|(Au_n^\pm, u_n^\pm)\right| &= \left|(Au_n, u_n^\pm)\right| \\ &= \left|(dJ(u_n) - dF(u_n), u_n^\pm)\right| \le C\left\|u_n^\pm\right\|. \end{aligned} \tag{3.9}$$

Since $A_\pm$ has a bounded inverse on $H_\pm$, (3.9) implies the boundedness of $u_n^\pm$. If u_n^0 is bounded, so is u_n, and then it has a weakly convergent subsequence. By the compactness of dF, and the finite dimensional conditions on H_0, we get a strong convergent subsequence. On the contrary, if

$\|u_n^0\| \to \infty$ (we ignore a subsequence) then $F(u_n) \to 0$, $dF(u_n) \to \theta$. From (3.8), $u_n^{\pm} \to \theta$. Finally, (3.7) implies that $J(u_n) \to 0$, i.e., $c = 0$.

A new idea is employed to avoid the difficulty arising from the lack of compactness (the $(\mathrm{PS})_0$ condition). We compactify the space H by adding some ∞ points, extend the critical point theory (in particular, the deformation theorem) to the enlarged space, and distinguish the genuine and the pseudo critical points that come from the ∞ points. We proceed as follows:

1. Compactify the subspace $H_0 = \ker A$. Let

$$\Sigma = H_0 \cup \{\infty\} \simeq S^{m_0}, \quad m_0 = \dim H_0,$$

and

$$E = H_0^{\perp} \times \Sigma \quad \text{where } H_0^{\perp} = H_+ \oplus H_-.$$

Define

$$\widehat{J}(x) = \begin{cases} J(v+s) & (v,s) \in H_0^{\perp} \times H_0 \\ \frac{1}{2}(Av, v) & (v,s) \in H_0^{\perp} \times \{\infty\}, \end{cases}$$

where $x = (v, s) \in E$.

The assumption $J(v+s) \to 0$ as $\|s\| \to \infty$ implies that $\widehat{J}$ is continuous (but not differentiable in general).

Although we cannot comment on the critical set of $\widehat{J}$, we call $\widehat{K} = K \cup \{(\theta, \infty)\}$ the pseudo critical set, where K is the critical set of J. Points in K are said to be genuine, and the single point (θ, ∞) to be pseudo . Thus

$$\widehat{K}_c := \widehat{K} \cap \widehat{J}^{-1}(c) = \begin{cases} K_c & \text{if } c \neq 0, \\ K_0 \cup \{(0, \infty)\} & \text{if } c = 0. \end{cases}$$

2. The first deformation theorem can be extended to study the pseudo critical set. Namely, we have

Lemma 3.2. *Under the assumptions* (H_1), (H_2) *of Theorem 5.2 of Chapter II, and* (H_3'), *if* K_0 *is bounded in* H, *then* $\forall\, c \in \mathbb{R}^1$, $\forall$ *closed neighborhoods* N *of* $\widehat{K}_c$ *in* E, *there exist* $\eta \in C([0,1] \times E, E)$ *and constants* $0 < \varepsilon < \bar{\varepsilon}$ *such that*

(1) $\eta(0, \cdot) = \mathrm{id}$,
(2) $\eta(t, \cdot)\,|_{C\widehat{J}^{-1}[c-\bar{\varepsilon}, c+\bar{\varepsilon}]} = \mathrm{id}\,|_{C\widehat{J}^{-1}[c-\bar{\varepsilon}, c+\bar{\varepsilon}]}$,
(3) $\eta(t, \cdot) : E \to E$ *is a homeomorphism* $\forall\, t \in [0,1]$,
(4) $\eta(1, \widehat{J}_{c+\varepsilon} \backslash N) \subset \widehat{J}_{c-\varepsilon}$, *and if* $\widehat{K}_c = \emptyset$ *then* $\eta(1, \widehat{J}_{c+\varepsilon}) \subset \widehat{J}_{c-\varepsilon}$.

Proof. The proof is similar to the standard one (cf. Theorem 3.3 of Chapter I). The main point is to construct a flow. Let $\xi(t, v, s)$ be the flow defined on $H_0^{\perp} \times H_0$ by the vector field $V(x)$ (cf. Theorem 3.4 of

Chapter I). Let the functions p, q have the same meaning as those in the referred theorem. Let W be a vector field on $H_0^\perp \times \{\infty\}$ defined by

$$W(v) = -p\left(\frac{1}{2}(Av, v)\right) q(\|Av\|)Av.$$

The associated flow $\dot{\zeta} = W(\zeta)$, $\zeta(0, v) = v$, is denoted by $\zeta(t, v)$. Set

$$\eta(t, v, s) = \begin{cases} \xi(t, v, s) & \text{if } (v, x) \in H_0^\perp \times H_0 \\ (\zeta(t, v), \infty) & \text{if } (v, s) \in H_0^\perp \times \{\infty\}. \end{cases}$$

Obviously this mapping satisfies all the conditions stated above. It remains to verify its continuity at points on $H_0^\perp \times \{\infty\}$.

Claim. First, we show that

$$\|V(v + s) - W(v)\| \to 0 \quad \text{as } v \text{ remains on a bounded set and } \|s\| \to \infty.$$

This is due to the facts

$$\left|J(v + s) - \frac{1}{2}(Av, v)\right| = |F(v + s)| \to 0, \quad \text{and}$$

$$\|X(v + s) - Av\| = \left\|\sum_\beta \eta_\beta(v + s)dF(x_\beta)\right\| \to 0,$$

where $\{\eta_\beta\}$ is the partition of unity in the construction of p.g.v.f. X, in combining with the following inequalities:

$$\left|p(J(v + s)) - p\left(\frac{1}{2}(Av, v)\right)\right| \le C\left|J(v + s) - \frac{1}{2}(Av, v)\right| \to 0$$
$$|q(\|X(x)\|) - q(\|Av\|)| \le C\|X(x) - Av\| \to 0.$$

Next, assume that $t_n \to t$, $v_n \to v$, $s_n \to \infty$, where $t_n \in [0, 1]$, $v_n \in H_0^\perp$, and $s_n \in H_0$. We want to show that

$$P_0\xi(t_n, v_n + s_n) \to \infty \quad \text{and} \quad (I - P_0)\xi(t_n, v_n + s_n) \to \zeta(t, v).$$

Consider the equation

$$\begin{cases} \dot{\xi} = V(\xi) \\ \xi(0, v_n + s_n) = v_n + s_n. \end{cases}$$

Since $\|V\| \le 1$, $\|\xi(t, v_n + s_n) - \xi(0, v_n + s_n)\| \le 1$, it follows $\|(I - P_0) \cdot \xi(t, v_n + s_n)\| \le \|v_n\| + 1 \le C$, $\|P_0\xi(t, v_n + s_n)\| \ge |s_n| - 1 \to \infty$. We conclude $\|V(\xi(t, v_n + s_n)) - W((I - P_0)\xi(t, v_n + s_n))\| \to 0$ from above.

We turn to the differential inequation:

$$\begin{aligned}\frac{d}{dt}&\|\zeta(t,v) - (I-P_0)\xi\,(t,v_n+s_n)\| \\ &\le \|W(\zeta(t,v)) - (I-P_0)V\,(\xi\,(t,v_n+s_n))\| \\ &= \|W\,(\zeta(t,v)) - W((I-P_0)\xi\,(t,v_n+s_n))\| + o(1) \\ &\le C\,\|\zeta(t,v) - (I-P_0)\xi\,(t,v_n+s_n)\| + o(1).\end{aligned}$$

According to Gronwall's inequality, we have

$$\begin{aligned}&\|\zeta(t,v) - (I-P_0)\,\xi\,(t,v_n+s_n)\| \\ &\quad\le C\,\|\zeta(0,v) - (I-P_0)\,\xi\,(0,v_n+s_n)\| + o(1) \\ &\quad= C\,\|v-v_n\| + o(1).\end{aligned}$$

On the other hand

$$\|\xi\,(t,v_n+s_n) - \xi\,(t_n,v_n+s_n)\| \le |t-t_n|$$

follows from $\|V\| \le 1$. Thus

$$\|\zeta\,(t,v_n) - (I-P_0)\,\xi\,(t_n,v_n+s_n)\| \to 0.$$

Theorem 3.4. *Under assumptions* (H_1), (H_2) *of Theorem 5.2 of Chapter II and* (H), *equation* (1.1) *possesses a solution. If, further, we assume that*

$$\varphi(x,0)=0, \quad \varphi\in C^1\left(\overline{\Omega}\times\mathbb{R}^1,\mathbb{R}^1\right)$$

and that either

(1) $\Phi(x,0)=0$ *or*

(2) $\Phi(x,0)>0$, *and* $\varphi_t'(x,0)\notin[-\widehat{\lambda}+\underline{\lambda},0]$, *where* $\underline{\lambda}=\max\{\lambda_j\ |\in \sigma(-\Delta),\ \lambda_j<\widehat{\lambda}\}$, *or*

(3) $\Phi(x,0)<0$, *and* $\varphi_t'(x,0)\notin[0,\overline{\lambda}-\widehat{\lambda}]$, *where* $\overline{\lambda}=\min\{\lambda_j\ |\in\sigma(-\Delta),\ \hat{\lambda}>\lambda_j\}$,

then there is a nontrivial solution.

Proof. One may assume that K_0 is bounded, since otherwise we would be done. Now Theorem 5.3 of Chapter II, is applied to assure at least $CL(S^{m_0})+1=2$ pseudo critical points. Therefore there is at least one genuine critical point, which is the desired solution.

Suppose $\varphi(x,0)=0$; then θ must be a critical point. However by the minimax principle for subordinate classes, $\exists$ a critical value $c\neq 0$. In case (1), $\Phi(x,0)=0$, θ is on the level $J^{-1}(0)$, so a point u_0 in K_c is a nontrivial solution.

If $c < 0$, we may choose u_0 to be the one, which corresponds to a m_- relative homology class. According to Theorem 1.5 of Chapter II, $C_{m_-}(J, u_0) \neq 0$, and if $c > 0$, we may choose it to be the one, which corresponds to a $m_- + m_0$ relative homology class. Therefore $C_{m_-+m_0}(J, u_0) \neq 0$.

However,

$$\begin{aligned} C_q(J,\theta) &= \delta_{q\underline{m}} \quad \text{if } \lambda_k < \widehat{\lambda} + \varphi_t'(x,0) < \lambda_{k+1} \\ &= 0 \qquad \text{if } \widehat{\lambda} + \varphi_t'(x,0) = \lambda_{k+1} \text{ and } q \notin [\underline{m}, \overline{m}] \end{aligned}$$

where $\lambda_k < \lambda_{k+1}$ is a pair of consecutive eigenvalues of $-\Delta$, and

$$\underline{m} = \sum_{j=1}^{k} \dim\ker(-\Delta - \lambda_j I), \quad \overline{m} = \sum_{j=1}^{k+1} \dim\ker(-\Delta - \lambda_j I).$$

Thus

$$C_{m_-}(J,\theta) = 0 \quad \text{if } \varphi_t'(x,0) < -\widehat{\lambda} + \underline{\lambda} \text{ or } \varphi_t'(x,0) > 0$$

and

$$C_{m_-+m_0}(J,\theta) = 0 \quad \text{if } \varphi_t'(x,0) < 0 \text{ or } \varphi_t'(x,0) > \overline{\lambda} - \widehat{\lambda}.$$

The proof is finished.

Several sufficient conditions can be given to assure (H). Namely,

(1) $\varphi(x,t) \to 0$ and $\Phi(x,t) \to 0$ as $|t| \to \infty$, if $\hat{\lambda}$ is simple and the nodal set of the associate eigenfunction has measure zero.

(2) $\varphi(x,t) = \varphi_0(t) + h(x)$, where $h \in \ker(-\Delta - \widehat{\lambda} I)^\perp$, and $\varphi_0(t)$ together with its primitive $\Phi_0(t) = \int_0^t \varphi_0(s)ds$ are bounded and uniformly continuous on $\mathbb{R}^1$. And $\dim(\Omega) = 1$

As a special case of (2), we assume that φ_0 is a T-periodic function, with $\int_0^T \varphi_0(t)dt = 0$. Indeed, the verification of (1) is trivial. (2) needs a little real analysis, so we refer to J. Mawhin [Maw1]. See also Solimini [Sol3] and Ward [War1].

Theorem 3.4 is due to Chang and Liu [ChL2].

3.3. A Bifurcation Problem

For simplicity, the function $g(x,t)$ in (1.1) is replaced by $g(t)$. We assume that

(1) $\lim_{|t|\to\infty} \frac{g(t)}{t} < \lambda_1$, the first eigenvalue of $-\Delta$ with 0-Dirichlet boundary value.

(2) $g(0) = 0$, and $g \in C^1(\mathbb{R}^1)$.

Theorem 3.5. *Let $\lambda = g'(0)$, then*

(i) *For $\lambda > \lambda_1$, the* BVP (1.1) *has at least two nontrivial solutions.*

(ii) *For $\lambda > \lambda_2$, or $\lambda = \lambda_2$ with $\frac{g(t)}{t} \geq \lambda$ in a neighborhood U of $t = 0$, (1.1) has at least three nontrivial solutions.*

(iii) *For $\lambda > \lambda_2$, we assume that if $\lambda \in \sigma(-\Delta)$ either*

$$\frac{g(t)}{t} \geq \lambda \quad \text{or} \quad \frac{g(t)}{t} \leq \lambda \ \textit{holds}$$

for $t \neq 0$ in a neighborhood U of 0, then (1.1) has at least four nontrivial solutions.

Proof. By condition (1), there exists an $\alpha \in (0, \lambda_1)$ and a constant $C_\alpha > 0$ such that $g(t) \leq \alpha t + C_\alpha$ if $t > 0$, and $g(t) \geq -\alpha t - C_\alpha$ if $t < 0$. Let φ_0 be the solution of the following equation

$$\begin{cases} -\Delta\varphi_0 = \alpha\varphi_0 + C_\alpha & \text{in } \Omega \\ \varphi_0\big|_{\partial\Omega} = 0. \end{cases}$$

Then, by the maximum principle, $\varphi_0 > 0$, and hence, $-\varphi_0 < \varphi_0$ is a pair of sub- and super-solutions of (1.1).

According to the cut-off technique, we may assume that $g(t)$ is bounded, and define the following functional

$$J(u) = \int_\Omega \left[\frac{(\nabla u)^2}{2} - G(u)\right] dx \quad \text{on} \quad H_0^1(\Omega),$$

which is bounded from below so that the P.S. condition is satisfied.

(i) Let φ_1 be the first eigenfunction, with $\max_{x\in\overline{\Omega}} \varphi_1 = 1$, and $\varphi_1 > 0$. We may choose $\varepsilon > 0$ so small that

$$-\varphi_0 < -\varepsilon\varphi_1 \quad \text{and} \quad \varepsilon\varphi_1 < \varphi_0$$

are two pairs of strict sub- and super-solutions of (1.1). According to Lemma 2.1, we have two distinct solutions $z_1, z_2 \in H_0^1(\Omega)$, satisfying

$$C_k(J, z_i) = \begin{cases} G & k = 0 \\ 0 & k \neq 0 \end{cases}, \quad i = 1, 2,$$

if they are isolated.

(ii) We may assume that there are at most finitely many solutions. The weak version of the mountain pass lemma is employed. We obtain a third solution z_3. According to the Kato-Hess Theorem and Theorem 1.6 in Chapter II, we have

$$C_k(J, z_3) = \begin{cases} G & k = 1 \\ 0 & k \neq 1. \end{cases}$$

We shall prove that $z_3 \neq \theta$. Indeed, if $\lambda > \lambda_2$, then

$$C_k(J,\theta) = C_{k-j}(\widetilde{J},\theta), \quad \text{where } j = \operatorname{ind} d^2J(\theta) \geq 2.$$

So $z_3 \neq \theta$. If $\lambda = \lambda_2$, with $\frac{g(t)}{t} \geq \lambda_2$ for $|t| \neq 0$ small, then $\operatorname{ind}(J,\theta) = 1$, θ is degenerate, and is a local maximum of $\widetilde{J}$ on the characteristic submanifold $\mathcal{N}$ at θ. Since $\dim \mathcal{N} = \dim \ker(-\Delta - \lambda I) = m_0$, we have

$$C_k(J,\theta) = \begin{cases} G & k = 1 + m_0, \\ 0 & k \neq 1 + m_0. \end{cases}$$

Again, the critical groups single out z_3 from θ.

(iii) In the case $\lambda \in \sigma(-\Delta)$, θ is a degenerate critical point. But, $g(t)/t \geq \lambda$ for $t \in U \setminus \{0\}$ implies that $\widetilde{J} = J|_{\mathcal{N}} \leq 0$, where $\mathcal{N}$ is a neighborhood in the characteristic submanifold at θ, $\dim \mathcal{N} = \dim \ker(-\Delta - \lambda I)$, equal to m_0. Let m_- be the Morse index of J at θ. We have $m \geq 2$, and

$$C_k(J,\theta) = \begin{cases} G & k = m_- + m_0 \\ 0 & k \neq m_- + m_0 \end{cases}$$

because θ is a local maximum of $\widetilde{J}$.

Similarly, in the cases either $\lambda \notin \sigma(-\Delta)$ or $\lambda \in \sigma(-\Delta)$ but $g(t)/t \leq \lambda$ for $t \in U \setminus \{0\}$, θ is a local minimum of $\widetilde{J}$; thus

$$C_k(J,\theta) = \begin{cases} G & k = m_- \\ 0 & k \neq m_-. \end{cases}$$

If there were no other critical points, then a contradiction would occur due to the Morse inequalities: $\beta_0 = 1$, $\beta_k = 0$, $k \neq 0$. In fact, for $k > j$, one would have

$$M_k - M_{k-1} + \cdots + (-1)^k M_0 = \beta_k - \beta_{k-1} + \cdots + (-1)^k \beta_0.$$

The LHS is even, but the RHS is odd. Therefore there are at least four nontrivial solutions. The theorem is proved.

A special case of this problem is due to the fact that the function is of the following form:

$$g(u) = \lambda u - h(u), \tag{3.10}$$

where λ is a real parameter and $h(u)$ satisfies the following conditions: $h \in C^1(\mathbb{R}^1)$, $h(0) = h'(0) = 0$ and $\lim_{|u|\to\infty} \frac{h(u)}{u} = +\infty$.

In this sense, we call it a bifurcation problem.

Example 1. The special form (3.10) of Theorem 3.5 has been studied by many authors. Cf. Ambrosetti [Amb1], Struwe [Str], for at least three solutions, Hofer [Hof], Tian [Tia1] and Dancer [Dan1] for at least four solutions in cases (i) and (iii).

3.4. Jumping Nonlinearities

Elliptic equations with jumping nonlinearities were first studied by Ambrosetti and Prodi [AmP1] and followed by many others: cf. Amann and Hess [AmH1], Berger and Podolak [BeP1], Fučik [Fuč1], Kazdan and Warner [KaW1], Hess [Hes1], Dancer [Dan1], H. Berestycki and P. L. Lions [BeL1] and author [Cha4]. After an observation due to Lazer and McKenna [LaM1], more solutions were obtained. In this respect, the reader is referred to Solimini [Sol1], Ambrosetti [Amb1], Hofer [Hof1] and Dancer [Dan1].

We consider the following BVP with a real parameter $t \in \mathbb{R}^1$.

$$(P_t)\begin{cases} -\Delta u = f(x,u) + t\varphi_1 & \text{in } \Omega \\ u\big|_{\partial\Omega} = 0 \end{cases}$$

where φ_1 is the first eigenfunction, with $\varphi_1(x) > 0\ \forall x \in \Omega$. Assume that $f \in C^1(\overline{\Omega} \times \mathbb{R}^1)$, satisfying the following conditions:

(1) $\lim_{\xi\to+\infty} f'_\xi(x,\xi) = \gamma$ uniformly in $x \in \overline{\Omega}$, and $\gamma \in (\lambda_j, \lambda_{j+1})$ for some $j \geq 1$, where $\{\lambda_j \mid j = 1, 2, \dots\} = \sigma(-\Delta)$.

(2) $\lim_{\xi\to-\infty} \frac{f(x,\xi)}{\xi} \leq \lambda_1 - \delta$, uniformly in $x \in \overline{\Omega}$, for some $\delta > 0$.

(3) There exists a constant M such that

$$\left|f'_\xi(x,\xi)\right| \leq M\left(1 + |\xi|^{\frac{4}{n-2}}\right).$$

We note that condition (1) implies that

$$\lim_{\xi\to+\infty} \frac{f(x,\xi)}{\xi} = \gamma.$$

Theorem 3.6. *Suppose that the conditions* (1)–(3) *are fulfilled. Then there exists $t^* \in \mathbb{R}^1$ such that (P_t) has*

(1) *no solution, if $t > t^*$;*
(2) *at least one solution, if $t = t^*$;*
(3) *at least two solutions, if $t < t^*$.*

If further, we assume $j \geq 2$, i.e., $\gamma > \lambda_2$, the second eigenvalue of $-\Delta$, then there exists $t^{**} < t^*$ such that (P_t) has at least four solutions if $t < t^{**}$.

The proof depends on the following lemmas.

Lemma 3.3. *Assume conditions* (1), (2), *and* $f \in C(\overline{\Omega} \times \mathbb{R}^1)$. *Let*

$$J_t(u) = \int_\Omega \left[\frac{|\nabla u|^2}{2} - F(x,u) - t\varphi_1 u \right] dx \quad u \in H_0^1(\Omega)$$

where $F(x,\xi) = \int_0^\xi f(x,s)ds$. *Then for all* $t \in \mathbb{R}^1$, J_t *satisfies the P.S. condition.*

Proof. For each function $u \in L^1_{loc}(\Omega)$ we denote $u^+ = \max\{u, 0\}$, and $u^- = u - u^+$. Assume that $\{u_n\} \subset H_0^1(\Omega)$ is a sequence satisfying

$$\int_\Omega (\nabla u_n \cdot \nabla v - f(x,u_n)v - t\varphi_1 v)\, dx = o(\|v\|) \quad v \in H_0^1(\Omega) \tag{3.11}$$

where $\|\cdot\|$ is the $H_0^1(\Omega)$ norm. Then we obtain

$$\int_\Omega [\nabla u_n^+ \nabla v - f(x,u_n)^+ v - (t\varphi_1)^+ v]\, dx = o(\|v\|) \quad \forall\, v \in H_0^1(\Omega).$$

Let $\rho_n = \rho_n^+ - \rho_n^-$, where

$$\rho_n^\pm = u_n^\pm - (-\Delta)^{-1} \left[f(x,u_n)^\pm + (t\varphi_1)^\pm \right] \to \theta \quad \text{in } H_0^1(\Omega).$$

By condition (1), $\|f(x,u_n)^+ - \gamma u_n^+\|_{L^2} = o(\|u_n^+\|_{L^2})$; but

$$u_n^+ = \left(id - \gamma(-\Delta)^{-1}\right)^{-1} \{(-\Delta)^{-1}[(f(x,u_n)^+ - \gamma u_n^+) + t^+\varphi_1] + \rho_n^+\};$$

it follows that $\{\|u_n^+\|\}$ is bounded. From conditions (1) and (2), we have $\delta > 0$ and $C > 0$ such that

$$f(x,\xi) - \lambda_1 \xi > \delta|\xi| - C. \tag{3.12}$$

Let us choose $\mu < \lambda_1$ such that $\lambda_1 - \mu < \delta$ then we have

$$\begin{aligned}(-\Delta - \mu)(u_n - \rho_n) &= f(x,u_n) + t\varphi_1 - \mu(u_n - \rho_n) \\ &\geq -C + t\varphi_1 + \mu\rho_n.\end{aligned}$$

By the weak Maximum Principle, one deduces

$$u_n - \rho_n \geq (-\Delta - \mu)^{-1}[-C + t\varphi_1 + \mu\rho_n] \tag{3.13}$$

noticing $\rho_n \to \theta$ $(H_0^1(\Omega))$. Combining (3.13) with the boundedness of u_n^+, we obtain that

$$\|u_n\|_{L^2} \quad \text{is bounded.}$$

Substituting this fact in (3.11), we get that $\{\|u_n\|\}$ is bounded. After a standard procedure, the (PS) condition is verified.

Lemma 3.4. *Under conditions* (1) *and* (2) *there exists a subsolution* $\underline{u}_t$ *for the* BVP (P_t) *such that for each solution* u_t *of* (P_t) *we have* $u_t > \underline{u}_t$.

Proof. According to (3.13), if we define $\underline{u}_t$ to be the solution of the following BVP:

$$\begin{cases} -\Delta u - \mu u = -C + t\varphi_1 & \text{in } \Omega \\ u\big|_{\partial\Omega} = 0 \end{cases}$$

the conclusion follows from the weak Maximum Principle, and the inequality (3.12).

Lemma 3.5. *Under conditions* (1) and (2) *there exists* $t_0 \in \mathbb{R}^1$ *such that if* (P_t) *is solvable, then* $t < t_0$.

Proof. By (3.12), we have $\delta > 0$, $C > 0$ such that

$$f(x,\xi) - \lambda_1 \xi \geq \delta|\xi| - C.$$

Thus, if u_t is a solution of (P_t), then multiplying by φ_1 on both sides of the equation, and by integration, we obtain

$$\lambda_1 \int_\Omega u_t \varphi_1 \, dx = \int_\Omega f(x, u_t)\varphi_1 \, dx + t \int_\Omega \varphi_1^2 \, dx.$$

From this one deduces

$$t \int_\Omega \varphi_1^2 \, dx + \delta \int_\Omega |u_t| \cdot \varphi_1 \, dx - C \int_\Omega \varphi_1 \, dx \leq 0,$$

that is

$$t < \left(\int_\Omega \varphi_1^2 \, dx \right)^{-1} C \int_\Omega \varphi_1 \, dx.$$

Lemma 3.6. *Under conditions* (1), (2) *and* (3), *there exists* $t_1 \in \mathbb{R}^1$ *such that* (P_t) *possesses a positive solution* $\overline{u}_t$ *which is a nondegenerate critical point of* J_t *with index* $d^2J_t(\overline{u}_t) = h_j$ *for* $t < t_1$, *where* $h_j = \sum_{k \leq j} \dim\ker(-\Delta - \lambda_k I)$.

Proof. Let

$$g(x,\xi) = -\gamma\xi + f(x,\xi) \quad \xi \geq 0.$$

We have $g(x,\xi) = o(|\xi|)$ uniformly in $x \in \overline{\Omega}$, as $\xi \to \infty$. We extend the function g to be a function $\widehat{g}$ such that $\widehat{g} \in C^1(\overline{\Omega} \times \mathbb{R}^1)$, with $|\widehat{g}(x,\xi)| = o(|\xi|)$ uniformly in $x \in \overline{\Omega}$. According to Theorem 3.1, the equation

$$\begin{cases} -\Delta u = \gamma u + \widehat{g}(x,u) + t\varphi_1 & \text{in } \Omega \\ u\big|_{\partial\Omega} = 0 \end{cases}$$

possesses a solution $\overline{u}_t$. Define

$$v_t = \overline{u}_t - \frac{t\varphi_1}{\lambda_1 - \gamma}. \tag{3.14}$$

We obtain

$$\begin{cases} -\Delta v_t = \gamma v_t + \widehat{g}(x, \overline{u}_t) & \text{in } \Omega, \\ v_t\big|_{\partial\Omega} = 0. \end{cases}$$

Thus, the L^p a priori bounds for v_t are employed to deduce

$$\|v_t\|_{C^1} = o(|t|) \quad \text{as} \quad |t| \to \infty.$$

Substituting the estimate in (3.14), we obtain

$$\overline{u}_t > 0 \quad \text{for} \quad t < t_1$$

where $-t_1$ is a real number large enough. This proves that $\overline{u}_t$ is a solution of the problem (P_t).

Again, we have

$$\overline{u}_t(x) = v_t(x) + \frac{t}{\lambda_1 - \gamma}\varphi_1(x) \to +\infty, \quad \text{a.e. as } t \to -\infty.$$

This implies that

$$f'_\xi(x, \overline{u}_t(x)) \to \gamma \quad \text{a.e. as } t \to -\infty.$$

Since

$$d^2J(\overline{u}_t) = id - (-\Delta)^{-1} f'_\xi(x, \overline{u}_t(x)),$$

by the Hölder inequality as well as the Sobolev embedding theorem, we see that

$$\|d^2J(\overline{u}_t) - (id - \gamma(-\Delta)^{-1})\|_{\mathcal{L}(H_0^1(\Omega))} \leq \left(\int_\Omega |f'_\xi(x, \overline{u}_t(x)) - \gamma|^p dx\right)^{1/p},$$

where $p = \frac{n}{2}$. Applying the Lebesgue dominance theorem, we arrive at

$$\|d^2J(\overline{u}_t) - (id - \gamma(-\Delta)^{-1})\|_{\mathcal{L}(H_0^1(\Omega))} = o(1).$$

Since $\gamma \notin \sigma(-\Delta)$ is assumed, $id - \gamma(-\Delta)^{-1}$ is invertible. Thus $\overline{u}_t$ is nondegenerate with Morse index h_j.

Proof of Theorem 3.6. Define $t^* = \sup\{t \in \mathbb{R}^1 \mid (P_t) \text{ is solvable}\}$. Combining Lemma 3.5 with Lemma 3.6, $t^* \in \mathbb{R}^1$, hence (1) is proved.

Now we come to look for a strict supersolution for $t < t^*$. In fact, choose $t' \in (t, t^*)$ and let $u_{t'}$ be a solution of $(P_{t'})$. Then $u_{t'}$ is a strict supersolution of (P_t) which satisfies $u_{t'} > \underline{u}_{t'} > \underline{u}_t$. Hence $[\underline{u}_t, u_{t'}]$ is a pair of strict sub- and supersolutions of (P_t). Lemma 2.1 is employed to deduce a solution u_t of (P_t) which is a local minimum of the functional $\widetilde{J}_t = J_t\big|_{C_0^1(\overline{\Omega})}$, so that

$$C_k(J_t, u_t) = \begin{cases} G & k = 0, \\ 0 & k \neq 0. \end{cases}$$

Noticing that the functional $\widetilde{J}_t$ is unbounded from below along the ray $u_s = s\varphi_1$, $s > 0$, Theorem 2.1 is applied. We find a second solution $\widehat{u}_t$ with critical groups

$$C_k(J_t, \widehat{u}_t) = \begin{cases} G & k = 1, \\ 0 & k \neq 1. \end{cases}$$

The conclusion (3) is proved.

As for conclusion (2), we prove, by the same method as in Lemma 3.3, that the set $\{u_t \mid t \in [t^* - 1, t^*]\}$, where u_t is the solution of (P_t) obtained by the previous sub- and supersolutions, is bounded in $H_0^1(\Omega)$. We obtain a sequence $t_i \to t^*$ such that u_{t_i} weakly converges in $H_0^1(\Omega)$, say to u^*. Then u^* is a solution of (P_{t^*}).

Finally, we assume $\gamma \in (\lambda_j, \lambda_{j+1})$, with $j \geq 2$. According to Lemma 3.6, there is a $t^{**} < t^*$ such that there exists a third solution $\bar{u}_t$ of (P_t) such that $\overline{u}_t$ is nondegenerate, with

$$C_k(J_t, \overline{u}_t) = \begin{cases} G & k = h_j, \\ 0 & k \neq h_j. \end{cases}$$

One more solution will then be obtained by a computation of the Leray–Schauder degree. In fact, by Lemma 3.3, we conclude that all solutions of the equation

$$u = (-\Delta)^{-1}(f(x, u) + t\varphi_1) \tag{3.15}$$

are bounded in an open ball $\overset{\circ}{B}_{R_t}$, where R_t, the radius, depends on t continuously. By the homotopy invariance of the Leray–Schauder degree, one has

$$\deg(\mathrm{id} - (-\Delta)^{-1}F_t, \overset{\circ}{B}_{R_t}, \theta) = \text{const.} \quad \forall\, t \in \mathbb{R}^1,$$

where

$$F_t u = f(x, u(x)) + t\varphi_1(x).$$

But, from conclusion (1), if $t > t^*$, (3.15) has no solution. It follows that

$$\deg(\mathrm{id} - (-\Delta)^{-1}F_t, \overset{\circ}{B}_t, \theta) = 0, \quad \forall\, t \in \mathbb{R}^1.$$

If $t < t^{**}$, suppose that there are only three solutions $\overline{u}_t$, $\widehat{u}_t$ and u_t, then by Theorem 3.2 of Chapter II, the Leray–Schauder degree would be

$$\deg(\mathrm{id} - (\Delta)^{-1}F_t, \overset{\circ}{B}_t, \theta) = (-1)^{h_j}.$$

This will be a contradiction.

Remark 3.2. Lemmas 3.4 and 3.5 are due to Kazdan and Warner [KaW1], and Lemma 3.6 is due to Ambrosetti [Amb1] and Lazer and McKenna

[LaM1]. The idea of the proof is taken from Hofer [Hof1], Dancer [Dan1] and Chang [Cha4]. An extension, in which $\lim_{\xi\to-\infty}\frac{f(x,\xi)}{\xi}<\lambda_i$, $i>1$, has been studied by many others. The reader is referred to the survey paper by Lazer [Laz1]; see also Lazer and McKenna [LaM2] and Dancer [Dan1].

3.5. Other examples

Suppose that $g\in C^1(\mathbb{R}^1)$ satisfies the following conditions:

(1) $g(0)=0$, $0\le g'(0)<\lambda_1$;
(2) $g'(t)>0$ and strictly increasing in t for $t>0$;
(3) $g'(\infty)=\lim_{|t|\to\infty}g'(t)$ exists and lies in (λ_1,λ_2).

Theorem 3.7. *Under conditions* (1), (2), (3) *the equation*

$$\begin{cases} -\Delta u = g(u) & \text{in } \Omega \\ u\big|_{\partial\Omega}=0 \end{cases} \tag{3.16}$$

has at least three distinct solutions.

Proof. 1. It is obvious that θ is a solution, which is also a strict local minimum of the functional

$$J(u)=\int_\Omega\left[\frac{1}{2}(\nabla u)^2-G(u)\right]dx \quad \text{on } H_0^1(\Omega),$$

where G is the primitive of g, with $G(0)=0$.

2. Modify g to be a new function

$$\widehat{g}(t)=\begin{cases} g(t) & t\ge 0 \\ 0 & t<0, \end{cases}$$

and consider a new functional

$$\widehat{J}(u)=\int_\Omega\left[\frac{1}{2}(\nabla u)^2-\widehat{G}(u)\right]dx,$$

where $\widehat{G}(t)=\int_0^t\widehat{g}(t)dt$. It is easily seen that θ is also a strict local minimum of $\widehat{J}$, which is a C^1-functional with a (PS) condition.

Since $\widehat{J}$ is unbounded from below, along the ray $u_s=s\varphi_1(x)$, $s>0$, Theorem 2.1 yields a critical point $u_0\ne\theta$ of $\widehat{J}$ which solves the equation

$$\begin{cases} -\Delta u=\widehat{g}(u) & x\in\Omega. \\ u\big|_{\partial\Omega}=0. \end{cases}$$

Since $\widehat{g}(u)\ge 0$, by the Maximum Principle, $u_0\ge 0$, hence u_0 is a solution of (3.16).

3. Now we shall prove that $-\Delta - g'(u_0(x))$ has a bounded inverse operator on $L^2(\Omega)$, which is equivalent to the fact that $\mathrm{Id}-(-\Delta)^{-1}g'(u_0(x))$ has a bounded inverse on $H_0^1(\Omega)$, i.e., u_0 is nondegenerate.

Since u_0 satisfies (3.16), it is also a solution of the equation

$$-\Delta u_0 - q(x)u_0(x) = 0, \quad u_0\big|_{\partial\Omega} = 0,$$

where

$$q(x) = \int_0^1 g'(tu_0(x))dt.$$

Let $\mu_1 < \mu_2 < \dots$ be eigenvalues of the problem

$$\begin{cases} -\Delta w - \mu g'(u_0(x))w = 0, \\ w\big|_{\partial\Omega} = 0. \end{cases}$$

We shall prove that $\mu_1 < 1 < \mu_2$. This implies the invertibility of the operator $-\Delta - g'(u_0(x))$.

In fact, according to assumption (2), we have

$$q(x) < g'(u_0(x)) \quad \forall\, x \in \Omega$$

so that

$$\mu_1 = \min \frac{\int (\nabla w)^2}{\int g'(u_0)w^2} < \min \frac{\int (\nabla w)^2}{\int q(x)w^2} \leq 1.$$

Again, by assumptions (2) and (3), we have

$$g'(u_0(x)) < \lambda_2 \quad \forall\, x \in \Omega.$$

According to the Rayleigh quotient characterization of the eigenvalues

$$\mu_2 = \sup_{E_1} \inf_{w \in E_1^\perp} \frac{\int (\nabla w)^2}{\int g'(u_0(x))w^2} > \frac{1}{\lambda_2} \sup_{E_1} \inf_{w\in E_1^\perp} \frac{\int (\nabla w)^2}{\int w^2} = 1$$

where E_1 is any one-dimensional subspace in $H_0^1(\Omega)$.

4. The Morse identity yields an odd number of critical points. Therefore there are at least three solutions of (3.16).

Finally, we turn to the following example.

Theorem 3.8. *Suppose that $g \in C^1(\mathbb{R}^1)$ satisfies the following conditions:*

(1) $g(0) = 0$, *and* $\lambda_2 < g'(0) < \lambda_3$;

(2) $g'(\infty) = \lim_{t\to\pm\infty} g'(t)$ *exists, and* $g'(\infty) \notin \sigma(-\Delta)$, *with* $g'(\infty) > \lambda_3$;

(3) $|g(t)| < 1$ *and* $0 \leq g'(t) < \lambda_3$ *in the interval* $[-c, c]$, *where* $c = \max_{x\in\overline{\Omega}} e(x)$, *and* $e(x)$ *is the solution of the* BVP:

$$\begin{cases} -\Delta e = 1 \quad in\ \Omega \\ e\big|_{\partial\Omega} = 0. \end{cases}$$

Then equation (3.16) possess at least five nontrivial solutions.

Proof. Define

$$\widetilde{g}(t) = \begin{cases} g(c) & \text{if } t > c \\ g(t) & \text{if } |t| \le c \\ g(-c) & \text{if } t < -c \end{cases}$$

and let

$$\widetilde{J}(u) = \int_\Omega \left[\frac{1}{2}(\nabla u)^2 - \widetilde{G}(u)\right] dx,$$

where $\widetilde{G}(t) = \int_0^t \widetilde{g}(s)ds$. The truncated equation

$$\begin{cases} -\Delta u = \widetilde{g}(u) & \text{in } \Omega \\ u\big|_{\partial\Omega} = 0 \end{cases} \tag{3.17}$$

possesses at least three solutions θ, u_1, u_2, because there are two pairs of sub- and supersolutions $[\varepsilon\varphi_1, e]$ and $[-e, -\varepsilon\varphi_1]$, where φ_1 is the first eigenfunction of $-\Delta$, with $\varphi_1(x) > 0$, and $\varepsilon > 0$ a small enough constant.

By the weak version of the Mountain Pass Lemma, there is a mountain pass point u_3. That $u_3 \neq \theta$ follows from the fact that

$$C_k(\widetilde{J}, u_3) = \begin{cases} G & k = 1 \\ 0 & k \neq 1. \end{cases}$$

But from condition (1)

$$C_k(\widetilde{J}, \theta) = \begin{cases} G & k = m_1 + m_2 \\ 0 & k \neq m_1 + m_2, \end{cases}$$

where $m_i = \dim\ker(-\Delta - \lambda_i I)$, $i = 1, 2, \ldots$. By Lemma 2.1, one has

$$C_k(\widetilde{J}, u_i) = \begin{cases} G & k = 0 \\ 0 & k \neq 0 \end{cases}, \quad i = 1, 2.$$

Noticing that $\widetilde{J}$ is bounded from below, we conclude that there is at least another critical point u_4.

Obviously, all these critical points u_i, $i = 1, 2, 3, 4$, are solutions of equation (3.17).

On account of the first condition in (3), in combination with the Maximum Principle, all solutions of (3.17) are bounded in the interval $[-c, c]$. Therefore they are solutions of (3.16); moreover, all these solutions u, because of their ranges, are included in $[-c, c]$, and we conclude:

$$\operatorname{ind}(J, u) + \dim\ker(d^2 J(u)) \le m := \dim \bigoplus_{j=1}^{2} (-\Delta - \lambda_j I),$$

provided by the second condition in (3).

Because of condition (2), we learned from Theorem 3.1, Theorem 5.2 of Chapter II is applicable, with $\gamma > m$, because $g'(\infty) > \lambda_3$. Therefore there exists another critical point u_5, which yields the fifth nontrivial solution for the equation (3.16).

Cf. Chang [Cha12].

4. Bounded Nonlinearities

4.1. Functionals Bounded from Below

The functionals J associated with equation (1.1) in this section are considered to be bounded from below. We shall study several cases which occurred in PDE about numbers of solutions.

First we assume

(g_6) $\exists\,\alpha < \lambda_1/2$, and $\beta > 0$ such that

$$G(x,t) = \int_0^t g(x,s)ds \le \alpha t^2 + \beta$$

where λ_1 is the first eigenvalue of $-\Delta$ with 0-Dirichlet data;

(g_7) $|g_t'(x,t)| \le C(1+|t|)^\gamma$, $\gamma < \frac{4}{n-2}$, if $n > 2$.

Theorem 4.1. *Under assumptions* (g_6) *and* (g_7), *suppose that*

$$g(x,0) = 0, \quad \text{and } \exists\, m \ge 1 \quad \text{such that} \quad \lambda_m < g_t'(x,0) < \lambda_{m+1}, \tag{4.1}$$

where $\{\lambda_1, \lambda_2, \dots\} = \sigma(-\Delta)$. *Then* (1.1) *has at least three solutions.*

Proof. Again, we consider the functional (1.3)

$$J(u) = \int_\Omega \left[\frac{1}{2}|\nabla u|^2 - G(x,u)\right] dx$$

which is well-defined and C^2 on $H_0^1(\Omega)$ provided by (g_7). (g_6) implies that J is bounded from below:

$$J(u) \ge \frac{1}{2}\left(1 - \frac{2\alpha}{\lambda_1}\right)\int_\Omega |\nabla u|^2 dx - \beta\,\mathrm{mes}(\Omega). \tag{4.2}$$

And θ is a nonminimum, nondegenerate critical point with finite Morse index of J provided by assumption (4.1). In order to apply Theorem 5.4 of Chapter II it suffices to verify the (PS) conditions. In fact, the coercive condition (4.2) in conjunction with the boundedness of $J(u_n)$ imply that $\{u_n\}$ is bounded, and hence is weakly compact. From (g_7), we see that

$$|g(x,t)| \le C_1(1+|t|)^\mu, \quad \mu < \frac{n+2}{n-2}, \quad \text{if } n > 2.$$

We use the same argument as in Lemma 2.2, from

$$dJ(u_n) = u_n - (-\Delta)^{-1} g(x, u_n(x)) \to \theta.$$

We conclude that the sequence u_n is subconvergent.

4.2. Oscillating Nonlinearity

The following technique is often used to reduce problems such as (1.1) to problems with bounded nonlinear terms.

Lemma 4.1. *Suppose that the function g in* (1.2) *satisfies*
(g_8) $\exists\, \xi \geq 0$ *such that* $g(x, \xi) \leq 0\ \forall\, x \in \overline{\Omega}$, *and let*

$$\widehat{g}(x,t) = \begin{cases} g(x,\xi) & \text{if } t > \xi \\ g(x,t) & \text{if } t \leq \xi. \end{cases}$$

Assume that $u \in H_0^1(\Omega)$ is a solution of the following equation:

$$\begin{cases} -\Delta u = \widehat{g}(x, u(x)) & x \in \Omega \\ u\big|_{\partial\Omega} = 0. \end{cases} \tag{4.3}$$

Then $u(x) \leq \xi$, so it is also a solution of (1.1).

Proof. By the standard regularity theorem $u \in W_p^2(\Omega)\ \forall\, p < +\infty$ so $u \in C^1(\overline{\Omega})$. Considering the domain $D = \{x \in \Omega \mid u(x) > \xi\}$, we have

$$\begin{cases} -\Delta u(x) \leq 0 & \forall\, x \in D, \\ u\big|_{\partial D} \leq \xi. \end{cases}$$

By the Maximum Principle, we have $u(x) < \xi$ in D, and hence $D = \emptyset$, i.e., $u(x) \leq \xi$.

Corollary 4.1. *If in the above lemma,* (g_8) *is replaced by*
(g_8)$'$ $\exists\, \xi \geq 0$ *such that* $g(x, \xi) < 0\ \forall\, x \in \overline{\Omega}$.
Then $u(x) < \xi\ \forall\, x \in \overline{\Omega}$.

Similarly, we introduce the assumption
(g_8)* $\exists\, \xi \leq 0$ such that $g(x, \xi) \geq 0$ (or > 0) $\forall\, x \in \overline{\Omega}$, and consider the truncation

$$\breve{g}(x,t) = \begin{cases} g(x,t) & \text{if } t \geq \xi, \\ g(x,\xi) & \text{if } t < \xi. \end{cases}$$

We have

Corollary 4.2. *Under (1.2) and $(g_8)^*$, if $u \in H_0^1(\Omega)$ is a solution of*

$$\begin{cases} -\Delta u(x) = \breve{g}(x, u(x)) & x \in \Omega \\ u\big|_{\partial\Omega} = 0, \end{cases}$$

then $u(x) \geq \xi\ \forall\, x \in \overline{\Omega}$, and is also a solution of (1.1).

By the same trick, if one looks for positive solutions, the function g defined on $\overline{\Omega} \times \mathbb{R}^1_+$, is extended continuously to be $\widetilde{g}: \overline{\Omega} \times \mathbb{R}^1 \to \mathbb{R}^1$, with $\widetilde{g}$ nonnegative for $t < 0$.

Keeping this in mind, we consider some examples. For the sake of simplicity, we assume $g(x,t) = g(t)$, and study the eigenvalue problem

$$\begin{cases} -\Delta u(x) = \lambda g(u(x)) & x \in \Omega \\ u\big|_{\partial\Omega} = 0. \end{cases} \tag{4.4}$$

Theorem 4.2. *Suppose $g \in C^1(\mathbb{R}^1_+)$ and $g(0) \geq 0$. Assume that*

(g_9) There exists $0 < a_1 < a_2 < \cdots < a_m$, such that $g(a_i) = 0$, $i = 1, 2, \ldots, m$.

(g_{10}) $G(t) = \int_0^t g(s)ds$ satisfies

$$\max\{G(t) \mid 0 \leq t \leq a_{i-1}\} < G(a_i), \quad i = 2, 3, \ldots, m \quad \text{and} \quad G(a_1) > 0.$$

Then $\exists\, \lambda_0 > 0$ such that for $\lambda > \lambda_0$, (4.4) possesses at least $(2m-1)$ nontrivial solutions. Furthermore, if $g(0) = 0$ and $g'(0) \leq 0$ then $\exists\, \lambda_1 > 0$ such that for $\lambda > \lambda_1$, there are at least $2m$ nontrivial solutions for (4.4).

Proof. By the truncation trick, we consider the functions

$$\widetilde{g}_i(t) = \begin{cases} g_i(0) & t < 0 \\ g_i(t) & 0 \leq t \leq a_i \\ 0 & t > a_i \end{cases}$$

$\widetilde{g}_i \in C(\mathbb{R}^1)$, and the functionals with parameter λ

$$J_i(u, \lambda) = \int_\Omega \left[\frac{1}{2}|\nabla u|^2 - \lambda G(u(x))\right] dx, \quad i = 1, 2, \ldots, m.$$

Given the above explanation, we know that the critical set $K_i(\lambda)$ of $J_i(\cdot, \lambda)$ must be a subset of the critical set $K_{i+1}(\lambda)$, $i = 1, 2, \ldots, m-1$, and by Lemma 4.1, $\forall\, u \in K_i(\lambda)$, $0 \leq u(x) \leq a_i$, $i = 1, 2, \ldots\ \forall\, x \in \overline{\Omega}$.

First, we see that $J_i(\cdot, \lambda)$ possesses a global minimum $u_i(\lambda)$, provided by the boundedness from below plus the (PS) condition. Thus $u_i(\lambda) \in K_i(\lambda)$, $i = 1, 2, \ldots, m$.

Second, $u_i(\lambda) \notin K_{i-1}(\lambda)$ for λ large, $i = 2, 3, \ldots, m$.

Indeed, we only want to show $\exists\, \lambda_i > 0$ and $w \in H^1_0(\Omega)$, with $0 \le w(x) \le a_i$, such that $J_i(w, \lambda) < J_{i-1}(u_{i-1}(\lambda), \lambda)\ \forall\, \lambda > \lambda_i$.

Let $\alpha = G(a_i) - \text{Max}\{G(t) \mid 0 \le t \le a_{i-1}\}$. By (g_{10}), $\alpha > 0$. Let $\Omega_\delta = \{x \in \Omega \mid \text{dist}(x, \partial\Omega) < \delta\}$ for $\delta > 0$, and let

$$w_\delta \in C_0^\infty(\Omega), \quad 0 \le w_\delta \le a_i, \quad w_\delta(x) = a_i \quad \forall\, x \in \Omega \setminus \Omega_\delta.$$

Thus

$$\begin{aligned}\int_\Omega G(w_\delta)dx &= \int_{\Omega\setminus\Omega_\delta} G(a_i)dx + \int_{\Omega_\delta} G(w_\delta(x))dx \\ &\ge G(a_i)|\Omega| - 2C_i|\Omega_\delta| \\ &\ge \int_\Omega G(u_{i-1}(\lambda))dx + \alpha|\Omega| - 2C_i|\Omega_\delta|,\end{aligned}$$

where $C_i = \max\{G(t) \mid 0 \le t \le a_i\}$. Hence

$$\begin{aligned}&J_i(w_\delta, \lambda) - J_{i-1}(u_{i-1}(\lambda), \lambda) \\ &\quad = \frac{1}{2}\int_\Omega \left[|\nabla w_\delta|^2 - |\nabla u_{i-1}(\lambda)|^2\right] dx - \lambda \int_\Omega [G(w_\delta) - G(u_{i-1}(\lambda))]\, dx \\ &\quad \le \frac{1}{2}\int_\Omega |\nabla w_\delta|^2 dx - \lambda(\alpha|\Omega| - 2C_i|\Omega_\delta|) < 0\end{aligned}$$

for $\delta = \delta_i > 0$ small, and $\lambda > \lambda_i$ large enough. The function $w = w_\delta$ is just what we need. One may assume $\lambda_1 \le \lambda_2 \le \cdots \le \lambda_m$.

Third, from $G(a_1) > 0$, we have

$$\text{Min}\ J_1(\cdot, \lambda) < 0 \quad \forall\, \lambda > \lambda_1,$$

using the above argument, so $u_1(\lambda) \ne \theta$. One may assume $\#K_m(\lambda) < +\infty$. Then the Morse equality is applied to the bounded from below function J_i. Noticing that $H^1_0(\Omega)$ is contractible, we have $\beta_0 = 1$, $\beta_1 = \beta_2 = \cdots = 0$, and

$$\sum_j (-1)^j M^i_j(\lambda) = 1,$$

where $M^i_j(\lambda)$, $j = 0, 1, \ldots$ are the Morse type numbers for $K_i(\lambda)$. But the Morse equality also holds for J_{i+1}, and we have known that $u_{i+1}(\lambda) \in K_{i+1}(\lambda) \setminus K_i(\lambda)$ for $\lambda > \lambda_{i+1}$, and that u_{i+1} is the global minimum of J_{i+1}, so $C_q(J_{i+1}, u_{i+1}) = \delta_{q0} G$, i.e., the contribution of $u_{i+1}(\lambda)$ in the alternative summation $\sum_j (-1)^j M^{i+1}_j(\lambda)$, is 1. If there were no other critical point in $K_{i+1}(\lambda) \setminus K_i(\lambda)$ for $\lambda > \lambda_{i+1}$, then the equality would lose the balance. Therefore, we concludes

$$\#\left(K_{i+1}(\lambda) \setminus K_i(\lambda)\right) \ge 2 \quad \text{if} \ \ \lambda > \lambda_{i+1}, \quad i = 1, 2, \ldots, m-1.$$

In the cases $g(0) = 0$ and $g'(0) > 0$, θ has no contribution in critical groups. This is proved by the standard perturbation technique in combining with the homotopy invariance property (cf [Cha16]). In summary, we have $\#(K_m(\lambda) \setminus \{\theta\}) \geq 2m - 1$, if $\lambda > \lambda_m$. The first conclusion is proved.

Assume that $g(0) = 0$, and $g'(0) \leq 0$; then $\forall\ \lambda$, $J_1(u, \lambda) \geq 0 = J_1(\theta, \lambda)$ for $\|u\|$ small. θ is a local minimum of $J_1(\cdot, \lambda)$, but not the global minimum $u_1(\lambda)\ \forall\ \lambda > \lambda_1$. We apply the Morse equality to J_1, that there must be one more point in $K_1(\lambda)$ for $\lambda > \lambda_1$, i.e., $\#(K_1(\lambda) \setminus \{\theta\}) \geq 2$, $\lambda > \lambda_1$ so is $\#(K_m(\lambda) \setminus \{\theta\}) \geq 2m$, if $\lambda > \lambda_m$.

4.3. Even Functionals

Theorem 4.3. *Suppose that $g(x,t)$ is of the form $a(x)t + p(x,t)$ where $a \in C(\overline{\Omega})$, and $p \in C^1(\overline{\Omega} \times \mathbb{R}^1, \mathbb{R}^1)$. Assume that $a > 0$ in $\overline{\Omega}$, and that* (g_8),

(g_{11}) *$p(x,t) = o(|t|)$ uniformly with respect to $x \in \overline{\Omega}$, and*

(g_{12}) *$p(x,t) = -p(x,-t)\ \forall\ (x,t) \in \overline{\Omega} \times \mathbb{R}^1$,*

hold. Then the equation

$$\begin{cases} -\Delta u(x) = \lambda g(x, u(x)) & \text{in } \Omega \\ u\big|_{\partial\Omega} = 0 \end{cases}$$

has at least k distinct pairs of solutions, if $\lambda > \lambda_k$, where λ_k is the kth eigenvalue of the eigenvalue problem

$$\begin{cases} -\Delta v(x) = \mu a(x) v(x) & \text{in } \Omega \\ v\big|_{\partial\Omega} = 0. \end{cases} \tag{4.5}$$

Proof. $\forall \lambda$, the functional is written as

$$J_\lambda(u) = \int_\Omega \frac{1}{2}|\nabla u|^2 - \lambda \left(\frac{au^2}{2} + P(x, u(x)) \right) dx$$

where $P(x,t) = \int_0^t p(x,s) ds$ is an even function with respect to t, provided by (g_{12}). Thus J_λ is an even functional.

According to (g_8) and Lemma 4.1, J_λ is bounded from below.

And $a > 0$ plus (g_{11}) imply that there exists $\rho > 0$, such that $J_\lambda\big|_{S_\rho \cap E_k} < 0$ for $\lambda > \lambda_k$, where S_ρ is the sphere with radius ρ centered at θ in $H_0^1(\Omega)$, and E_k is the direct sum of eigenspaces with eigenvalues $\leq \lambda_k$ of the problem (4.5). The verification of the (PS) condition is omitted.

Now we apply Theorem 4.1 of Chapter II. There are at least k pairs of distinct solutions.

4.4. Variational Inequalities

A variety of variational problems with side constraints arising from mechanics and physics are called variational inequalities. They have been extensively studied since the 1960s. See, for instance, Duvaut and J. L. Lions [DuL1].

A typical example is as follows:

Given a closed convex set C in $H_0^1(\Omega)$, a continuous $g: \Omega \times \mathbb{R}^1 \to \mathbb{R}^1$ and $h \in L^{\frac{2n}{n-2}}(\Omega)$, find $u_0 \in C$ such that

(4.6)

$$\int_\Omega [\nabla u_0 \cdot \nabla(u - u_0) - (g(x, u_0(x)) - h(x))\,(u - u_0)(x)]\,dx \geq 0 \;\; \forall\, u \in C.$$

In fact, the variational inequality is attached to the following variational problem: to find $u_0 \in C$, which is the critical point of the functional

$$J(u) = \frac{1}{2}\int_\Omega \left[|\nabla u|^2 - G(x, u(x)) + h(x)u(x)\right] dx$$

with respect to the closed convex set C (cf. Definition 6.4 of Chapter I).

In this sense, all the critical point theories, including the Morse inequalities on closed convex sets, are suitable for the applications. In contrast with the well-developed variational inequality theory, in which g is assumed to be nonincreasing in t so that the solution is a minimum of the functional J, the restriction on g is avoided in this subsection.

Indeed, one can find minimax points.

We are satisfied to study the following two examples mainly by explaining the differences.

Example 1. Assume that g satisfies (g_1), (g_2) and (g_3). Let $C = P$ be the positive cone in $H_0^1(\Omega)$; then there are at least two solutions of (4.6), if $g(x, 0) = 0$ and $g(x, t) \geq 0 \;\forall\, (x, t) \in \Omega \times \mathbb{R}^1_+$, and if $\exists\, t_0 > 0$, such that $g(x, t_0) > 0$.

Claim. We follow Lemma 2.2 step by step to verify the (PS) condition with respect to P. Note that

$$\|- dJ(u_k)\|_{u_k} := \sup\left\{(-dJ(u_k), v - u_k) \mid v \in P, \|v - u_k\|_{H_0^1} < 1\right\} \to 0,$$

where (,) is the inner product in $H_0^1(\Omega)$. It implies $\forall\, \varepsilon > 0 \;\exists\, k_0 \in \mathbb{Z}_+$ such that

$$(dJ(u_k), u_k) \leq \varepsilon \|u_k\|, \quad \forall\, k \geq k_0.$$

(WARNING: This is only a one side inequality! Not like that in Lemma 2.2 in which we got $|(dJ(u_k), u_k)| \leq \varepsilon\|u_k\|$). This is enough to assure the boundedness of $\|u_k\|$, as shown in Lemma 2.2.

Now we prove the subconvergence of u_k. As shown in Lemma 2.2, we obtain a subsequence, which we still write in u_k, such that

$$(-\Delta)^{-1} \circ i^* \circ g(\cdot, u_k) \to u^*.$$

Since g is positive in $t \geq 0$, $u_k \in P$, and $(-\Delta)^{-1}$ preserves the positive cone (Maximum Principle), $u^* \in P$.

Again, from $\| - dJ(u_k)\|_{u_k} \to 0$, it follows that $\forall\, \varepsilon > 0\ \exists\, k_0 \in \mathbb{Z}_+$, such that $(-u_k + (-\Delta)^{-1} \circ i^* \circ g(\cdot, u_k), v - u_k) \leq \frac{\varepsilon}{2}\|v - u_k\|$, $\forall\, v \in P$, $\forall\, k \geq k_0$. Consequently, $\exists\, k_1 \in \mathbb{Z}_+$, such that

$$(-u_k + u^*, v - u_k) \leq \varepsilon\|v - u_k\| \quad \forall\, v \in P, \ \forall\, k \geq k_1.$$

In particular, set $v = u^*$, this proves $u_k \to u^*$.

To study the multiple solutions, it is easily seen that θ is a local minimum for J in P. Since the first eigenvector $\varphi_1 \in P$, J is unbounded from below in P. A weak version of the Mountain Pass theorem with respect to P is applied to obtain the second solution.

For the same functional J, but we change to a different closed convex set, one has

Example 2. Suppose $\psi \in H^1(\Omega)$, and $C = \{u \in H_0^1(\Omega) \mid 0 \leq u(x) \leq \psi(x) \text{ a.e.}\}$. Under the same assumptions on g and h in the Example 1, assume that

$$\inf\{J(u) \mid u \in C\} < 0. \tag{4.7}$$

Then the variational inequality (4.7) possesses at least three solutions.

Claim. The (PS) condition with respect to C can be verified as above. Now, $u_1 = \theta$ is a local minimum, and J has a global minimum u_2. Assumption (4.7) implies $u_1 \neq u_2$. We apply the Morse equality which provides the third critical point of J.

Remark 4.1. Theorem 4.1 is taken from K.C. Chang [Cha1]. For an extension of it see K.C. Chang [Cha2]. Theorem 4.2 is an extension of the results due to K.J. Brown and H. Budin [BrB1] and P. Hess [Hes2], in which only the case $g(0) > 0$ was discussed. Section 4.4 was studied in K.C. Chang [Cha7].

CHAPTER IV

Multiple Periodic Solutions of Hamiltonian Systems

0. Introduction

In this chapter, we shall apply Morse theory to estimate the numbers of solutions of Hamiltonian systems.

Let $H(t,z)$ be a C^1 function defined on $\mathbb{R}^1 \times \mathbb{R}^{2n}$ which is 2π-periodic with respect to the first variable t. We are interested in the existence and multiplicity of the 1-periodic solutions of the following Hamiltonian system:

$$\begin{cases} \dot{q} = -H_p(t;q,p) \\ \dot{p} = H_q(t;q,p), \end{cases} \tag{0.1}$$

where $q, p \in \mathbb{R}^n$, $z = (q,p)$. The function H then is called the Hamiltonian function. Letting J be the standard symplectic structure on $\mathbb{R}^{2n}$, i.e.,

$$J = \begin{pmatrix} 0 & -I_n \\ I_n & 0 \end{pmatrix},$$

where I_n is the $n \times n$ identity matrix, the equation (0.1) can be written in a compact version

$$-J\dot{z} = H_z(t,z). \tag{0.2}$$

Equation (0.2) is very similar to the operator equation considered in Chapter II, Section 5. Indeed, let $X = L^2\left((0,1), \mathbb{R}^{2n}\right)$, and let

$$A: z(t) \longmapsto -J\dot{z}(t)$$

with domain

$$D(A) = H^1_*\left([0,2\pi], \mathbb{R}^{2n}\right) = \left\{ z(t) \in H^1([0,2\pi], \mathbb{R}^{2n}) \mid z(0) = z(2\pi) \right\}.$$

For the sake of convenience, we make the real space $\mathbb{R}^{2n}$ complex. Let $\mathbb{C}^n = \mathbb{R}^n + i\mathbb{R}^n$, and let $\{e_1, e_2, \dots, e_{2n}\}$ be an orthonormal basis in $\mathbb{R}^{2n}$. Let

$$\varphi_j = e_j + ie_{j+n}, \qquad j = 1, 2, \dots, n,$$

which defines a basis in $\mathbb{C}^n$. The linear isomorphism $\mathbb{R}^{2n} \to \mathbb{C}^n$

$$z = \sum_{j=1}^{2n} z_j e_j \longmapsto \widehat{z} = \sum_{j=1}^{n} (z_j - iz_{j+n})\varphi_j,$$

is called the complexification of $\mathbb{R}^{2n}$, which preserves the inner product. Namely,

$$\begin{aligned}[\widehat{z}, \widehat{w}] &= \operatorname{Re} \sum_{j=1}^{n} (z_j - iz_{j+n})\overline{(w_j - iw_{j+n})} \\ &= \sum_{j=1}^{n} (z_j w_j + z_{j+n} w_{j+n}) = \sum_{j=1}^{2n} z_j w_j = \langle z, w \rangle ,\end{aligned}$$

where $[\ ,\ ]$ is the inner product on $\mathbb{C}^n$.

We introduce the complex Hilbert space $L^2([0, 2\pi], \mathbb{C}^n)$ to replace the real Hilbert space $L^2([0, 2\pi], \mathbb{R}^{2n})$, whose inner product reads as

$$\begin{aligned}(\widehat{z}, \widehat{w}) &= \int_0^{2\pi} [\widehat{z}(t), \widehat{w}(t)] dt \\ &= \int_0^{2\pi} \langle z(t), w(t) \rangle dt = (z, w).\end{aligned}$$

From now on, if there is no confusion, we shall not distinguish between these two. Sometime, we only write z but not $\widehat{z}$. One important thing is that $Jz \mapsto i\widehat{z}$. Thus, if we expand $z \in L^2([0, 2\pi], \mathbb{C}^n)$ in Fourier series:

$$z(t) = \sum_{j=1}^{n} \left(\sum_{m=-\infty}^{\infty} c_{jm} e^{-imt} \right) \varphi_j,$$

$$D(A) = \left\{ z \in L^2([0, 2\pi], \mathbb{C}^n) \mid \sum_{j=1}^{n} \sum_{m=-\infty}^{\infty} (1 + |m|)^2 |c_{jm}|^2 < +\infty \right\},$$

and A is self-adjoint with the following spectral decomposition:

$$L^2([0, \pi], \mathbb{C}^n) = \bigoplus_{m \in \mathbb{Z}} M(m),$$

where $M(m) = \operatorname{span}\left\{e^{-imt}\varphi_1, e^{-imt}\varphi_2, \ldots, e^{-imt}\varphi_n\right\}$ is the eigenspace of A with eigenvalue m, $\dim M(m) = n$, $m = 0, \pm 1, \pm 2, \ldots$.

Let E^*, where $*$ is $+$, 0, or $-$, be the positive, zero, and negative subspaces respectively according to the spectral decomposition, and let P^*, where $*$ is $+$, 0, or $-$, be the associated orthogonal projections. Set

$$A^+ = AP^+, \quad A^- = -AP^-, \quad A^0 = AP^0, \quad A = A^+ - A^-.$$

Let

$$H = D\left(|A|^{1/2}\right)$$
$$= \left\{ z \in L^2([0,2\pi],\mathbb{C}^n) \mid \sum_{j=1}^{n} \sum_{m=-\infty}^{\infty} (1+|m|)|c_{j\cdot m}|^2 < +\infty \right\}$$

with the following scalar product:

$$(z,w)_H = (z,w)_{L^2} + \left(|A|^{1/2}z, |A|^{1/2}w\right)_{L^2}.$$

Then H is a Hilbert space.

There are few simple properties often used:

(1) The embedding $H \hookrightarrow L^p([0,2\pi],\mathbb{C}^n)$ is compact, $\forall\, p < +\infty$.

Claim. Since $H^{1/2}([0,2\pi],\mathbb{C}^n) \hookrightarrow L^p([0,2\pi],\mathbb{C}^n)$ is compact and $D(|A|^{1/2}) \hookrightarrow H^{1/2}([0,2\pi],\mathbb{C}^n)$ is continuous.

(2) $\exists$ a compact linear operator $K: L^{p'}([0,2\pi],\mathbb{C}^n) \to H$, such that

$$(Kf,z)_H = \int_0^{2\pi} \langle f(t), z(t)\rangle dt, \qquad \frac{1}{p} + \frac{1}{p'} = 1.$$

Claim. $K = (I+|A|)^{-1}$, the rest follows from (1).

(3) Let $H^* = P^*H$, where $*$ is $+$, $-$, or 0. There exists a bounded linear operator $\widetilde{A} \in \mathcal{L}(H,H)$ satisfying

$$(\widetilde{A}z,w)_H = \left((A^+)^{1/2}z, (A^+)^{1/2}w\right)_{L^2} - \left((A^-)^{1/2}z, (A^-)^{1/2}w\right).$$

Moreover, $\widetilde{A}\big|_{H^\pm}$ has bounded inverse with $\ker(\widetilde{A}) = H^0$.

Claim. $\widetilde{A} = A(I+|A|)^{-1}$.

For a given polynomial growth Hamiltonian function $H(t,z)$, the functional

$$g(z) = -\int_0^{2\pi} H(t,z(t))\,dt$$

is well-defined and is in $C^1(H,\mathbb{R}^1)$.

Then the Hamiltonian system (0.2) is rewritten as

$$\widetilde{A}z + dg(z) = \theta. \tag{0.3}$$

This is the problem we studied in Chapter I, Section 5 and Chapter II except now we assume $\dim H^- = +\infty$. We look for critical points of the functional on H:

$$J(z) = \frac{1}{2}(\widetilde{A}z,z) + g(z), \tag{0.4}$$

where $dg = K \circ d_z H(t, \cdot) \circ i : H \to H$ is compact, with

$$H \xrightarrow{i} L^p \xrightarrow{d_z H(t,\cdot)} L^2 \xrightarrow{K} H.$$

1. Asymptotically Linear Systems

The Morse index plays an important role in distinguishing critical points. For functionals defined by Hamiltonian systems (0.4),

$$d^2 J(z)(w, w) = (\tilde{A}w, w)_H - \int_0^{2\pi} d_z^2 H(t, z) w^2(t) dt.$$

The Morse index $\text{ind}(J, z)$ depends on $\widetilde{A}$ or A and the matrix $d_z^2 H\,(t, z(t))$, but since $\dim H^- = \infty$, it is very possible that $\text{ind}(J, z) = \infty$. So, it is necessary to introduce some replacement of the Morse index.

First of all, let us study the meaning of nondegeneracy.

Definition 1.1. Let $\text{Sym}(2n, \mathbb{R})$ be the set of all symmetric $2n \times 2n$ matrices. Let $B \in C\,([0, 2\pi], \text{Sym}(2n, \mathbb{R}))$, with $B(0) = B(2\pi)$. We say that λ is a Floquet multiplier of

$$-J\dot{w}(t) = B(t)w(t) \tag{1.1}$$

if λ is an eigenvalue of $\overline{W}(2\pi)$, where $\overline{W}(t)$ is the fundamental solution matrix of (1.1).

By definition, z is nondegenerate for $J(z)$ (see (0.4)) if and only if the linear system

$$-J\dot{w}(t) = d_z^2 H\,(t, z(t))\, w(t) \tag{1.2}$$

has no nontrivial 2π-periodic solution, or equivalently, 1 is not a Floquet multiplier of (1.2).

In order to classify this continuous loop of symmetric matrices without a Floquet multiplier 1, we study the associated fundamental solution matrices $\{\overline{W}(t)\}$. Noticing that, if $B(t)$: $[0, 2\pi] \to \text{Sym}(2n, \mathbb{R})$ is continuous, then

$$\begin{aligned} \frac{d}{dt}\left(\overline{W}^T(t) J \overline{W}(t)\right) &= \dot{\overline{W}}^T J \overline{W} + \overline{W}^T J \dot{\overline{W}} \\ &= \overline{W}^T B \overline{W} - \overline{W}^T B \overline{W} = 0, \end{aligned}$$

and from $\overline{W}^T(0) J \overline{W}(0) = J$, it follows that $\overline{W}^T J W = J$. Conversely, if $\overline{W}^T(t) J \overline{W}(t) = J$, then $\overline{W}(t)$ is invertible.

Let $B = -J\dot{\overline{W}}\overline{W}^{-1}$; then $B^T = (\overline{W}^{-1})^T \dot{\overline{W}}^T J = (\overline{W}^{-1})^T J \dot{\overline{W}}^{-1} = J\overline{W}\dot{\overline{W}}^{-1} = B$. Therefore, $B \in C\,([0, 2\pi], \text{Sym}(2n, \mathbb{R}))$.

Recall that a matrix M is called symplectic if $M^T JM = J$. We denote by $\mathrm{Sp}(n, \mathbb{R})$ the set of all real $2n \times 2n$ symplectic matrices.

Let $P = \{\gamma \in C([0, 2\pi], \mathrm{Sp}(n, \mathbb{R})) \mid \gamma(0) = id,\ \gamma(2\pi)\text{has no eigenvalue } 1\}$.

Definition 1.2. Let $\gamma_0, \gamma_1 \in P$. We say that γ_0 is equivalent to γ_1, denoted by $\gamma_0 \sim \gamma_1$, if $\exists\, \delta \in C([0,1] \times [0, 2\pi], \mathrm{Sp}(n, \mathbb{R}))$ such that $\delta(i, t) = \gamma_i(t)$, $i = 0, 1$, $\delta(s, 0) = id$, and $1 \notin \sigma(\delta(s, 2\pi))\ \forall\, s \in [0, 1]$.

The following classifying theorem is applied [CoZ2], [LoZ1].

Theorem 1.1 (Conley–Zehnder–Long). *Every path $Y \in P$ belongs to one and only one equivalent class P_k, $k \in \mathbb{Z}$.*

Moreover, $\forall\, k \in \mathbb{Z}$, one finds a distinguished "standard path" $\beta_k(t)$ as follows.

In the case $n \geq 2$, let

$$\begin{aligned} X_k &= diag\,(0, (k - n + 2)/2, 1/2, \dots, 1/2)\,, \\ Y &= diag\,(\ln 2, 0, 0, \dots, 0)\,, \\ Z_k &= diag\,((k - n + 1)/2, 1/2, 1/2, \dots, 1/2)\,, \end{aligned}$$

be $n \times n$ matrices, and $\beta_k(t) = \exp(tJB_k)$, $0 \leq t \leq 2\pi$, where

$$B_k = \begin{cases} \begin{pmatrix} X_k & Y \\ Y & X_k \end{pmatrix} & \text{if } (-1)^{n+k} = -1, \\ \begin{pmatrix} Z_k & 0 \\ 0 & Z_k \end{pmatrix} & \text{if } (-1)^{n+k} = 1. \end{cases}$$

In the case $n = 1$, set $\omega(t) = \frac{1}{2}(1 + \cos t)$,

$$\omega_1(t) = \begin{cases} \omega(t), & 0 \leq t \leq \pi \\ 0, & \pi \leq t \leq 2\pi \end{cases}, \text{ and } \omega_2(t) = \begin{cases} 0, & 0 \leq t \leq \pi \\ \omega(t), & \pi \leq t \leq 2\pi; \end{cases}$$

and let

$$K = \begin{pmatrix} 0 & 1 \\ 1 & 0 \end{pmatrix}, \ R(\theta) = \begin{pmatrix} \cos\theta & -\sin\theta \\ \sin\theta & \cos\theta \end{pmatrix}, \ H(a) = \begin{pmatrix} a^{-1} & 0 \\ 0 & a \end{pmatrix}, \ a \neq 0.$$

If k is odd, then $\beta_k(t) = R(tk/2)$, $0 \leq t \leq 2\pi$, and $B_k = k/2$. If $k = 0$, then $\beta_0(t) = H\left(2^{t/2\pi}\right)$, $0 \leq t \leq 2\pi$, and $B_k = \frac{1}{2\pi}(\ln 2)K$; if k even, but not 0, then $\beta_k(t) = H\left(2^{\omega_2(t)}\right) R(-\omega_1(t)k\pi)$, $0 \leq t \leq 2\pi$, and

$$B_k(t) = -\frac{k}{2}\dot{\omega}_1(t)\mathrm{id} + \frac{\ln 2}{2\pi}\dot{\omega}_2(t)K.$$

Definition 1.3. We define a map $j \colon P \to \mathbb{Z}$ as follows: If $\gamma \in P_k$, then $j(\gamma) = k$. The number k is called the Maslov index of γ and is denoted by $j(\gamma)$ or $j(B)$, if $\gamma(t)$ is the fundamental solution matrix of (1.1).

In the following, we shall show that the Maslov index of $\gamma = \overline{W}(t)$, which is associated with $d_z^2 H\,(t, z(t))$, is a candidate for the replacement of the Morse index.

Let us denote $E_m = \bigoplus_{|j| \leq m} M(j)$, $m = 0, 1, 2, \dots$, and let P_m be the orthogonal projection onto E_m, $m = 0, 1, 2, \dots$, in which $\dim E_m = 2n(2m + 1)$.

Theorem 1.2. *Assume that $B \in C([0,2\pi], \mathrm{Sym}(2n,\mathbb{R}))$ has no Floquet multiplier 1 and $k = j(B)$. Let $Q_m = A - P_m B P_m \in \mathcal{L}(E_m\ E_m)$, and m^*, $* = -, 0$ be the Morse index and the nullity of Q_m. Then $\exists\, N > 0$ such that*

$$m^- = \frac{1}{2}\dim E_m + k \text{ and } m^0 = 0, \quad \forall\, m \geq N. \tag{1.3}$$

Proof. We have seen the one-to-one correspondence between the set of loops in $\mathrm{Sym}(2n,\mathbb{R})$ without the Floquet multiplier 1 and $P\colon B(t) = -J\dot{\gamma}(t)\gamma(t)^{-1}$. In order to transfer the deformation on γ to B, we claim that the above equivalent classification may be restricted to a smaller class

$$\begin{aligned} P^* = \{\gamma \in C^1([0,2\pi], \mathrm{Sp}(n,\mathbb{R})) \mid \dot{\gamma}(2\pi) \\ = \dot{\gamma}(0)\gamma(2\pi), \gamma(0) = \mathrm{id},\ 1 \notin \sigma(\gamma(2\pi))\}. \end{aligned}$$

Claim. Given $\gamma_0, \gamma_1 \in P^*$, $\gamma_0 \sim \gamma_1$ in P, i.e. $\exists\ \delta\colon [0,1] \times [0,2\pi] \to \mathrm{Sp}(n,\mathbb{R})$, continuous, such that $\delta(i,\tau) = \gamma_i(t)$, $i = 0,1$, $\delta(s,0) = \mathrm{id}$ and $1 \notin \sigma(\delta(s,t))\ \forall\, s \in [0,1]$. By reparametrization of s, we may assume that $\delta(s,t) = \gamma_0(t)$ for $0 \leq s \leq \varepsilon$, and $\delta(s,t) = \gamma_1(t)$ for $1-\varepsilon \leq s \leq 1$. Moreover, by reparametrizing in the t variable along $s = 0$ and $s = 1$, we can achieve that $\delta(s,t)$ is differentiable in t in the neighborhood of the boundary of $[0,1] \times [0,2\pi]$ and satisfies, in addition, the condition $\dot{\gamma}(2\pi) = \dot{\gamma}(0) \cdot \gamma(2\pi)$. It remains to smooth out ε in the interior. This is readily done by smoothing the generating functions of the symplectic matrices in the parameter t.

Then we obtain a continuous family $B_s(t)$, $s \in [0,1]$, such that $B_0(t) = B(t)$, $B_1(t)$ is the standard path in the above theorem, and $A - B_s(t)$ is invertible in $\mathcal{L}\left(L^2([0,2\pi], \mathbb{R}^{2n})\right)$ (or equivalently, $\widetilde{A} - KB_s(t)$ is invertible in $\mathcal{L}(H)$) $\forall\, s \in [0,1]$. Therefore (by small perturbation), $\exists\, \varepsilon > 0$, $\exists\, N_1 > 0$ such that, for $m > N_1$,

$$\|(AP_m - P_m B_s P_m)x\| \geq \varepsilon \|P_m x\| \quad \forall\, s \in [0,1],\ \forall\, x \in E_m.$$

This enables us to figure out the Morse index and nullity of $B(t)$ by the counterparts of $B_1(t)$, the standard path. Obviously, we have $m^0 = 0$, because $Q_m = AP_m - P_m B_1 P_m$ is invertible. As to m^-, we only want to verify the standard paths, which are constant matrices except for $n = 1$ and k even but not 0. It satisfies us to verify the simplest case and leave the other cases as exercises because they are elementary linear algebra, cf. Long–Zehnder [LoZ1]. Indeed, the representation of Q_m on the invariant subspaces $M(j) \oplus M(-j)$ under the basis (real) $\{\cos jte_\ell, \sin jte_\ell \mid \ell = 1,2,\ldots,2n\}$ (and $M(0)$ under the basis $\{e_1, e_2, \ldots, e_{2n}\}$), reads as follows:

$$\begin{pmatrix} -B_1 & -jJ \\ jJ & -B_1 \end{pmatrix} \text{ as } j = 1,2,\ldots,m, \text{ and } \ -B_1 \text{ as } j = 0.$$

For the $4n\times 4n$ matrix $\begin{pmatrix} -B_1 & -jJ \\ jJ & -B_1 \end{pmatrix}$, we split them again according to the standard path. In the simplest case ($n \geq 2$, $n+k$ even), $B_1 = \begin{pmatrix} Z_k & 0 \\ 0 & Z_k \end{pmatrix}$, and $Z_k = \operatorname{diag}\{(k-n+1)/2, 1/2, \dots, 1/2\}$, where $k = j(B_1) = j(B)$. To those invariant subspaces, in which the elements in $\mathbb{Z}_k$ equal to $1/2$, we have

$$\operatorname{ind}\begin{pmatrix} -1/2 & & & -j \\ & -1/2 & j & \\ & j & -1/2 & \\ -j & & & -1/2 \end{pmatrix} = 2 \quad \text{for} \quad j = 1, 2, \dots, m,$$

and

$$\operatorname{ind}\begin{pmatrix} -1/2 & 0 \\ 0 & -1/2 \end{pmatrix} = 2 \quad \text{for} \quad j = 0.$$

There are $(n-1)$ such elements, and the total contribution in the index is $2(m+1)(n-1)$. The remaining element in $\mathbb{Z}_k$ is $(k-n+1)/2$. In the case $k \geq n$,

$$\operatorname{ind}\begin{pmatrix} -(k-n+1)/2 & & & -j \\ & -(k-n+1)/2 & j & \\ & j & -(k-n+1)/2 & \\ -j & & & -(k-n+1)/2 \end{pmatrix}$$

$$= \begin{cases} 2, & \text{if } j > \frac{k-n+1}{2} \\ 4, & \text{if } j < \frac{k-n+1}{2} \end{cases},$$

and

$$\operatorname{ind}\begin{pmatrix} -(k-n+1)/2 & \\ & -(k-n+1)/2 \end{pmatrix} = 2 \quad \text{if} \quad j = 0.$$

The contribution to the index is $2\left(m - \frac{k-n}{2}\right) + 4\left(\frac{k-n}{2}\right) + 2 = 2m + k - n + 2$. Therefore $\operatorname{ind}(Q_m) = 2(m+1)(n-1) + 2m + k - n + 2 = (2m+1)n + k$, i.e.,

$$m^- = \frac{1}{2}\dim E_m + k.$$

Similarly, we verify the case $k < n$.

This finishes the proof.

We turn to study the existence and multiplicity of solutions of asymptotically linear Hamiltonian systems. A Hamiltonian system is called asymptotically linear if

(H_∞) $\exists\, B_\infty \in C([0, 2\pi], \mathrm{Sym}(2n, \mathbb{R}))$ such that $B_\infty(0) = B_\infty(2\pi)$ and

$$\|d_z H(t,z) - B_\infty(t) z\|_{\mathbb{R}^{2n}} = o(\,\|z\|)_{\mathbb{R}^{2n}}$$

uniformly in t, as $\|z\|_{\mathbb{R}^{2n}} \to \infty$.

Lemma 1.1. *Assume that* $H \in C^1\left([0,2\pi] \times \mathbb{R}^{2n}, \mathbb{R}^1\right)$ *satisfies* (H_∞), *where* B_∞ *has no Floquet multiplier* 1. *Then the functional (0.4)* J *satisfies the* $(PS)^*$ *condition with respect to* $\{E_m \mid m = 1, 2, \dots\}$, *and* $J_m = J|_{E_m}$ *satisfies the* (PS) *condition* $\forall\, m$ *large.*

Proof. 1. We claim that $\|dJ_m(z) - Q_m^\infty z\|_H = o(\|z\|_H)$ as $\|z\| \to \infty$, for $z \in E_m$, where $Q_m^\infty = AP_m - P_m B_\infty P_m$. Indeed,

$$\begin{aligned}\|dJ_m(z) - Q_m^\infty z\|_H &= \|P_m\left(d_z H(t,z) - B_\infty z\right)\|_H \\ &\le \|d_z H(t,z) - B_\infty z\|_H = o(\|z\|_H).\end{aligned}$$

2. Since KB_∞ is compact, and P_m strongly converges to the identity, we have $\|K(P_m B_\infty - B_\infty)\| \to 0$ as $n \to \infty$. The operator $K(A - B_\infty) = \widetilde{A} - KB_\infty$ is invertible, so by the Banach inverse operator theorem, $\widetilde{A}P_m - P_m K B_\infty P_m$ has also a bounded inverse. Moreover, there exists a constant C such that

$$\|\left(\widetilde{A}P_m - P_m K B_\infty P_m\right)^{-1}\| \le C \quad \text{for } m \text{ large enough.}$$

Combining the above two conclusions, we obtain from $dJ_m(z_j) \to \theta$ that $\|z_j\|_{E_m}$ is bounded. Thus the (PS) condition for J_m holds.

3. Assume $\|dJ_m(z_m)\|_H \to \theta$, as $m \to \infty$ for $z_m \in E_m$. Then $\|z_m\|_H$ is bounded, provided by the same reason. Noticing that P_0 is of finite rank, we have a subsequence of z_m, which is still written by z_m, such that $Kd_z H(t, z_m) \to u$ and $P_0 z_m \to v$ as $m \to \infty$. Consequently,

$$\widetilde{A}z_m + P_0 z_m = dJ_m(z_m) + P_m K d_z H(t, z_m) + P_0 z_m \to u + v,$$

i.e.,

$$z_m \to (\widetilde{A} + P_0)^{-1}(u + v) \text{ in } H \text{ as } m \to \infty,$$

since $\widetilde{A} + P_0$ is invertible in H. This proves the $(PS)^*$ condition.

Theorem 1.3. *Suppose that* $H \in C^1\left([0,2\pi] \times \mathbb{R}^{2n}, \mathbb{R}^1\right)$ *being* 2π *periodic in* t *satisfies* (H_∞), *where* B_∞ *has no Floquet multiplier* 1. *Then the Hamiltonian system* (0.2) *possesses a* 2π*-periodic solution.*

Further, we assume

(H_0) $\exists\, B_0 \in C\left([0,2\pi], Sym(2n, \mathbb{R})\right)$ *such that* $B_0(0) = B_0(2\pi)$ *and* $\|d_z H(t,z) - B_0(t)z\|_{\mathbb{R}^{2n}} = o(\|z\|_{\mathbb{R}^{2n}})$ *uniformly in* t, *as* $\|z\|_{\mathbb{R}^{2n}} \to 0$.

If B_0 *has no Floquet multiplier* 1, *and if*

$$j(B_0) \ne j(B_\infty), \tag{1.4}$$

then (0.2) *possesses at least a nontrivial* 2π*-periodic solution.*

Proof. 1. First, we use the Galerkin method and study the restriction J_m of J on E_m, $m = 1, 2, \dots$. For large m, we learn from Theorem 1.2

and Lemma 5.1 in Chapter II that $\dim H_q(E_m, (J_m)_{a_m}) = \delta_{m_\infty^- q}$ for some a_m, where $m_\infty^- = \mathrm{ind}(B_\infty)$, so there exists a critical point $z_m \in E_m$ of J_m. According to (H_0), H is C^2 at $z = \theta$ on $\mathbb{R}^{2n}$; therefore, the functional $\int_0^{2\pi} H(t, z(t))dt$ is C^2 at $z = \theta$ under the topology $C([0, 2\pi], \mathbb{R}^n)$, and then it is C^2 at $z = \theta$ on each finite dimensional subspace E_m. One concludes $z_m \neq \theta$ by the fact that θ is a nondegenerate critical point of J_m with Morse index $m_0^- = n(2m+1) + j(B_0) \neq m_\infty^-$ (cf. Theorem 5.2 from Chapter II), due to (1.4).

We want to give a lower bound for $\|z_m\|$, in order to distinguish the limit, if it exists, from θ.

Since both $\widetilde{A} - KB_\infty$ and $\widetilde{A} - KB_0$ have bounded inverse, we have $R > 0$ and $N > 0$ such that

$$\left.\begin{aligned} \| \left(\widetilde{A}P_m - KP_mB_\infty P_m\right)^{-1} \|_{\mathcal{L}(H)} < R, \\ \| \left(\widetilde{A}P_m - KP_mB_0 P_m\right)^{-1} \|_{\mathcal{L}(H)} < R, \end{aligned}\right\} \quad \text{for } m \geq N.$$

Take $\delta > 0$ so small that

$$\begin{aligned} \|dJ(z) - (\widetilde{A} - KB_0)z\|_H &= \|K(d_zH(t,z) - B_0 z)\|_H \\ &< \frac{1}{2R}\|z\|_H \quad \forall\, \|z\|_H < \delta. \end{aligned}$$

Then, for large m, we have

$$\begin{aligned} \|dJ(z) - (\widetilde{A}P_m - KP_mB_0P_m)z\|_H &\leq \frac{1}{2R}\|z\|_H \\ &\leq \frac{1}{2}\|(\widetilde{A}P_m - KP_mB_0P_m)^{-1}\|^{-1}_{\mathcal{L}(H)}\|z\|_H \quad \forall\, \|z\|_H < \delta. \end{aligned}$$

But if we write $Q_m^0 = \widetilde{A}P_m - KP_mB_0P_m$, then

$$\begin{aligned} \|dJ(z)\| &\geq \|Q_m^0 z\|_H - \frac{1}{2}\|(Q_m^0)^{-1}\|^{-1}_{\mathcal{L}(H)}\|z\|_H \\ &\geq \|(Q_m^0)^{-1}\|^{-1}_{\mathcal{L}(H)}\|z\|_H - \frac{1}{2}\|(Q_m^0)^{-1}\|^{-1}_{\mathcal{L}(H)}\|z\|_H \\ &= \frac{1}{2}\|(Q_m^0)^{-1}\|^{-1}_{\mathcal{L}(H)} \cdot \|z\|. \end{aligned}$$

Consequently, if z_m is a critical point of J_m with $\|z_m\|_H < \delta$, then $z_m = \theta$. Thus $\|z_m\|_H \geq \delta$ for m large enough.

2. Applying the $(PS)^*$ condition, there exists a limit point z^* of $\{z_m\}$. Therefore, z^* is a critical point of J, with $\|z^*\| \geq \delta$. This is the nontrivial solution.

The proof is finished.

If one wishes to extend the above result to the degenerate case the Maslov index should be extended. In this respect we refer the reader to Y. Long [Lon], Li Liu [LiL1] and Ding Liu [DiLi].

2. Reductions and Periodic Nonlinearities

We have seen that for indefinite functionals, the Morse indices of critical points could be infinite (e.g., the functionals arising from the Hamiltonian systems), the Galerkin method plays an important role. Nevertheless, there is a kind of Lyapunov–Schmidt procedure, called the saddle point reduction, which reduces the infinite dimensional problems to finite dimensional ones. The later method has the advantage of simplicity. We shall introduce this method in Section 2.1, and apply it to the study of (0.1) in Section 2.2.

We shall also investigate a class of Hamiltonian systems in which the Hamiltonian functions are periodic in some of their variables. It is interesting to note that it causes multiple periodic solutions. This is Section 2.3.

2.1. Saddle Point Reduction

Let H be a real Hilbert space, and let A be a self-adjoint operator with domain $D(A) \subset H$. Let F be a potential operator with $\Phi \in C^1(H, \mathbb{R}^1)$, $F = d\Phi$, $\Phi(\theta) = 0$. Assume that

(A) There exist real numbers $\alpha < \beta$ such that $\alpha, \beta \notin \sigma(A)$, and that $\sigma(A) \cap [\alpha, \beta]$ consists of at most finitely many eigenvalues of finite multiplicities.

(F) F is Gateaux differentiable in H, which satisfies

$$\left\| dF(u) - \frac{\alpha+\beta}{2} I \right\| \le \frac{\beta-\alpha}{2} \quad \forall\, u \in H.$$

The problem is to find the solutions of the following equation:

$$Ax = F(x) \quad x \in D(A). \tag{2.1}$$

With no loss of generality, we may assume $\alpha = -\beta$, $\beta > 0$.

A Lyapunov–Schmidt procedure is applied for a finite dimensional reduction.

Let

$$P_0 = \int_{-\beta}^{\beta} dE_\lambda, \quad P_+ = \int_{\beta}^{+\infty} dE_\lambda, \quad P_- = \int_{-\infty}^{-\beta} dE_\lambda,$$

where $\{E_\lambda\}$ is the spectral resolution of A, and let

$$H_0 = P_0 H, \quad H_\pm = P_\pm H.$$

According to (A), there exists $\varepsilon > 0$ small, such that $-\varepsilon \notin \sigma(A)$. We assume further the following condition:

(D) $\Phi \in C^2(V, \mathbb{R}^1)$, where $V = D(|A|^{1/2})$, with the graph norm

$$\|x\|_V = \left(\| \, |A|^{1/2} x\|_H^2 + \varepsilon^2 \|x\|_H^2\right)^{1/2}.$$

We decompose the space V as follows:

$$V = V_0 \oplus V_- \oplus V_+,$$

where

$$V_0 = |A_\varepsilon|^{-1/2} H_0, \quad V_\pm = |A_\varepsilon|^{-1/2} H_\pm, \text{ and } A_\varepsilon = \varepsilon I + A.$$

For each $u \in H$, we have the decomposition

$$u = u_+ + u_0 + u_-,$$

where $u_0 \in H_0$, $u_\pm \in H_\pm$; let $x = x_+ + x_0 + x_- \in V$, where

$$x_0 = |A_\varepsilon|^{-1/2} u_0, \quad x_\pm = |A_\varepsilon|^{-1/2} u_\pm.$$

Thus we have

$$\|x_\pm\|_{V_\pm} = \|u_\pm\|_{H_\pm}, \quad \|x_0\|_{V_0} = \|u_0\|_{H_0},$$

and that $V_\pm, V_0$ are isomorphic to $H_\pm$ and H_0, respectively.

Now we define a functional on H as follows:

$$f(u) = \frac{1}{2}\left(\|u_+\|^2 + \|E_+ u_0\|^2 - \|E_- u_0\|^2 - \|u_-\|^2\right) - \Phi_\varepsilon(x),$$

where $E_+ = \int_0^\infty dE_\lambda$, $E_- = \int_{-\infty}^0 dE_\lambda$, and $\Phi_\varepsilon(x) = \frac{\varepsilon}{2}\|x\|_H^2 + \Phi(x)$. Let $F_\varepsilon = \varepsilon I + F$.

The Euler equation of this functional is the system

$$u_\pm = \pm |A_\varepsilon|^{-1/2} P_\pm F_\varepsilon(x), \tag{2.2}$$

$$E_\pm u_0 = \pm |A_\varepsilon|^{-1/2} E_\pm P_0 F_\varepsilon(x). \tag{2.3}$$

Thus $x = x_+ + x_0 + x_-$ is a solution of (2.1) if and only if $u = u_+ + u_0 + u_-$ is a critical point of f. However, the system (2.2) is reduced to

$$A_\varepsilon x_\pm = P_\pm F_\varepsilon(x_+ + x_- + x_0),$$

which is equivalent to

$$x_\pm = A_\varepsilon^{-1} P_\pm F_\varepsilon(x_+ + x_- + x_0). \tag{2.4}$$

By assumption (D), $F \in C^1(V, V)$, and by assumption (F)

$$\|F_\varepsilon(u) - F_\varepsilon(v)\|_H \le (\varepsilon + \beta)\|u - v\|_H \quad \forall\, u, v \in H.$$

Furthermore, there is a $\gamma > \beta + \varepsilon$ such that

$$\|A_\varepsilon^{-1}|_{H_+ \oplus H_-}\| \le \frac{1}{\gamma}$$

by assumption (A). We shall prove that the operator $\mathcal{F} = A_\varepsilon^{-1}(P_+ + P_-)F_\varepsilon \in C^1(V, V)$ is contractible with respect to variables in $V_+ \oplus V_-$. In fact, $\forall\, x = x_+ + x_- + z$, $y = y_+ + y_- + z$, for fixed $z \in V_0$,

$$\begin{aligned}\|\mathcal{F}(x) - \mathcal{F}(y)\|_V &= \|\, |A_\varepsilon|^{-1/2}(P_+ + P_-)(F_\varepsilon(x) - F_\varepsilon(y))\|_H \\ &\le \|\, |A_\varepsilon|^{-1/2}(P_+ - P_-)\|_{\mathcal{B}(H)}\|F_\varepsilon(x) - F_\varepsilon(y)\|_H \\ &\le (\varepsilon + \beta)\|\, |A_\varepsilon|^{-1/2}(P_+ + P_-)\|_{\mathcal{B}(H)}\|(x_+ + x_-) \\ &\qquad - (y_+ + y_-)\|_H\,.\end{aligned}$$

Since

$$\|x_\pm\|_H \;\|\, |A_\varepsilon|^{-1/2}u_\pm\|_H \le \frac{1}{\sqrt{\gamma}}\|u_\pm\|_H = \frac{1}{\sqrt{\gamma}}\|x_\pm\|_V,$$

and

$$\|\, |A_\varepsilon|^{-1/2}(P_+ + P_-)\|_{\mathcal{B}(H)} \le \frac{1}{\sqrt{\gamma}},$$

we obtain

$$\|\mathcal{F}(x) - \mathcal{F}(y)\|_V \le \frac{\varepsilon + \beta}{\gamma}\|x - y\|_V.$$

The implicit function theorem can be applied, yielding a solution $x_\pm(x_0)$, for fixed $x_0 \in V_0$, such that $x_\pm \in C^1(V_0, V_\pm)$. Since $\dim V_0$ is finite, all topologies on V_0 are equivalent. We have

$$u_\pm(x_0) = |A_\varepsilon|^{1/2}x_\pm(x_0) \in C^1(H_0, H),$$

which solves the system (2.2).

Let

$$a(x_0) = f\left(u_+(x_0) + u_-(x_0) + u_0(x_0)\right),$$

where $u_0(x_0) = |A_\varepsilon|^{1/2}x_0$ and let $z = x_0$. We have

$$\begin{aligned}a(z) &= \frac{1}{2}\left(\|A_\varepsilon^{1/2}x_+\|^2 + \|A_\varepsilon^{1/2}E_+z\|^2 - \|(-A_\varepsilon)^{1/2}x_-\|^2 - \|(-A_\varepsilon)^{1/2}E_-z\|^2\right) \\ &\qquad - \Phi_\varepsilon(x) \\ &= \frac{1}{2}\left(Ax(z), x(z)\right) - \Phi(x(z)),\end{aligned}$$

where $x(z) = \xi(z) + z$, $\xi(z) = x_+(z) + x_-(z) \in D(A)$. Noticing that

$$d\xi(z) = A_\varepsilon^{-1}(P_+ + P_-)F'_\varepsilon(x(z))dx(z)$$

by (2.4), one sees that $d\xi(z) \in D(A)$ and that

$$Ad\xi(z) = (I - P_0)F'(x(z))dx(z).$$

Thus

$$\begin{aligned} da(z) &= (dx(z))^*[Ax(z) - F(x(z))] = Az - P_0F(x(z)) \\ &= Ax(z) - F(x(z)) \end{aligned} \tag{2.5 }$$

and

$$\begin{aligned} d^2a(z) &= [A - F'(x(z))]dx(z) \\ &= A|_{H_0} - P_0F'(x(z))dx(z). \end{aligned}$$

In summary, we have proven

Theorem 2.1. *Under the assumptions* (A), (F) *and* (D), *there is a one-one correspondence:*

$$z \mapsto x = x(z) = x_+(z) + x_-(z) + z,$$

between the critical points of the C^2-function $a \in C^2(H_0, \mathbb{R}^1)$ with the solutions of the operator equation

$$Ax = F(x) \quad x \in D(A).$$

Now we turn to the asymptotic behavior of the function a.

Lemma 2.1. *Under the assumptions* (A), (F) *and* (D), *we assume further that there is a bounded self-adjoint operator F_∞ satisfying*

$$(\mathrm{F}_\infty)\begin{cases} \text{(i)} & P_0F_\infty = F_\infty P_0; \\ \text{(ii)} & \|F(u) - F_\infty u\| = o(\|u\|) \ \text{ as } \ \|u\| \to \infty; \\ \text{(iii)} & 0 \notin \sigma(A - F_\infty). \end{cases}$$

Then we have that

(1) $\xi(z) = o(\|z\|)$ *as* $\|z\| \to \infty$, *and*
(2) *the function $a(z)$ is asymptotically quadratic with asymptotics $A - F_\infty|_{H_0}$, i.e.,*

$$\|da(z) - (A - F_\infty)z\| = o(\|z\|) \ \text{ as } \ \|z\| \to \infty.$$

Proof. By (2.4), we obtain

$$A\xi(z) = (I - P_0)F(x(z)).$$

Since P_0 commutes with F_∞, we have

$$(A - F_\infty)\xi(z) = (I - P_0)[F(x(z)) - F_\infty x(z)].$$

Hence, $\forall\, \varepsilon > 0$ there exists $R > 0$, such that

$$\begin{aligned}\|\xi(z)\| &\le \|(A - F_\infty)^{-1}\|\,\|F(x(z)) - F_\infty x(z)\|\\ &< \varepsilon C\left(\|z\| + \|\xi(z)\|\right), \text{ if } \|z\| > R,\end{aligned}$$

where $C = \|(A - F_\infty)^{-1}\|$; it follows that

$$\|\xi(z)\| = o(\|z\|).$$

By (2.5) we have

$$\begin{aligned}\|da(z) - (A - F_\infty)z\| &= \|Az - P_0F(x(z)) - (A - F_\infty)z\|\\ &\le \|F(x(z)) - F_\infty x(z)\| + \|F_\infty x(z) - F_\infty z\|\\ &= o(\|x(z)\|) = o(\|z\|) \text{ as } \|z\| \to \infty.\end{aligned}$$

Lemma 2.2. *Under the assumptions* (A), (F) *and* (D), *we assume that* $F(\theta) = \theta$.

(1) *If there is a self-adjoint operator* $C_0^- \in \mathcal{L}(H, H)$ *which commutes with* P_0 *and* P_-, *such that*

$$\min(\sigma(A) \cap [\alpha, \beta])I \le C_0^- \le F'(\theta),$$

then

$$a(z) \le \frac{1}{2}\left((A - C_0^-)z, z\right) + o(\|z\|^2) \text{ as } \|z\| \to 0.$$

(2) *If there is a self-adjoint operator* $C_0^+ \in \mathcal{L}(H, H)$ *which commutes with* P_0 *and* P_+, *such that*

$$F'(\theta) \le C_0^+ \le \max\left(\sigma(A) \cap [\alpha, \beta]\right) I,$$

then

$$a(z) \ge \frac{1}{2}\left((A - C_0^+)z, z\right) + o(\|z\|^2) \text{ as } \|z\| \to 0.$$

Proof. By the definition and (2.5),

$$\begin{aligned}a(z) &= \frac{1}{2}\left(Ax(z), x(z)\right) - \Phi(x(z))\\ &= \frac{1}{2}(Aq, q) - \Phi(q) + \frac{1}{2}\left(Ax_+(z), x_+(z)\right) - \left(\Phi(x(z)) - \Phi(q)\right),\end{aligned}$$

where $q = x_-(z) + z$. We shall prove that

$$\frac{1}{2}(Ax_+(z), x_+(z)) - (\Phi(x(z)) - \Phi(q)) \leq 0,$$

that is,

$$a(z) \leq \frac{1}{2}(Aq, q) - \Phi(q). \tag{2.6}$$

In fact,

$$\begin{aligned}
&\Phi(x(z)) - \Phi(q) - \frac{1}{2}(Ax_+(z), x_+(z)) \\
&\quad = \frac{1}{2}(Ax_+(z), x_+(z)) + \int_0^1 (F(tx_+(z) + q) - F(x_+(z) + q), x_+(z))\, dt \\
&\quad \geq \frac{\beta}{2}\|x_+(z)\|^2 - \int_0^1 \beta(1-t)dt\|x_+(z)\|^2 = 0.
\end{aligned}$$

However,

$$\begin{aligned}
\left|\Phi(q) - \frac{1}{2}(F'(\theta)q, q)\right| &= \left|\int_0^1 (F(tq) - F'(\theta)tq, q)dt\right| \\
&\leq \frac{1}{2}\sup_{0<t<1}\|F'(tq) - F'(\theta)\|_{\mathcal{L}(V,V)}\|q\|_V^2,
\end{aligned}$$

that is,

$$-\Phi(q) \leq -\frac{1}{2}(F'(\theta)q, q) + o(\|q\|_V^2). \tag{2.7}$$

Note that $x_- \in C^1(V_0, V_-)$; this implies that if $\|z\| \to 0$, then $\|x_-(z)\|_{V_-} = O(\|z\|_{V_0})$ because $x_-(\theta) = \theta$. Thus

$$\|q\|_V = O(\|z\|). \tag{2.8}$$

Substituting (2.7) and (2.8) into (2.6), we obtain

$$\begin{aligned}
a(z) &\leq \frac{1}{2}\left((A - F'(\theta))q, q\right) + o(\|z\|^2) \\
&\leq \frac{1}{2}\left((A - C_0^-)q, q\right) + o(\|z\|^2) \\
&= \frac{1}{2}\left((A - C_0^-)x_-(z), x_-(z)\right) + \frac{1}{2}\left((A - C_0^-)z, z\right) + o(\|z\|^2)
\end{aligned}$$

as $\|z\| \to 0$. Let

$$\alpha_- = \min\{\sigma(A) \cap [\alpha, \beta]\},$$

and, by the assumption,

$$\alpha_- I \leq C_0^-.$$

This implies

$$((A - C_0^-)x_-(z), x_-(z)) \le ((A - \alpha_- I)x_-(z), x_-(z)) \le 0;$$

therefore

$$a(z) \le \frac{1}{2}\left((A - C_0^-)z, z\right) + o(\|z\|^2) \text{ as } z \to \theta.$$

We prove the second assumption in a similar way .

Finally, we apply Theorem 5.2 of Chapter II to solve the operator equation (2.1) under conditions (A), (F), (D) and (F_∞).

For a symmetric matrix B, let $m^\pm(B)$ be the dimension of the maximal positive/negative subspace.

Theorem 2.2. *Under assumptions* (A), (F), (D) *and* (F_∞), *we assume* $F(\theta) = \theta$. *If one of the following conditions holds:*

(1) *There exists a bounded self-adjoint* C_0^-, *commuting with* P_0 *and* P_-, *such that*

$$\min\{\sigma(A) \cap [\alpha, \beta]\}I \le C_0^- \le F'(\theta)$$

and

$$m^-\left(A - C_0^-|_{H_0}\right) > m^-\left(A - F_\infty|_{H_0}\right);$$

(2) *There exists a bounded self-adjoint* C_0^+, *commuting with* P_0 *and* P_+, *such that*

$$F'(\theta) \le C_0^+ \le \max\{\sigma(A) \cap [\alpha, \beta]\}I$$

and

$$m^+\left(A - C_0^+|_{H_0}\right) > m^+\left(A - F_\infty|_{H_0}\right);$$

then there exists at least one nontrivial solution of the equation (2.1).

Proof. By Theorem 2.1, problem (2.1) is reduced to finding critical points of the function $a \in C^2(H_0, \mathbb{R}^1)$. According to Lemma 2.1, a is an asymptotically quadratic function with a nonsingular symmetric matrix $A - F_\infty|_{H_0}$ as asymptotics. By Lemma 2.2, condition (1) means that $d^2a(\theta)$ is negative on the subspace Z_- on which $A - C_0^-$ is negative. Thus

$$m^-(d^2a(\theta)) \ge m^-\left(A - C_0^-|_{H_0}\right) > m^-\left(A - F_\infty|_{H_0}\right).$$

Similarly, condition (2) means that

$$m^+(d^2a(\theta)) \ge m^+\left(A - C_0^+\ |_{H_0}\right) > m^+\left(A - F_\infty|_{H_0}\right).$$

In this case,

$$\begin{aligned} m^-\left(A - F_\infty|_{H_0}\right) &= \dim H_0 - m^+\left(A - F_\infty|_{H_0}\right) \\ &> \dim H_0 - m^+(d^2a(\theta)) \\ &= m^-(d^2a(\theta)) + \dim\ker(d^2a(\theta)). \end{aligned}$$

Both cases imply that

$$m^-\left(A - F_\infty|_{H_0}\right) \notin \left[m^-(d^2a(\theta)), m^-(d^2a(\theta)) + \dim\ker(d^2a(\theta))\right].$$

The conclusion follows from Theorem 5.2 of Chapter II.

Remark 2.1. The finite dimensional reduction method presented here is a modification of a method due to Amann [Ama1] and Amann and Zehnder [AmZ1]. Avoiding the use of monotone operators and a dull verification of the implicit function theorem, we change a few of the assumptions and gain a considerable simplification of the reduction theory.

2.2. A Multiple Solution Theorem

We apply the saddle point reduction to Hamiltonian systems. Let $H = L^2\left([0,2\pi], \mathbb{R}^{2n}\right)$, $A = -J\frac{d}{dt}$, with $D(A)$ given in the preliminary. For $H \in C^2(\mathbb{R}^1 \times \mathbb{R}^{2n}, \mathbb{R}^1)$ being 2π-periodic in t, we define

$$F(z) = d_z H(t, z(t)).$$

Suppose that there is a constant $C > 0$ such that

$$\|d_z^2 H(t,z)\|_{\mathcal{L}(\mathbb{R}^{2n})} \le C; \tag{2.9}$$

then

$$\Phi(z) = \int_0^{2\pi} H(t, z(t))dt \in C^1(H, \mathbb{R}^1).$$

Again, the derivative $F(z) = d\Phi(z)$ is Gateaux differentiable with

$$\|dF(z)\|_{\mathcal{L}(H)} \le C \quad \forall\, z \in H,$$

so conditions (A) and (F) are satisfied. By observing the continuous imbeddings

$$D\left(|A|^{1/2}\right) \to H^{1/2}\left([0,2\pi], \mathbb{R}^{2n}\right) \to L^p\left([0,2\pi], \mathbb{R}^{2n}\right), \quad \forall\, p < +\infty,$$

condition (D) is also easily verified.

When we study Hamiltonian systems under condition (2.9), the equation is reduced to

$$da(z) = 0,$$

where

$$a(z) = \frac{1}{2}\left(Ax(z), x(z)\right) - \Phi(x, (z)).$$

Lemma 2.3. *Suppose that x_0 is a nondegenerate 2π-periodic solution of (0.2), i.e., the linearized equation*

$$-J\dot{z} = d^2H(t, x_0(t))z, \quad z(0) = z(2\pi), \tag{2.10}$$

has no Floquet multiplier 1, then the correspondence $z_0 \in H_0$ is a nondegenerate critical point of $a(z)$.

Proof. Since

$$dF(x_0)z = d_z^2 H\left(t, x_0(t)\right) z(t) \quad \forall\, z \in H,$$

$0 \notin \sigma\left(A - dF(x_0)\right)$, because 1 is not a Floquet multiplier of (2.10). And since $\sigma\left(A - dF(x_0)\right)$ consists of eigenvalues, $\left(A - dF(u_0)\right)^{-1}$ exists and is bounded. However,

$$d^2a(z_0) = [A - dF(x_0)]dx(z_0)$$

where

$$dx(z_0) = id_{H_0} + d\xi(z_0),$$

hence $d^2a(z_0)$ must be invertible, i.e., z_0 is nondegenerate.

Now we turn to a result which is concerned with the existence of at least two nontrivial periodic solutions.

Theorem 2.3. *Suppose that $H \in C^2(\mathbb{R}^1 \times \mathbb{R}^{2n}, \mathbb{R}^1)$ satisfies the following conditions:*

(1) *There exist constants $\alpha < \beta$ such that*

$$\alpha I \le d_z^2 H(t, z) \le \beta I \quad \forall\, (t, z) \in \mathbb{R}^1 \times \mathbb{R}^{2n}.$$

(2) *Let $j_0, j_0 + 1, \ldots, j_1$ be all integers within $[\alpha, \beta]$ (without loss of generality, we may assume $\alpha, \beta \notin \mathbb{Z}$). Suppose that there exist γ and C, such that $j_1 < \gamma < \beta$ and*

$$H(t, z) \ge \frac{1}{2}\gamma\|z\|^2 - C \quad \forall\, (t, z) \in \mathbb{R}^1 \times \mathbb{R}^{2n}.$$

(3) *$H_x(t, \theta) = \theta$. $\exists\, j \in [j_0, j_1) \cap \mathbb{Z}$ such that*

$$jI < d_z^2 H(t, \theta) < (j + 1)I \quad \forall\, t \in \mathbb{R}^1.$$

(4) *H is 2π-periodic with respect to t.*

Then the Hamiltonian system (0.2) possesses at least two nontrivial periodic solutions.

Proof. According to the finite dimensional reduction, we turn to the function

$$a(z) = \frac{1}{2}\,(Au(z), u(z)) - \int_0^{2\pi} H(t, u(z))dt,$$

where $u(z) = z + u_+(z) + u_-(z)$, $z \in Z \triangleq H_0$, and $u_\pm(z) \in H_\pm$. Since

$$\begin{aligned} a(z) = \frac{1}{2}(Aw, w) - \int_0^{2\pi} H(t, w)dt \\ + \left\{\frac{1}{2}[(Au(z), u(z)) - (Aw, w)] - \int_0^{2\pi} [H(t, u(z)) - H(t, w)]dt\right\}, \end{aligned}$$

where $w = z + u_-(z)$, and the terms in the bracket are equal to

$$\begin{aligned} &- \int_0^{2\pi} (F(su_+ - w), u_+)\, ds + \frac{1}{2}(Au(z), u_+) \\ &= \int_0^{2\pi}\int_0^{2\pi} (d_z^2 H(t, su_+ + w)u_+, u_+)\, s\, ds\, dt - \frac{1}{2}(Au_+, u_+) \le 0 \end{aligned}$$

by condition (1), we obtain

$$\begin{aligned} a(z) &\le \frac{1}{2}(Aw, w) - \int_0^{2\pi} H(t, w)dt \\ &\le \frac{1}{2}(2\pi j_1 - \gamma)\|w\|^2 + C \to -\infty \quad \text{as} \quad \|z\| \to \infty \end{aligned}$$

using condition (2).

Thus the function $-a(z)$ is bounded from below and satisfies the (PS) condition.

In order to apply the three critical points theorem we claim that θ is neither a minimum nor degenerate. In fact, using condition (3), it follows from Lemma 2.2 that

$$\frac{1}{2}\left((A - \overline{\lambda}I)z, z\right) + o(\|z\|^2) \le a(z) \le \frac{1}{2}\left((A - \widehat{\lambda}I)z, z\right) + o(\|z\|^2),$$

where $(\overline{\lambda}, \widehat{\lambda}) \subset (j, j+1)$, as $\|z\| \to 0$. The theorem is proved.

Remark 2.2. Saddle point reduction was first applied to Hamiltonian systems by Amann Zehnder [AmZ1]. Theorem 2.3 is due to Chang [Cha1].

2.3. Periodic Nonlinearity

A special class of Hamiltonian systems (0.2), in which the function $H(t,q;p)$, $q=(q_1,q_2,\dots,q_n)$, $p=(p_1,p_2,\dots,p_n)$, is periodic in some of its variables, say $q_1,\dots,q_s,p_1,\dots,p_r,p_{s+1},\dots$, $1\le r\le s\le T$, provides a possible way of gaining more solutions. This is due to the fact that the Hamiltonian function H, and then the functional J, is invariant under certain translation groups; therefore, the quotient space contains certain tori.

We start with an abstract theorem:

Theorem 2.4. *Under assumptions* (A), (F) *and* (D), *we assume that*

(P) $\exists\, e_1,\dots,e_r\in\ker A$, *they are linearly independent, and* $\exists\,(T_1,\dots,T_r)\in R^1$, *such that*

$$\Phi\left(x+\sum_{j=1}^{r}m_jT_je_j\right)=\Phi(x),\quad \forall\,(m_1,\dots,m_r)\in\mathbb{Z}^r,\quad \forall\,x\in H.$$

(LL) $\Phi(x)\to\pm\infty$ *if* $|x|\to\infty$ $\forall\,x\in\ker A\cap\operatorname{span}\{e_1,\dots,e_r\}^{\perp}$.

Then the equation

$$Ax-\Phi'(x)=0$$

has at least $r+1$ *distinct solutions; and if all solutions are nondegenerate, then there are at least* 2^r *distinct solutions.*

Proof. A saddle point reduction procedure is applied.

Consider the function on the finite dimensional space V_0 defined below:

$$a(z)=\frac{1}{2}\,(Ax(z),x(z))-\Phi(x(z)).$$

We shall prove that

1. $x_\pm\left(z+\sum_{j=1}^{r}T_je_j\right)=x_\pm(z),\quad \forall\,z\in V_0.$

In fact,

$$P_\pm F_\varepsilon\left(x_++x_-+z+\sum_{j=1}^{r}T_je_j\right)=P_\pm F_\varepsilon(x_++x_-+z),$$

and therefore

$$x_\pm(z)=x_\pm\left(z+\sum_{j=1}^{r}T_je_j\right).$$

2. $a\left(z+\sum_{j=1}^{r}T_je_j\right)=a(z).$

Claim.

$$\begin{aligned}
a\left(z+\sum_{j=1}^{r} T_j e_j\right) &= \frac{1}{2}\left(Ax\left(z+\sum_{j=1}^{r} T_j e_j\right), x\left(z+\sum_{j=1}^{r} T_j e_j\right)\right) \\
&\quad - \Phi\left(x\left(z+\sum_{j=1}^{r} T_j e_j\right)\right) \\
&= \frac{1}{2}\left(Ax(z), x(z)+\sum_{j=1}^{r} T_j e_j\right) - \Phi\left(x(z)+\sum_{j=1}^{r} T_j e_j\right) \\
&= \frac{1}{2}\left(Ax(z), x(z)\right) - \Phi(x(z)) \\
&= a(z).
\end{aligned}$$

3. a satisfies the (PS) condition on $T^r \times (Y \oplus N(A)^\perp) \cap V_0$ where $Y = N(A) \cap \operatorname{span}\{e_1, \ldots, e_r\}^\perp$, and T^r is the r-torus defined by

$$\mathbb{R}^r/(T_1 Z^1 \times \cdots \times T_r Z^1).$$

Claim. Suppose that $\{z^k\}$ is a sequence along which $\{a(z^k)\}$ is bounded, and $\|da(z^k)\| = o(1)$.

According to (2.5),

$$|Ax(z^k) - F(x(z^k))|_H = o(1).$$

Let Q be the orthogonal projection onto Y, which is considered to be a subspace of the Hilbert space $\mathcal{K} = Y \oplus N(A)^\perp$. Thus on the space $\mathcal{K}$,

$$(I-Q)x(z^k) = A^{-1}(I-Q)F(x(z^k)) + o(1),$$

and since F is bounded, $\|(I-Q)x(z^k)\|$ is bounded. Noticing

$$\begin{aligned}
\Phi(Qx(z^k)) &= \Phi(x(z^k)) - \int_0^1 (F(x_t(z^k)), (I-Q)x(z^k))dt \\
&= a(z^k) - \frac{1}{2}\left(Ax(z^k), x(z^k)\right) - \int_0^1 (F(x_t(z^k)), (I-Q)x(z^k))dt,
\end{aligned}$$

where

$$x_t(z) = ((1-t)I + tQ)\, x(z),$$

and

$$\begin{aligned}
\left(Ax(z^k), x(z^k)\right) &= \left(Ax(z^k), (I-Q)x(z^k)\right) \\
&= (F(x(z^k)) + o(1), (I-Q)x(z^k)),
\end{aligned}$$

$\Phi(Qx(z^k))$ must be bounded. According to condition (LL), $Qx(z^k)$ is bounded. The compactness of z^k now follows from the boundedness of $x(z^k)$ and the finiteness of the dimension of V_0.

4. If we decompose V_0 into $\text{span}\{e_1, \ldots, e_r\} \oplus (Y \oplus N(A)^\perp) \cap V_0$,

$$z = v + w, \quad (v, w) \in \text{span}\{e_1, \ldots, e_r\} \oplus (Y \oplus N(A)^\perp) \cap V_0,$$

and let

$$g(w, v) = \frac{1}{2}\left(A\xi(w + v), \xi(w + v)\right) - \Phi(x(w + v)),$$

where

$$\xi(z) = x_+(z) + x_-(z),$$

then g is well-defined on $T^r \times (Y \oplus N(A)^\perp) \cap V_0$, and

$$dg(w, v) = P_0 F(x(w + v)),$$

which is bounded and then is compact on finite dimensional manifold. The function $a(z)$ now is written in the following form:

$$a(w, v) = \frac{1}{2}(Aw, w) - g(w, v).$$

Noticing that F is bounded, $\|\xi(z)\|$ is always bounded. If we denote by y the projection of w onto Y, we have

$$g(y, v) = \frac{1}{2}\left(A\xi(y + v), \xi(y + v)\right) - \Phi(y) - \left[\Phi(\xi(y + v) + y + v) - \Phi(y)\right].$$

The first term and the third term are bounded, therefore

$$g(y, v) \to \pm\infty \quad \text{as } \|y\| \to \infty.$$

The function $a(w, v)$ satisfies all the assumptions of Theorem 5.3 of Chapter II. Theorem 2.4 is proved, provided cuplength $(T^r) = r$, and the sum of the Betti numbers of T^r is 2^r.

Now we study the periodic solutions of the Hamiltonian systems in which the Hamiltonian functions are periodic in some of the variables.

We use the following notations: $p, q \in R^n$,

$$p = (p_1, \ldots, p_n), \quad q = (q_1, \ldots, q_n), \quad 1 \le r \le s \le t \le n,$$

$$\begin{aligned} \overline{p} &= (p_1, \ldots, p_r), & \overline{q} &= (q_1, \ldots, q_r), \\ \widetilde{p} &= (p_{r+1}, \ldots, p_s), & \widetilde{q} &= (q_{r+1}, \ldots, q_s), \\ \widehat{p} &= (p_{s+1}, \ldots, p_T), & \widehat{q} &= (q_{s+1}, \ldots, q_T), \\ \check{p} &+ (p_{T+1}, \ldots, p_n), & \check{q} &= (q_{T+1}, \ldots, q_n). \end{aligned}$$

We make the following assumptions:

(I) $A(t)$, $B(t)$, $C(t)$ and $D(t)$ are symmetric continuous matrix functions on S^1, of order $(s-r)\times(s-r)$, $(T-s)\times(T-s)$, $(n-T)\times(n-T)$ and $(n-T)\times(n-T)$ respectively. Let $\widetilde{A}=\int_{S^1}A(t)$, and $\widehat{B}=\int_{S^1}B(t)$ be invertible.

(II) $\widehat{H}\in C^2(S^1\times R^{2n},R^1)$ is periodic in the following variables $\overline{p}$, $\overline{q}$, $\widehat{p}$, $\widetilde{q}$, and $d^2\widehat{H}$ is bounded.

(III) Let $\operatorname{span}\{\varphi_1,\dots,\varphi_m\}=\ker\left(-\check{J}\frac{d}{dt}-(C(t)\oplus D(t))\right)$ where

$$\check{J}=\begin{pmatrix}0 & -I_{n-T}\\ I_{n-T} & 0\end{pmatrix},$$

and $\varphi_1,\dots,\varphi_m$ are linearly independent, and

$$\widehat{H}\left(t,\sum_{j=1}^{m}\tau_j\varphi_j\right)\to\pm\infty\quad\text{as } |\tau|=(\tau_1^2+\cdots+\tau_m^2)^{1/2}\to\infty.$$

(IV) $c,d\in C(S^1,R^T)$, with $c=(c_1,\dots,c_T)$, $d=(d_1,\dots,d_T)$ and

$$\int_{S^1}c_i(t)=\int_{S^1}d_j(t)=0,$$

$i=1,\dots,r,\ s+1,\dots,T,\ j=1,\dots,s.$

We define a Hamiltonian function as follows:

$$\begin{aligned}H(t,p,q)&=\frac{1}{2}A(t)\widetilde{p}\cdot\widetilde{p}+\frac{1}{2}B(t)\widehat{q}\cdot\widehat{q}+\frac{1}{2}(C(t)\check{p}\cdot\check{p}+D(t)\check{q}\cdot\check{q})\\&\quad+\sum_{i=1}^{T}(c_i(t)p_i+d_i(t)q_i)+\widehat{H}(t,p,q).\end{aligned}$$

Theorem 2.5. *Under conditions* (I)–(IV), *the Hamiltonian system* (0.2)

$$-J\frac{d}{dt}z=H_z(t,z),\quad t\in S^1,$$

has at least $r+T+1$ periodic solutions, and if all solutions have no Floquet multiplier 1, then (0.2) *has at least 2^{r+T} periodic solutions.*

Proof. Let

$$\Lambda(t)=\begin{pmatrix}0 &&&&&&&\\ & A(t) &&&&&&\\ && 0 &&&&&\\ &&& C(t) &&&&\\ &&&& 0 &&&\\ &&&&& 0 &&\\ &&&&&& B(t) &\\ &&&&&&& D(t)\end{pmatrix},$$

and let

$$\begin{aligned}\mathcal{A} &= \left(-J\frac{d}{dt} - \Lambda(t)\right)\\ &= \left(-\overline{J}\frac{d}{dt}\right) \oplus \left(-\widetilde{J}\frac{d}{dt} - \begin{pmatrix} A(t) & \\ & 0\end{pmatrix}\right) \oplus \left(-\widehat{J}\frac{d}{dt} - \begin{pmatrix} 0 & \\ & B(t)\end{pmatrix}\right)\\ &\quad \oplus \left(-\check{J}\frac{d}{dt} - \begin{pmatrix} C(t) & \\ & D(t)\end{pmatrix}\right),\end{aligned}$$

where the subscripts on J coincide with those on p. We have

$$\begin{aligned}(\widetilde{p},\widetilde{q}) &\in \ker\left(-\widetilde{J}\frac{d}{dt} - \begin{pmatrix} A(t) & \\ & 0\end{pmatrix}\right),\\ &\Leftrightarrow \begin{cases}\dot{\widetilde{q}} = A(t)\widetilde{p}\\ \dot{\widetilde{p}} = 0,\end{cases}\\ &\Leftrightarrow \begin{cases}\widetilde{q} = \int_0^t A(s)ds \cdot \widetilde{c} + \widetilde{d}, & \text{with } \widetilde{q}(2\pi) = \widetilde{q}(0),\\ \widetilde{p} = \widetilde{c}\end{cases}\end{aligned}$$

(i.e., with $\widetilde{A}\cdot\widetilde{c} = \theta$). According to assumption I, $\widetilde{c} = \theta$. We have

$$\ker\left(-\widetilde{J}\frac{d}{dt} - \begin{pmatrix} A(t) & \\ & 0\end{pmatrix}\right) = \{(\theta,\widetilde{d}) \mid \widetilde{d} \in R^{s-r}\} \cong R^{s-r}.$$

Similarly,

$$\ker(-\widehat{J}\frac{d}{dt} - \begin{pmatrix} 0 & \\ & B(t)\end{pmatrix}) = \{(\hat{c},\theta) \mid \hat{c} \in \mathbb{R}^{T-s}\} \cong \mathbb{R}^{T-s}.$$

Thus

$$\ker(\mathcal{A}) = R^{2r} \oplus R^{s-r} \oplus R^{T-s} \oplus \operatorname{span}\{\varphi_1,\dots,\varphi_m\}.$$

Let

$$\Phi(z) = \int_{S^1}\left\{\widehat{H}(t,z(t)) + \sum_{i=1}^{T}[c_i(t)p_i(t) + d_i(t)q_i(t)]\right\}dt.$$

Then all the assumptions (A), (F), (D), (P) and (LL) are satisfied. The proof is complete.

Example 2.1. If the Hamiltonian function $H \in C^2(S^1 \times R^{2n}, R^1)$ is periodic in each variable, then (HS) has at least $2n+1$ periodic solutions.

This is the case where $r = s = T = n$.

This result, related to the Arnold conjecture (cf. Section 5), was first obtained by Conley and Zehnder [CoZ1], see also Chang [Cha5].

Example 2.2. If $H \in C^2(S^1 \times R^{2n}, R^1)$, where H is periodic in the components of q, and that there is an $R > 0$ such that for $|p| \geq R$,

$$H(t,p,q) = \frac{1}{2}Mp \cdot p + a \cdot p$$

where $a \in R^n$, and M is a symmetric nonsingular time independent matrix, then the corresponding (HS) possesses at least $N+1$ distinct periodic solutions.

This is the case where $r = 0$, $s = T = n$.

This is a result obtained by Conley and Zehnder [CoZ1]; see also P.H. Rabinowitz [Rab6].

Remark 2.3. Periodic nonlinearity has been studied by many authors: Conley–Zehnder [CoZ1], Franks [Fra1], Mawhin [Maw2], Mawhin–Willems [MaW1], Li [Li1], Rabinowitz [Rab6], Pucci–Serrin [PS1–2]. Fonda–Mawhin [FoM1] and Chang [Cha9]. Theorems 2.4 and 2.5 are due to Chang. The condition $H \in C^2$ can be weakened to $H \in C^1$, cf. Liu [Liu4].

3. Singular Potentials

Most Hamiltonian systems interested in mechanics have singularities in their potentials.

Let Ω be an open subset in $\mathbb{R}^n$ with compact complement $C = \mathbb{R}^n \setminus \Omega$, $n \geq 2$. Find $x(\cdot) \in C^2([0, 2\pi], \Omega)$ satisfying

$$\begin{cases} \ddot{x}(t) = \text{grad}_x V(t, x(t)), & \\ x(0) = x(2\pi), & \dot{x}(0) = \dot{x}(2\pi), \end{cases} \tag{3.1}$$

where $V \in C^2\left([0, 2\pi] \times \Omega, \mathbb{R}^1\right)$ is assumed to be 2π-periodic in t, with additional conditions:

(A_1) There exists $R_0 > 0$ such that

$$\sup\{|V(t,x)| + \|V_x'(t,x)\|_{\mathbb{R}^n} \mid (t,x) \in [0,2\pi] \times (\mathbb{R}^n \setminus B_{R_0})\} < +\infty.$$

(A_2) There is a neighborhood U of C in $\mathbb{R}^n$ such that

$$V(t,x) \geq \frac{A}{d^2(x,C)} \quad \text{for} \quad (t,x) \in [0,2\pi] \times U,$$

where $d(x, C)$ is the distance function to C, and $A > 0$ is a constant.

The condition (A_2) is called the strong force condition, according to W. B. Gordon [Gor1].

For the sake of simplicity, from now on we shall denote the subset of $C^2([0,2\pi],\Omega)$, satisfying the 2π-periodic condition, by $C^2(S^1,\Omega)$. Similar notations will be used for other 2π-periodic function spaces.

We shall study the problem (3.1) by critical point theory.

Let us introduce an open set of the Hilbert space $H^1(S^1, \mathbb{R}^n)$ as follows:

$$\Lambda^1\Omega = \left\{x \in H^1(S^1, \mathbb{R}^n) \mid x(t) \in \Omega,\ \forall\, t \in S^1\right\}.$$

This is the loop space on Ω. Let us define

$$J(x) = \int_0^{2\pi} \left\{\frac{1}{2}\|\dot{x}(t)\|^2_{\mathbb{R}^n} + V(t, x(t))\right\} dt \tag{3.2}$$

on $\Lambda^1\Omega$, the Euler equation for J is (3.1).

In order to apply critical point theory on the open set $\Lambda^1\Omega$, one should take care of the boundary behavior of J, i.e., we should know what happens if x tends to $\partial(\Lambda^1\Omega)$.

Lemma 3.1. *Assume* (A_1) *and* (A_2). *Let* $\{x_k\} \subset \Lambda^1\Omega$ *and* $x_k \rightharpoonup x$ *weakly in* $H^1(S^1, \mathbb{R}^n)$, *with* $x \in \partial(\Lambda^1\Omega)$. *Then* $J(x_k) \to +\infty$.

Proof. It suffices to prove

$$\int_0^{2\pi} V(t, x_k(t))\, dt \to +\infty.$$

Moreover, since $V(t, x)$ is bounded from below, it remains to prove that there is an interval $[a, b] \subset [0, 2\pi]$ such that $\int_a^b V(t, x_k(t))dt \to +\infty$.

By definition, $x \in \partial(\Lambda^1\Omega)$ means that there is $t^* \in [0, 2\pi]$ such that $x(t^*) \in \partial\Omega$. According to (A_1) and (A_2), there is a constant $B > 0$ such that

$$V(t, x) \geq \frac{A}{d^2(x, C)} - B;$$

hence

$$\int_{t^*}^{t^*+\delta} V(t, x(t))dt \geq \int_{t^*}^{t^*+\delta} \left(\frac{A}{\|x(t) - x(t^*)\|^2_{\mathbb{R}^n}} - B\right) dt$$

$\forall\, \delta > 0$. However, we have

$$\|x(t) - x(t^*)\|_{\mathbb{R}^n} \leq |t - t^*|^{1/2} \left(\int_0^{2\pi} \|\dot{x}(t)\|^2_{\mathbb{R}^n} dt\right)^{1/2}$$

from the Schwarz's inequality; thus

$$\int_{t^*}^{t^*+\delta} V(t, x(t))\, dt = +\infty. \tag{3.3}$$

Since the embedding $H^1(S^1, \mathbb{R}^n) \hookrightarrow C(S^1, \mathbb{R}^n)$ is compact, we have

$$\operatorname{Max}\left\{\|x(t) - x_k(t)\|_{\mathbb{R}^n} \mid t \in S^1\right\} \to 0 \quad \text{as } k \to \infty,$$

after omitting a subsequence. Consequently,

$$\int_{t^*}^{t^*+\delta} V(t, x_k(t))\, dt \to +\infty,$$

provided by Fatou's Lemma and (3.3).

Lemma 3.2. *Assume* (A_1) *and* (A_2); *then there is a constant* c_0 *depending on the* C^1 *norm of the function* V *on* $S^1 \times (\mathbb{R}^n \setminus B_{R_0})$, *such that* J *satisfies the* $(PS)_c$ *condition for* $c > c_0$.

Proof. Assume that $\{x_k\} \subset \Lambda^1\Omega$ satisfies

$$J(x_k) \to c, \tag{3.4}$$

and

$$dJ(x_k) = x_k + \mathbb{K}\,(\mathrm{grad}_x V(\cdot, x_k(\cdot)) - x_k) \to \theta \text{ in } H^1(S^1, \mathbb{R}^n), \tag{3.5}$$

where

$$\mathbb{K} = \left(\mathrm{id} - \frac{d^2}{dt^2}\right)^{-1} : L^2(S^1, \mathbb{R}^n) \to H^1(S^1, \mathbb{R}^n) \text{ is compact.}$$

We shall prove the subconvergence of $\{x_k\}$ in $\Lambda^1\Omega$. Since V is bounded below, (3.4) implies a constant $C_1 > 0$ such that

$$\int_0^{2\pi} |\dot{x}_k|^2 dt \leq C_1. \tag{3.6}$$

Let $\xi_k = \frac{1}{2\pi}\int_0^{2\pi} x_k(t)dt$. If we can prove that $\{\xi_k\}$ is bounded, then $\{x_k\}$ is bounded in $H^1(S^1, \mathbb{R}^n)$. Hence, there is a subsequence $x_k \rightharpoonup x$ (weakly in H^1). Applying Lemma 3.1, we have that $x \in \Lambda^1\Omega$ and that $\|\mathrm{grad}_x V(\cdot, x_k)\|_{\mathbb{R}^n}$ is bounded. Hence, the strong convergence of $\{x_k\}$ follows from the compactness of $\mathbb{K}$ and (3.5).

It remains to prove the boundedness of $\{\xi_k\}$. If not, we may assume that $|\xi_k| \to \infty$; then for large k, we have

$$\|x_k(t)\|_{\mathbb{R}^n} \geq \|\xi_k\|_{\mathbb{R}^n} - \left(2\pi\int_0^{2\pi} \|\dot{x}_k(t)\|^2 dt\right)^{1/2} \geq R_0,$$

which implies

$$\left|\int_0^{2\pi} V(t, x_k(t))dt\right| \leq 2\pi \sup\left\{|V(t,x)| \mid (t,x) \in S^1 \times (\mathbb{R}^n \setminus B_{R_0})\right\}. \tag{3.7}$$

From (3.5), we obtain

$$\int_0^{2\pi} \left[\|\dot{x}_k(t)\|^2 + V_x'(t, x_k(t))y_k(t)\right] dt \leq \|y_k\|_{H^1},$$

where $y_k = x_k - \xi_k$, for k large.

Since $\int_0^{2\pi} y_k(t)dt = 0$, we have

$$\|y_k\|_{H^1} = \|\dot{x}_k\|_{L^2}, \quad \text{and} \quad \|y_k\|_{L^2} \le \|y_k\|_{H^1};$$

hence

$$\int_0^{2\pi} \|\dot{x}_k(t)\|^2_{\mathbb{R}^n} dt \le \|\dot{x}_k\|_{L^2} + \|V'_x(t, x_k(t))\|_{L^2}\|\dot{x}_k\|_{L^2}.$$

It follows that

$$\begin{aligned} \|\dot{x}_k\|_{L^2} &\le 1 + \|V'_x(\cdot, x_k(\cdot))\|_{L^2} \\ &\le 1 + 2\pi \sup_{(t,x)\in S^1\times(\mathbb{R}^n\setminus B_{R_0})} \|V'_x(t,x)\|_{\mathbb{R}^n}. \end{aligned} \tag{3.8}$$

Substituting (3.7) and (3.8) into (3.2), and letting

$$\begin{aligned} c_0 = \frac{1}{2}\left(1 + 2\pi \sup_{(t,x)\in S^1\times(\mathbb{R}^n\setminus B_{R_0})} \|V'_x(t,x)\|_{\mathbb{R}^n}\right)^2 \\ + 2\pi \sup_{(t,x)\in S^1\times(\mathbb{R}^n\setminus B_{R_0})} |V(t,x)|, \end{aligned}$$

we have $J(x_k) \le c_0$. This is a contradiction.

Lemma 3.3. *There exists a sequence of integers*

$$q_1 < q_2 < \cdots < q_j < \cdots, \quad q_j \nearrow +\infty,$$

such that $H_{q_j}(\Lambda^1\Omega) \ne 0$.

Proof. Pick a point $p_0 \in C$ and choose $R > 0$ such that $C \subset \mathring{B}_R$, we have

$$\mathbb{R}^n \setminus B_R \subset \Omega \subset \mathbb{R}^n \setminus \{p_0\}.$$

Since $\mathbb{R}^n \setminus B_R$ is a deformation retract of $\mathbb{R}^n \setminus \{p_0\}$, $\Lambda^1(\mathbb{R}^n \setminus B_R)$ is a deformation retract of $\Lambda^1(\mathbb{R}^n \setminus \{p_0\})$, and then $\Lambda^1(\mathbb{R}^n \setminus B_R)$ is a retract of $\Lambda^1\Omega$. We obtain

$$\begin{aligned} H_*(\Lambda^1\Omega) &\cong H_*\left(\Lambda^1(\mathbb{R}^n \setminus B_R)\right) \oplus H_*\left(\Lambda^1\Omega, \Lambda^1(\mathbb{R}^n \setminus B_R)\right) \\ &\cong H_*(\Lambda^1 S^{n-1}) \oplus H_*(\Lambda^1\Omega, \Lambda^1 S^{n-1}) \end{aligned}$$

from Chapter I, Section 10. According to Bott [Bot], the Poincaré series of $\Lambda^1 S^{n-1}$ is written as

$$P_t(\Lambda^1 S^{n-1}) = (1 + t^n) + \frac{t^{n-1}}{1 - t^{2(n-1)}}(1 + t^n)(1 + t^{n-1})$$

with $\mathbb{Z}_2$ coefficients. Our conclusion follows.

Lemma 3.4. *For each $b > 0$, there exists a finite dimensional singular complex $M = M_b$ such that the level set $J_b = \{x \in \Lambda^1\Omega \mid J(x) \leq b\}$ is deformed into M.*

Proof. According to (A_1) and (A_2), we have $b_1 > 0$ such that

$$\int_0^{2\pi} \|\dot{x}(t)\|^2 dt \leq b_1 \quad \forall\, x \in J_b.$$

From Lemma 3.1, there exists $\varepsilon_0 = \varepsilon(b, b_1) > 0$ such that

$$d(x(t), C) \geq \varepsilon_0 \quad \forall\, x \in J_b \quad \forall\, t \in S^1.$$

Let us choose an integer $N = N_b > 2\pi \frac{b_1^{1/2}}{\varepsilon_0}$, and let

$$t_i = \frac{2\pi i}{N}, \quad i = 0, 1, 2, \ldots, N.$$

Define a broken line

$$\overline{x}(t) = \left(1 - \frac{1}{2\pi} N(t - t_{i-1})\right) x(t_{i-1}) + \frac{1}{2\pi} N(t - t_{i-1}) x(t_i),$$

$\forall\, t = [t_{i-1}, t_i]$, $i = 1, 2, \ldots, N$, for any $x \in J_b$, and let $M = \{\overline{x}(t) \mid x \in J_b\}$. The correspondence $\overline{x} \mapsto (x(t_1), x(t_2), \ldots, x(t_N))$ defines a homeomorphism between M and a certain open subset of the N-fold product $\Omega \times \Omega \times \cdots \times \Omega$. We shall verify the following.

(1) $M \subset \Lambda^1\Omega$.

Indeed, $\forall\, x \in J_b$, $\forall\, t_1 > t_2$,

$$\begin{aligned} \|x(t_1) - x(t_2)\|_{\mathbb{R}^n} &\leq \int_{t_2}^{t_1} \|\dot{x}(t)\| dt \\ &\leq \left(\int_0^{2\pi} \|\dot{x}(t)\|^2 dt\right)^{1/2} |t_1 - t_2|^{1/2} \\ &\leq b_1^{1/2} |t_1 - t_2|. \end{aligned}$$

Therefore,

$$\begin{aligned} d(\overline{x}(t), C) &\geq d(x(t_i), C) - \left(1 - \frac{1}{2\pi} N(t - t_{i-1})\right) \|x(t_i) - x(t_{i-1})\|_{\mathbb{R}^n} \\ &\geq \varepsilon_0 - 2\pi N^{-1} b_1^{1/2} > 0 \end{aligned}$$

$\forall\, t \in [t_{i-1}, t_i]$, $i = 0, 1, 2, \ldots, N$.

(2) There exists $\eta \in C\left([0,1] \times J_b, \Lambda^1\Omega\right)$ such that $\eta(0,\cdot) = \mathrm{id}$, and $\eta(1, J_b) = M$.

We define η as follows:

$$\eta(s,x)(t) = \begin{cases} x(t) & \text{for } t \geq 2\pi s \\ \left(1 - \frac{t-t_{i-1}}{2\pi s - t_{i-1}}\right) x(t_{i-1}) & \\ \quad + \frac{t-t_{i-1}}{2\pi s - t_{i-1}} x(2\pi s) & \text{for } t_{i-1} < t < 2\pi s \\ \overline{x}(t) & \text{for } t \leq t_{i-1} \leq 2\pi s \leq t_i \end{cases}$$

then η is the required deformation.

We have proven that J_b is deformed into M in the loop space $\Lambda^1\Omega$. The proof is finished.

Lemma 3.5. *For each $q > nN$, where $N = N_{c_0}$ is as defined in Lemmas 3.2 and 3.4, set*

$$c = \inf_{z\in\alpha} \max_{x\in|z|} J(x),$$

where $\alpha \in H_q(\Lambda^1\Omega)$ is nontrivial. Then $c > c_0$ and then c is a critical value of J.

Proof. If not, $c \leq c_0$, then there is a $[z] \in \alpha$ such that $|z| \subset J_{c_0+1}$. According to Lemma 3.4, there exists a deformation η: $[0,1]\times J_{c_0+1} \to \Lambda^1\Omega$, such that $\eta(1, J_{c_0+1}) \subset M_{c_0+1}$, with $\dim M_{c_0+1} \leq nN_{c_0}$. This implies that $\eta(1,|z|) \subset M_{c_0+1}$. But $\eta(1,[z]) \in \alpha$, and $\alpha \in H_q(\Lambda^1\Omega)$, with $q > nN_{c_0}$. This is impossible.

Theorem 3.1. *Under assumptions* (A_1) *and* (A_2), (3.1) *possesses infinitely many 2π-periodic solutions.*

Proof. We prove the theorem by contradiction. Assume that there are only finitely many solutions: $K = \{x_1, x_2, \ldots, x_\ell\}$. Noticing that the nullity $\dim\ker(d^2J(x_j)) \leq 2n$, $\forall\, j$, let

$$q^* > \max\left\{nN_{c_0}, \mathrm{ind}(J,x_j) + \dim\ker(d^2J(x_j)) \mid 1 \leq j \leq \ell\right\},$$

and

$$b > \max\{c_0, J(x_j) \mid 1 \leq j \leq \ell\}.$$

It follows that

$$C_q(J, x_j) = 0 \quad \forall\, q \geq q^*,\ j = 1, 2, \ldots, \ell, \tag{3.9}$$

and

$$H_*(\Lambda^1\Omega, J_{c_0}) = H_*(J_b, J_{c_0}).$$

Consequently,

$$H_q(\Lambda^1\Omega, J_{c_0}) = 0 \quad \forall\, q \geq q^*, \tag{3.10}$$

provided by the Morse inequalities. But

$$i_*: H_q(\Lambda^1\Omega) \to H_q(\Lambda^1\Omega, J_{c_0})$$

is an injection for $q \geq q^*$, and the conclusion of Lemma 3.3 contradicts (3.10). The proof is finished.

For autonomous systems, i.e., the potential V is independent of t, in order to single out nonconstant 2π-periodic solutions, we have

Corollary 3.1. *Under the assumptions of Theorem 3.1, if, further, V is independent of t, then (3.1) possesses infinitely many 2π-periodic nonconstant solutions, if V'' is bounded from below on the critical set $\mathbf{K}$ of V.*

Proof. For any constant solution $x(t) = x_0$, the Hessian of J at x_0 reads as

$$d^2J(x_0)x = -\ddot{x} + V''(x_0)x$$

with periodic boundary conditions, and hence, the Morse index and the nullity must be bounded by a constant depending on a, where $V''(x) \geq aI_{n\times n}$ $\forall\, x \in \mathbf{K}$. We conclude that all constant solutions have a bounded order of critical groups. Therefore there must be infinitely many nonconstant solutions.

Remark 3.1. Problem (3.1) was studied by Gordon [Gor1]. The critical point approach was given by Ambrosetti–Coti–Zelati [AmZ1–2] and Coti–Zelati [Cot1]. Theorem 3.1 improves the results in [AmZ1–2] considerably, where assumption (A_1) was replaced by much stronger conditions:

$$|V(t,x)|, \|\text{grad}_x V(t,x)\| \to 0$$

uniformly in t, as $\|x\| \to +\infty$; and there exists $R_1 > 0$ such that

$$V(t,x) > 0 \quad \forall\, x, \quad \|x\| \geq R_1.$$

Theorem 3.1 is due to M.Y. Jiang [Jia1–2].

Some related problems of the three body type were recently studied by A. Bahri and P.H. Rabinowitz [BaR1].

By avoiding condition (A_2), Bahri and Rabinowitz introduced the concept of generalized solutions. The existence and multiplicity results for generalized solutions were studied in [BaR1]. A most important problem is to ask when the generalized solution is a regular solution.

4. The Multiple Pendulum Equation

The Problem. The simple mechanical system consists of double mathematical pendula having lengths $\ell_1, \ell_2 > 0$ and masses $m_1, m_2 > 0$, as illustrated in the following figure.

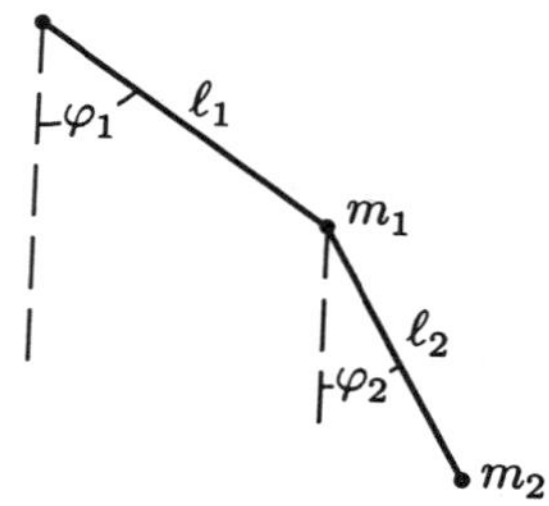

The positions of the system are described by two angle variables φ_1, φ_2 (mod 2π), so that the configuration space of the system can be viewed as a two-dimensional torus T^2. The motion of this system is governed by the Lagrangian function

$$L(\varphi,\dot\varphi) = \frac{1}{2}(m_1+m_2)\ell_1^2\dot\varphi_1^2 + (m_1+m_2)\ell_1\ell_2\cos(\varphi_1-\varphi_2)\dot\varphi_1\dot\varphi_2 + \frac{1}{2}m_2\ell_2^2\dot\varphi_2^2 + V(\varphi),$$

where $\varphi = (\varphi_1,\varphi_2)$, and the potential energy $-V(\theta)$ is given by

$$V(\varphi) = g\left((m_1+m_2)\ell_1\cos\varphi_1 + m_2\ell_2\cos\varphi_2\right).$$

Let 2λ be the smaller eigenvalue of the matrix

$$\begin{pmatrix} (m_1+m_2)\ell_1^2 & (m_1+m_2)\ell_1\ell_2 \\ (m_1+m_2)\ell_1\ell_2 & m_2\ell_2^2 \end{pmatrix}, \qquad \lambda > 0.$$

For the sake of simplicity, we assume the constant of gravitation $g = 1$. We shall add an additional forcing term $f = f(t) \in \mathbb{R}^2$, which is assumed to depend periodically on time t with period $T > 0$, and which, moreover, has mean value zero, i.e.,

$$f(t+T) = f(t) \quad \forall\, t$$

$$f \in L^2([0,T],\mathbb{R}^2), \quad \text{and} \quad \int_0^T f(t)dt = 0. \tag{4.1}$$

We look for periodic solutions $\varphi(t) = (\varphi_1(t),\varphi_2(t))$, having period T, of the Euler Lagrange equation

$$\frac{d}{dt}L_{\dot\varphi}(\varphi,\dot\varphi) - L_\varphi(\varphi,\dot\varphi) = f(t), \tag{4.2}$$

or, equivalently, the critical point in $H^1_*([0,T],\mathbb{R}^2)$ of the functional

$$J(\varphi) = \int_0^T \left[\frac{1}{2}L(\varphi,\dot\varphi) + f(t)\varphi\right] dt,$$

where

$$H^1_*([0,T],\mathbb{R}^2) = \left\{\varphi \in H^1([0,T],\mathbb{R}^2) \mid \varphi(0) = \varphi(t)\right\}.$$

We shall prove the following

Theorem 4.1. *Under (4.1), equation (4.2) possesses at least three periodic solutions having period T. Furthermore, if $\gamma_1 := (m_1+m_2)\ell_1 - m_2\ell_2 > 0$, and if $\gamma_2 := (m_1+m_2)\ell_1$ satisfies*

$$T\left\{\sqrt{T}\,\gamma_2 + \|f\|_{L^2}\right\}^2 < 16\pi^2\lambda\gamma_1,$$

then (4.2) possesses at least four distinct periodic solutions.

The Solution. The first conclusion is not surprising: it follows from the following simple observation:

$$J(\varphi_1 + 2k\pi, \varphi_2 + 2\ell\pi) = J(\varphi_1, \varphi_2) \quad \forall\,(k,\ell) \in \mathbb{Z}^2.$$

Let $\varphi_i = (\overline{\varphi}_i, \widehat{\varphi}_i)$, where

$$\begin{aligned}\overline{\varphi}_i &= \frac{1}{T}\int_0^T \varphi_i(t)dt, \quad \text{and}\\ \widehat{\varphi}_i &= \varphi_i - \overline{\varphi}_i,\end{aligned}$$

$i = 1,2$. Then J is well-defined on $M := T^2 \times \widehat{H}^1([0,T],\mathbb{R}^2)$, where

$$\widehat{H}^1([0,T],\mathbb{R}^2) = \left\{\varphi \in H^1_*([0,T],\mathbb{R}^2) \mid \frac{1}{T}\int_0^T \varphi(t)dt = \theta\right\}.$$

Lemma 4.1. *The functional J is bounded from below.*

Proof. Indeed,

$$\begin{aligned}\frac{1}{2}(m_1+m_2)\ell_1^2\dot\varphi_1^2 + (m_1+m_2)\ell_1\ell_2\cos(\varphi_1-\varphi_2)\dot\varphi_1\dot\varphi_2 + \frac{1}{2}m_2\ell_2^2\dot\varphi_2^2\\ \geq \lambda(\dot\varphi_1^2 + \dot\varphi_2^2).\end{aligned}$$

It follows that

$$J(\varphi) \geq \lambda\int_0^T |\dot\varphi|^2 dt - CT - \int_0^T \widehat{\varphi}\cdot f\,dt,$$

where $C = (m_1+m_2)\ell_1 + m_2\ell_2$, and $\widehat{\varphi}\cdot f = \widehat{\varphi}_1 f_1 + \widehat{\varphi}_2 f_2$, $f = (f_1, f_2)$. The first eigenvalue characterization provides the following estimate:

$$\inf\frac{\|\dot\varphi\|^2_{L^2}}{\|\widehat{\varphi}\|^2_{L^2}} = \lambda_1 = \left(\frac{2\pi}{T}\right)^2,$$

so

$$\|\widehat{\varphi}\|_{L^2} \leq \frac{T}{2\pi}\|\dot\varphi\|_{L^2}. \tag{4.3}$$

Consequently,

$$J(\varphi) \geq \lambda\|\dot\varphi\|^2_{L^2} - CT - \frac{T}{2\pi}\|f\|_{L^2}\|\dot\varphi\|_{L^2}. \tag{4.4}$$

This finishes the proof.

Lemma 4.2. *The functional J satisfies the (PS) condition.*

Proof. Let $\{\varphi_n\}$ be a sequence in M such that $J(\varphi_n)$ is bounded, and $dJ(\varphi_n) \to \theta$ as $n \to \infty$.

From (4.3) and (4.4), $\{\|\widehat{\varphi}_n\|_{H^1}\}$ is bounded, and since T^2 is compact, $\{\varphi_n\}$ is also bounded in M. By passing to a subsequence, one may assume that $\varphi_n \rightharpoonup \varphi_0$ in H^1_*, and $\varphi_n \to \varphi_0$ in $C\left([0,T],\mathbb{R}^2\right)$, as $n \to \infty$. Now

$$\begin{aligned}
&\langle dJ(\varphi_n), \varphi_n - \varphi_0\rangle - \langle dJ(\varphi_0), \varphi_n - \varphi_0\rangle \\
&\quad = (A(\varphi_n)\dot{\varphi}_n, \dot{\varphi}_n - \dot{\varphi}_0) + ((A'(\varphi_n)\dot{\varphi}_n)\dot{\varphi}_n, \varphi_n - \varphi_0) \\
&\qquad + (V'(\varphi_n), \varphi_n - \varphi_0) \\
&\qquad + (f, \varphi_n - \varphi_0) - (A(\varphi_0)\dot{\varphi}_0, \dot{\varphi}_n - \varphi_0) - ((A'(\varphi_0)\dot{\varphi}_0)\dot{\varphi}_0, \varphi_n - \varphi_0) \\
&\qquad - (V'(\varphi_0), \varphi_n - \varphi_0) - (f, \varphi_n - \varphi_0) \\
&\quad \geq (A(\varphi_0)(\dot{\varphi}_n - \dot{\varphi}_0), \dot{\varphi}_n - \dot{\varphi}_0) + ((A(\varphi_n) - A(\varphi_0))\dot{\varphi}_n, \dot{\varphi}_n - \dot{\varphi}_0) \\
&\qquad - o\left(\|\varphi_n - \varphi_0\|_{L^2}\right) \\
&\quad \geq \lambda\|\dot{\varphi}_n - \dot{\varphi}_0\|^2_{L^2} - O\left(|A(\varphi_n) - A(\varphi_0)|\right) - O\left(\|\varphi_n - \varphi_0\|_{L^2}\right)
\end{aligned}$$

where $\langle\ ,\ \rangle$ and $(\ ,\ )$ denote the duality on $T(M)$ and the inner product on L^2 respectively, and

$$A(\varphi) = \begin{pmatrix} (m_1+m_2)\ell_1^2, & (m_1+m_2)\ell_1\ell_2\cos(\varphi_1-\varphi_2) \\ (m_1+m_2)\ell_1\ell_2\cos(\varphi_1-\varphi_2), & m_2\ell_2^2 \end{pmatrix}.$$

Since $dJ(\varphi_n) \to \theta$, $\varphi_n - \varphi_0 \rightharpoonup \theta$ in H^1_* and $\varphi_n \to \varphi_0$ in C, as $n \to \infty$, we conclude from the inequality that

$$\|\dot{\varphi}_n - \dot{\varphi}_0\|_{L^2} \to 0.$$

Therefore, φ_n is convergent in M. This completes the proof.

Lemma 4.3. *The circle*

$$S_1 = \left\{\varphi \in M \mid \widehat{\varphi}_1 = \widehat{\varphi}_2 = \theta, \overline{\varphi}_1 = \pi, \overline{\varphi}_2 \in \frac{\mathbb{R}^1}{(\frac{T}{2\pi}\mathbb{Z})}\right\} \subset J_{-\gamma_1 T}.$$

Proof. Directly compute, $\forall \varphi \in S_1$, we have

$$J(\varphi) = \int_0^T [-(m_1+m_2)\ell_1 + m_2\ell_2\cos\overline{\varphi}_2]\,dt \leq -\gamma_1 T.$$

Lemma 4.4. $\exists\, s \in (0,1)$ *such that*

$$S_2 = \left\{\varphi \in M \mid \overline{\varphi}_1 = 0, \overline{\varphi}_2 \in \frac{\mathbb{R}^1}{(\frac{T}{2\pi}\mathbb{Z})}\right\} \cap J_{-s\gamma_1 T} = \emptyset.$$

Proof. $\forall\, \varphi \in S_2$,

$$\begin{aligned} J(\varphi) &\geq \lambda \|\dot{\varphi}\|_{L^2}^2 + \int_0^T (m_1+m_2)\ell_1(\cos\varphi_1 - \cos\overline{\varphi}_1)dt \\ &\quad + T\gamma_1 - \|f\|_{L^2}\cdot\|\widehat{\varphi}\|_{L^2} \\ &\geq \lambda\|\dot{\varphi}\|_{L^2}^2 + T\gamma_1 - \gamma_2\int_0^T |\widehat{\varphi}_1(t)|dt - \|f\|_{L^2}\|\widehat{\varphi}\|_{L^2} \\ &\geq \lambda\|\dot{\varphi}\|_{L^2}^2 + T\gamma_1 - \left(\sqrt{T}\,\gamma_2 + \|f\|_{L^2}\right)\|\widehat{\varphi}\|_{L^2} \\ &\geq \lambda\|\dot{\varphi}\|_{L^2}^2 + T\gamma_1 - \frac{T}{2\pi}\left(\sqrt{T}\,\gamma_2 + \|f\|_{L^2}\right)\|\dot{\varphi}\|_{L^2}. \end{aligned}$$

Hence, if there is a $\varphi \in J_{-s\gamma_1 T} \cap S_2$, then

$$\lambda\|\dot{\varphi}\|_{L^2}^2 + T\gamma_1 - \frac{T}{2\pi}\left(\sqrt{T}\,\gamma_2 + \|f\|_{L^2}\right)|\dot{\varphi}\|_{L^2} \leq -s\gamma_1 T.$$

But, by the assumption on γ_2 in the theorem, it cannot be true $\forall\, s \in (0,1)$. The contradiction proves the lemma.

Note that $\gamma: M = T^2 \times \widehat{H} \to T^2 \times \{\theta\}$ defined by

$$\gamma: (\varphi_1, \varphi_2) \mapsto (\overline{\varphi}_1, \overline{\varphi}_2)$$

is a deformation retract. We need

Lemma 4.5. *Let X be a topological space which contains subsets satisfying*

$$\begin{array}{ccccc} X & \supset & Y & \supset & Z \\ \cup & & \cup & & \cup \\ X' & \supset & Y' & \supset & Z'. \end{array}$$

If Z is a strong deformation retract of X, and if Z' is a strong deformation retract of X', then the inclusion map $j: (Z, Z') \to (Y, Y')$ induces a monomorphism

$$j_*: H_*(Z,Z') \to H_*(Y,Y') \tag{4.5}$$

in homology and an epimorphism

$$j^*: H^*(Y) \to H^*(Z) \tag{4.6}$$

in the cohomology ring.

Proof. We consider the commutative diagram

$$\begin{array}{ccccccc} & & & & H^*(Y',Z') & \xrightarrow{\xi} & H_*(X',Z') \\ & & & & \eta\downarrow & \swarrow & \xi \\ H_{*+1}(X,Z) & \to & H_*(Z,Z') & \xrightarrow{\beta} & H_*(X,Z') & \longrightarrow & H_*(X,Z) \\ & & j_*\downarrow & & \gamma\downarrow & & \\ & & H_*(Y,Y') & \longrightarrow & H_*(X,Y') & & \end{array} \tag{4.7}$$

where the longest row is the exact sequence of the triple (X, Z, Z') and the longest column is the exact sequence for the triple (X, Y', Z'). The indicated maps are induced by inclusions. By the assumptions

$$H_*(X, Z) = 0 \quad \text{and} \quad H_*(X', Z') = 0.$$

Therefore β is an isomorphism and ξ is a zero map. To prove that j_* is injective, assume $\alpha \in H_*(Z, Z')$ satisfies $j_*(\alpha) = 0$. Then by the commutativity of the rectangle in diagram (4.7), $\gamma \circ \beta(\alpha) = \delta \circ j_*(\alpha) = 0$. Therefore, by the exactness of the longest column in (4.7) there exists an $\overline{\alpha} \in H_*(Y', Z')$ such that $\eta(\overline{\alpha}) = \beta(\alpha)$. By the commutativity of the triangle in (4.7) and the property of ξ, we have $\eta(\overline{\alpha}) = \zeta \circ \xi(\overline{\alpha}) = 0$, and since β is an isomorphism, we conclude $\alpha = \beta^{-1} \circ \eta(\overline{\alpha}) = 0$ as claimed.

In order to prove (4.6) we consider the commutative diagram

$$\begin{array}{ccccccc} H^*(X,Z) & \longrightarrow & H^*(X) & \xrightarrow{\beta} & H^*(Z) & \longrightarrow & H^{*+1}(X,Z) \\ & & & \overset{\gamma}{\searrow} & \uparrow j^* & & \\ & & & & H^*(Y) & & \end{array} \tag{4.8}$$

where the longest row is the exact sequence for the pair (X, Z) and β and γ are homomorphisms induced by inclusions. Since $H^*(X, Z) = 0$, β is an isomorphism. If $\alpha \in H^*(Z)$, then by the commutativity of the triangle in (4.8) $j^*(\gamma \circ \beta^{-1}(\alpha)) = \alpha$, so that j^* is indeed surjective.

Proof of Theorem 4.1.

1. The first conclusion follows directly from Corollary 3.4 of Chapter II, because

$$\mathrm{CL}(T^2 \times \widehat{H}) = 2.$$

2. As to the second conclusion, we consider two separate pairs: $(M, J_{-s\gamma_1 T})$ and $(J_{-\gamma_1 T}, \emptyset)$, and we want to prove that there are at least two distinct critical points in each pair.

For the pair $(J_{-\gamma_1 T}, \emptyset)$, Lemmas 4.3 and 4.4 yield

$$(S^1 \setminus \{0\}) \times S^1 \times \widehat{H} \supset J_{-\gamma_1 T} \supset \{\pi\} \times S^1 \times \{\theta\}.$$

Construct a strong deformation retraction

$$\eta\colon [0,1] \times (S^1 \setminus \{0\}) \times S^1 \times \widehat{H} \to \{\pi\} \times S^1 \times \{\theta\},$$

by

$$\eta\,(t, (\overline{\varphi}_1, \overline{\varphi}_2, \widehat{\varphi}_1, \widehat{\varphi}_2)) = ((1-t)\overline{\varphi}_1 + t\pi, \overline{\varphi}_2, (1-t)\widehat{\varphi}_1, (1-t)\widehat{\varphi}_2)\,.$$

Apply Lemma 4.5. Then there are a monomorphism $j_*\colon H_*(S^1) \to H_*(J_{-\gamma_1 T})$ and an epimorphism $j^*\colon H^*(J_{-\gamma_1 T}) \to H^*(S^1)$.

We pick two homology classes, $0 \neq [\alpha_i] \in H_i(S^1)$, $i = 0, 1$, and a cohomology class, $0 \neq \beta \in H^1(S^1)$, such that

$$[\alpha_0] = [\alpha_1] \cap \beta.$$

Let $[z_i] = j_*[\alpha_i]$, $i = 0, 1$, and $\omega = j^{*-1}(\beta)$. Then $0 \neq [z_i] \in H_i(J_{-\gamma_1 T})$ for $i = 0, 1$ and $0 \neq \omega \in H^1(J_{-\gamma_1 T})$. Since

$$\begin{aligned}[z_2] \cap \omega = (j_*[\alpha_2]) \cap \omega &= j_*([\alpha_2] \cap j^*\omega) \\ &= j_*([\alpha_2] \cap \beta) = j_*[\alpha_1] = [z_1],\end{aligned}$$

Corollary 3.4 of Chapter II is used to give at least two critical points in $J_{-\gamma_1 T}$.

To the pair $(M, J_{-s\gamma_1 T})$, we observe that $M \supset M \supset T^2 \times \{\theta\}$, and

$$(S^1 \setminus \{\theta\}) \times S^1 \times \widehat{H} \supset J_{-s\gamma_1 T} \supset \{\pi\} \times S^1 \times \{\theta\}.$$

Again, applying Lemma 4.5, there are a monomorphism $k_*: H_*(T^2, S^1) \to H_*(M, J_{-s\gamma_1 T})$ and an epimorphism $k^*: H^*(M) \to H^*(T^2)$. Similarly, we pick two relative homology classes $[\delta_i] \in H_i(T^2, S^1)$, $i = 1, 2$, and a cohomology class $0 \neq \beta' \in H^1(T^2)$, and $[\delta_1] = [\delta_2] \cap \beta'$. Similarly, let $[w_i] = k_*[\delta_i]$, $i = 1, 2$, and $\omega' = k^{*-1}(\beta')$; we have $[w_1] = [w_2] \cap \omega'$. Then we use Theorem 3.4 of Chapter II to obtain at least two critical points in $(M, J_{-s\gamma_1 T})$.

In summary, we have proven that there are at least four distinct solutions.

Remark 4.1. The conclusion of Theorem 4.1 was first obtained by Fournier and Willem [FoW1] by a relative category method.

The above method enabled Chang, Long and Zehnder [CLZ1] to extend Theorem 4.1 to a n-pendulum problem. Under suitable parameters, they obtained 2^n solutions.

For a more general consideration, cf. Felmer [Fel1].

5. Some Results on Arnold Conjectures

5.1. The Conjectures

Let M be a compact symplectic manifold with a symplectic form ω, i.e., a closed nondegenerate 2-form. Let $h: \mathbb{R}^1 \times M \to \mathbb{R}^1$ be a time dependent smooth function. We call it the Hamiltonian. Supposing h is 2π-periodic in t, we associate a family of vector fields X_t on M, defined by

$$\omega(\cdot, X_t) = dh_t,$$

where X_t is called the Hamiltonian vector field associated with h_t. We consider the Hamiltonian system

$$\dot{\varphi}_t = X_t(\varphi_t), \quad \varphi_0 = \text{id}, \tag{5.1}$$

which defines a family of symplectic diffeomorphisms.

Arnold's first conjecture is concerned with the fixed point of the symplectic diffeomorphism $\varphi_{2\pi}$. Namely,

(AC_1) $\varphi_{2\pi}$ has at least as many fixed points as a function on M has critical points.

Let $CR(M)$ be the minimum number of critical points that a function on M must have, and $CRN(M)$ the minimum number if all are nondegenerate. Clearly,

$$CR(M) \geq CL(M) + 1$$

and

$$CRN(M) \geq SB(M),$$

where CL is the abbreviation of cuplength, and SB the sum of Betti numbers.

According to Conley–Zehnder [CoZ1], (AC_1) is rewritten as

$$\#\,\text{Fix}(\varphi_{2\pi}) \geq \begin{cases} CL(M) + 1 & \\ SB(M) & \text{if } \varphi_{2\pi}(M) \pitchfork M \text{ at } \text{Fix}(\varphi_{2\pi}). \end{cases}$$

This conjecture is somewhat related to Poincaré's last theorem: Let $\varphi\colon D^1 \times S^1 \to D^1 \times S^1$ be an area preserving homeomorphism such that

$$\varphi(p,q) = (f(p,q), q + g(p,q)) \quad (p,q) \in D^1 \times S^1 \tag{5.2}$$

where f, g are 2π-periodic in q, and for all $q \in S^1$, $f(\pm 1, q) = \pm 1$, $g(1,q) > 0$ and $g(-1,q) < 0$. Then φ has at least two fixed points, or, geometrically speaking, for an area preserving homeomorphism on an annulus, if it twists on the boundary, then it has at least two fixed points.

Indeed, the symplectic diffeomorphism $\varphi_{2\pi}$ preserves area (in the case $M = D^1 \times S^1$). If $\varphi_{2\pi}$ is written as (5.2), then the condition

$$h_q(t, \pm 1, q) = 0 \tag{5.3}$$

implies $f(\pm 1, q) = \pm 1$, and

$$ph_p > 0 \quad \text{for} \quad p = \pm 1 \tag{5.4}$$

implies $g(-1,q) < 0$ and $g(+1,q) > 0 \ \forall\, q$.

The symplectic diffeomorphism $\varphi_{2\pi}$ induced from a Hamiltonian $h \in C^2(\mathbb{R}^1 \times D^1 \times S^1, \mathbb{R}^1)$, 2π-periodic in t, satisfying (5.3) and (5.4), satisfies the hypotheses of Poincaré's last theorem.

On the other hand, there is a one-to-one correspondence between the fixed points of $\varphi_{2\pi}$ and the periodic solutions of the Hamiltonian system

$$\begin{cases} \dot{z}(t) = X_t(z(t)) \\ z(0) = z(2\pi) \end{cases} \tag{5.5}$$

provided that $\varphi_{2\pi}$ is the Poincaré map $z(2\pi) = \varphi_{2\pi}(z(0))$. This relationship enables us to reduce our study of (AC_1) to an estimate of the number of periodic solutions of Hamiltonian systems.

We turn to the second conjecture. A submanifold $L \subset M$ is called a Lagrangian submanifold if $\omega_x(\xi, \eta) = 0 \ \forall\, x \in L, \ \forall\, \xi, \eta \in T_x(L)$, the tangent space of L at x, and $\dim L = \frac{1}{2} \dim M$ (a symplectic manifold M is of even dimension). Arnold conjectured:

(AC_2) For any Lagrangian submanifold L,

$$\#\left(L \cap \varphi_{2\pi}(L)\right) \geq \begin{cases} CL(L) + 1 \\ SB(L) & \text{if } L \pitchfork \varphi_{2\pi}(L) \end{cases}$$

where $\varphi_{2\pi}$ is defined in (5.1).

We now give some examples of the Lagrangian submanifolds.

Example 5.1. $M = \mathbb{R}^{2n}$,

$$\omega_0 = \sum_{j=1}^{n} dx_j \wedge dx_{n+j}. \tag{5.6}$$

Then (M, ω_0) is a symplectic manifold, and

$$L = \left\{(x_1, x_2, \dots, x_{2n}) \in \mathbb{R}^{2n} \mid x_{n+1} = x_{n+2} = \cdots = x_{2n} = 0\right\}$$

is a Lagrangian submanifold.

Example 5.2. $M = T^{2n}$, the $2n$ torus, with the canonical symplectic form (5.6). Then (M, ω_0) is a compact symplectic manifold, and

$$L = \left\{(x_1, x_2, \dots, x_{2n}) \in \mathbb{R}^{2n}/\mathbb{Z}^{2n} \mid x_{n+1} = x_{n+2} = \cdots = x_{2n} = 0\right\}$$

($= T^n$) the n-torus, is a Lagrangian submanifold.

Example 5.3. $M = \mathbb{C}P^n = S^{2n+1}/S^1$, the complex projective space. It is defined as follows: First, we imbed S^{2n+1} into the complex space $\mathbb{C}^{n+1}$.

$$S^{2n+1} \cong \left\{(z_1, z_2, \dots, z_{n+1}) \in \mathbb{C}^{n+1} \mid |z_1|^2 + |z_2|^2 + \cdots + |z_{n+1}|^2 = 1\right\}.$$

A group action S^1 on $\mathbb{C}^{n+1}$ is defined:

$$S_\mu : (z_1, z_2, \ldots, z_{n+1}) \longmapsto e^{i\mu}(z_1, z_2, \ldots, z_{n+1})$$

$\forall\, \mu \in \mathbb{R}^1/2\pi\mathbb{Z}^1 \cong S^1$. The complex projective space is just the quotient space of S^{2n+1} under the group action S^1.

However, $\mathbb{C}^{n+1}$ has the canonical Hermite form

$$\langle z, w\rangle = \sum_{j=1}^{n+1} z_j \overline{w}_j \quad \forall\, z, w \in \mathbb{C}^{n+1}.$$

It induces a symplectic form

$$\omega_0(z, w) = -\mathrm{Im}\langle z, w\rangle.$$

In real coordinates $x, y, u, v \in \mathbb{R}^{n+1}$, $z = x + iy$, $w = u + iv$, where ω_0 is just the canonical symplectic form on $\mathbb{R}^{2(n+1)}$.

Noticing that

$$\begin{array}{ccc} S^{2n+1} & \xrightarrow{\ i\ } & \mathbb{C}^{n+1} \\ \downarrow{\scriptstyle \pi} & & \\ \mathbb{C}P^n & & \end{array}$$

where $\pi : S^{2n+1} \to \mathbb{C}P^n$ is the projection $z \mapsto [z]$, the equivalence class under the group action S^1, and $i : S^{2n+1} \to \mathbb{C}^{n+1}$ is the imbedding, we define a symplectic form ω on $\mathbb{C}P^n$ as follows: $\pi^*\omega = i^*\omega_0$.

It is well defined, because ω_0 is equivariant under the group action S^1.

Looking at the symplectic manifold $(\mathbb{C}P^n, \omega)$ in this way, the submanifold

$$L = \{[z] \in \mathbb{C}P^n \mid z \in [z], z = x + iy, y = \theta\}$$

is diffeomorphic to the real projective space $\mathbb{R}P^n$, and is easily verified to be a Lagrangian submanifold.

5.2. The Fixed Point Conjecture on (T^{2n}, ω_0)

Theorem 5.1. (AC$_1$) *is true for* (T^{2n}, ω_0), *i.e., there are at least* $2n+1$ *fixed points for* $\varphi_{2\pi}$, *and at least* 2^{2n} *fixed points if* $\varphi_{2\pi}(T^{2n})$ *is transversal to* T^{2n} *at all its fixed points.*

Proof. As we mentioned before, the problem is reduced to finding the number of 2π-periodic solutions of equation (5.5). Since ω_0 is canonical, and $\mathbb{R}^{2n}$ is the covering space of T^{2n}, one may extend the Hamiltonian function $h(t, \cdot)$ from $T^{2n} = \mathbb{R}^{2n}/2\pi\mathbb{Z}^n$ to $\mathbb{R}^{2n}$ by $H \in C^2(\mathbb{R}^1 \times \mathbb{R}^{2n}, \mathbb{R}^1)$, satisfying

$$\begin{cases} H(t, z) = H(t, z + 2\pi e_j), & j = 1, 2, \ldots, 2n, \\ H(t, z) = h(t, z) & \forall\, (t, z) \in \mathbb{R}^1 \times T^{2n}, \end{cases}$$

where $\{e_j \mid 1 \le j \le 2n\}$ is the orthonormal basis in $\mathbb{R}^{2n}$.

Noticing that the Hamiltonian system induced by H and the canonical symplectic form ω_0 reads as

$$-J\dot{z} = H_z(t,z), \tag{5.7}$$

this is the equation we have studied so far. Each solution of (5.7) with the boundary condition

$$z(2\pi) = z(0) + 2\pi\vec{k}_0 \quad \text{for some } \vec{k}_0 \in \mathbb{Z}^{2n},$$

corresponds to a 2π-periodic solution $[z]$ of (5.5) on T^{2n}. Moreover, two such solutions z_1, z_2 are in the same class $[z]$ if and only if there exists $\vec{k} \in \mathbb{Z}^{2n}$ such that

$$z_2(t) = z_1(t) + 2\pi\vec{k}.$$

Therefore, if there are two distinct 2π-periodic solutions z_j of (5.7) satisfying $z_j(0) = z_j(2\pi)$, $j = 1, 2$, then they must correspond to distinct classes $[z_j]$, $j = 1, 2$, in T^{2n}.

Now since H is of periodic nonlinearity, we apply Theorem 2.5 and conclude that there are at least $(2n+1)$ (or 2^{2n}) distinct 2π periodic solutions of (5.7) (if all these solutions do not have Floquet multiplier 1 respectively).

The proof is complete.

Let us return to the extension of Poincaré's last theorem.

Theorem 5.2. *Let $h \in C^2(\mathbb{R}^1 \times B^n \times \mathbb{R}^n, \mathbb{R}^1)$ be 2π-periodic in t and $q \in \mathbb{R}^n$. Assume that $h_q(t,p,q) = 0$ and $\langle p, h_p(t,p,q)\rangle > 0$ whenever $p \in \partial B^n$, where B^n is the unit ball in $\mathbb{R}^n$. Then the Hamiltonian system*

$$-Jz = h_z(t,z), \quad z = (p,q) \tag{5.8}$$

possesses at least $n+1$ (or 2^n) distinct 2π-periodic solutions (if all these solutions do not have Floquet multiplier 1 respectively).

Proof. Since h is 2π-periodic in t and q, we may restrict ourselves on the compact set $(t,q) \in S^1 \times T^n$. From $\langle p, h_p(t,p,q)\rangle > 0$ whenever $p \in \partial B^n$, we have $0 < \delta < \varepsilon/2$, such that $\langle p_1, h_p(t,p_2,q)\rangle > 0$, and $\langle p_1, p_2\rangle > 0$ for $1 - \varepsilon \le |p_i| \le 1$, $i = 1,2$ and $|p_1 - p_2| < \varepsilon$, and that $|h_q(t,p,q)| < \frac{\delta}{2\pi}$ for $1 - 2\delta \le |p| < 1$ and $(t,q) \in S^1 \times T^n$. Let us define a Hamiltonian $H \in C^2(\mathbb{R}^1 \times \mathbb{R}^{2n}, \mathbb{R}^1)$ being 2π-periodic in t and q, as follows:

$$H(t,p,q) = (1 - \rho(|p|))\, h(t,p,q) + \rho(|p|)\alpha|p|^2,$$

where $\rho \in C^\infty(\mathbb{R}^1_+)$ satisfies $0 \le \rho \le 1$,

$$\rho(s) = \begin{cases} 1 & \text{if } s \ge 1 \\ 0 & \text{if } s \le 1 - \delta, \end{cases}$$

and $0 \le \rho'(s) \le \frac{2}{\delta}$, and $\alpha \notin \mathbb{Z}$ is chosen such that $\alpha > \frac{1}{1-\varepsilon}\operatorname{Max} h(t,p,q)$. Consequently,

$$\begin{aligned}\langle p_1, H_p(t,p_2,q)\rangle &= \langle p_1, (1-\rho(|p_2|)h_p(t,p_2,q) \\ &\quad + \left[\rho'(|p_2|)\left(\alpha|p_2|^2 - h(t,p_2,q)\right) + 2\alpha|p_2|^2\rho(|p_2|)\right]\frac{p_2}{|p_2|}\rangle > 0\end{aligned}$$

$\forall\ 1-\varepsilon \le |p_i| \le 1$, $i = 1, 2$, $|p_1 - p_2| < \varepsilon$ and $\forall\ (t,q) \in S^1 \times T^n$. We turn to consider the new Hamiltonian system

$$-J\dot{z}(t) = H_z\left(t, z(t)\right), \quad z = (p,q), \tag{5.9}$$

and claim that (5.8) and (5.9) have the same 2π-periodic solutions. Indeed, we conclude that

1. (5.9) has no 2π-periodic solutions $z(t) = (p(t), q(t))$ such that $z(t) \notin \overline{B}^n \times \mathbb{R}^n$ for some t. If not, with no loss of generality, we may assume $p(0) \notin \overline{B}^n$. In a neighborhood of $(p(0), q(0))$ we have

$$\begin{cases} \dot{p} = -H_q = 0 \\ \dot{q} = H_p = 2\alpha p \end{cases},$$

so the solution must be $(p(0), q(0) + 2\alpha pt)$, which cannot be a periodic solution. Moreover,

2. (5.9) has no 2π-periodic solution $z(t) = (p(t), q(t))$ such that $z(t) \notin B^n_{1-\delta} \times \mathbb{R}^n$ for some t. Otherwise, from $1-\delta \le |p(0)| \le 1$ and $|\dot{p}| = |H_q| \le \frac{\delta}{2\pi}$, it follows that $1 - 2\delta \le |p(t)| \le 1$ and $|p(t_1) - p(t_2)| < \varepsilon$. Thus

$$0 < \left\langle p(0), \int_0^{2\pi} H_p(t, p(t), q(t))dt \right\rangle = \langle p(0), q(2\pi) - q(0)\rangle = 0.$$

This is a contradiction. However, according to Example 2.2, (5.9) has at least $n+1$ (or 2^n) distinct 2π-periodic solutions (if all solutions do not have Floquet multiplier 1 respectively).

This proves the theorem.

5.3. The Lagrange Intersections for $(\mathbb{C}P^n, \mathbb{R}P^n)$

We turn to (AC_2), where the symplectic manifold $M = \mathbb{C}P^n$ and the Lagrange submanifold L is taken to be $\mathbb{R}P^n$, as in Example 5.3. Since

$$CL(\mathbb{R}P^n) = SB(\mathbb{R}P^n) = n+1,$$

it is not necessary to consider the transversal case. We have

Theorem 5.3. (AC_2) *is true for* $(\mathbb{C}P^n, \mathbb{R}P^n)$, *i.e., there are at least* $(n+1)$ *intersections of* $\varphi_{2\pi}(\mathbb{R}P^n) \cap \mathbb{R}P^n$.

We reduce the intersection problem to a critical point problem by several steps:

Step 1. Reduction to a boundary value problem. By definition, $p_1 \in \mathbb{R}P^n \cap \varphi_{2\pi}(\mathbb{R}P^n)$ if and only if $\exists\, p_0 \in \mathbb{R}P^n$ such that $p_1 = \varphi_{2\pi}(p_0) \in \mathbb{R}P^n$, i.e., the equation

$$\begin{cases} \dot{w}(t) = X_t(w(t)) \\ w(0) = p_0,\ w(2\pi) = p_1, \quad p_0, p_1 \in \mathbb{R}P^n \end{cases} \tag{5.10}$$

possesses a solution $w(t)$.

Obviously, there is a one-to-one correspondence between the intersection points and the solutions of (5.10): $w(t) = \varphi_t p_0 = \varphi_t \circ \varphi_{2\pi}^{-1} p_1$.

The problem is transferred to find the number of distinct solutions of the BVP (5.10).

Step 2. Reduction to Hamiltonian systems on $\mathbb{C}^{n+1}$. Note that

$$C^{n+1} \xrightarrow{i} S^{2n+1} \xrightarrow{\pi} S^{2n+1}/S^1 = \mathbb{C}P^n\,,$$

where π is the Hopf fibration. We can associate with $h: \mathbb{R}^1 \times \mathbb{C}P^n \to \mathbb{R}^1$ a function $H: \mathbb{R}^1 \times \mathbb{C}^{n+1} \to \mathbb{R}^1$ satisfying

(1) $H(t, e^{i\mu} z) = H(t, z) \qquad \forall\, z = (z_1, z_2, \ldots, z_{n+1}) \in \mathbb{C}^{n+1}$;
(2) $H(t, \cdot)\big|_{S^{2n+1}} = h(t, \cdot) \circ \pi$;
(3) $H(t, \cdot)$ is positively 2-homogeneous in a neighborhood of the unit ball;
(4) $H(t, \cdot)$ is C^1 and C^1-bounded.

With no loss of generality, we may assume that $h(t, \cdot) > 0$, so that $H(t, z) > 0\ \forall\, z \neq \theta$, and also that $H(t, \theta) = 0$ and $H_z(t, \theta) = \theta$.

We turn to consider a new Hamiltonian system:

$$\begin{cases} -J\dot{z}(t) = H_z(t, z(t)) + \lambda z(t) \\ z(0), z(2\pi) \in \mathbb{R}^{n+1} \cap S^{2n+1}, \end{cases} \tag{5.11}$$

where

$$J = \begin{pmatrix} 0 & -I_{n+1} \\ I_{n+1} & 0 \end{pmatrix},$$

and λ is a Lagrange multiplier. It plays a role here as an eigenvalue.

Lemma 5.1. *Let* z *be a solution of* (5.11). *Then* $z(t) \in S^{2n+1}$, *and* $\pi(e^{i\lambda t} z(t))$ *solves* (5.10).

Proof. We consider the derivative of the norm:

$$\begin{aligned} \frac{d}{dt}|z(t)|^2 &= 2\langle z(t), \dot{z}(t)\rangle \\ &= 2\langle z(t), JH_z(t, z(t))\rangle, \end{aligned}$$

where $|\cdot|$ and $\langle\ ,\ \rangle$ are the norm and the inner product on $\mathbb{C}^{n+1} = \mathbb{R}^{2(n+1)}$ respectively. But $0 = \frac{d}{d\mu} H(t, e^{i\mu} z)\big|_{\mu=0} = \langle H_z(t), z(t)), -Jz(t)\rangle$ because H is S^1 invariant. Therefore $|z(t)| =$ const. Here and in the following, we write either $z = (x, y) \in \mathbb{R}^{n+1} \times \mathbb{R}^{n+1}$ or $z = x + iy \in \mathbb{C}^{n+1}$, if there is no confusion.

As to the second conclusion, we observe that the symplectic structure ω on $\mathbb{C}P^n$ is defined by

$$\pi^*\omega = i^*\omega_0,$$

where ω_0 is the canonical symplectic structure on $\mathbb{C}^{n+1}$. Thus $\omega(\cdot, X_t) = dh_t$, and

$$\begin{aligned} dh_t \circ \pi = dH_t \circ i &= i^*\omega_0(\cdot, Y_t) \\ &= \pi^*\omega(\cdot, Y_t) = \omega(\cdot, \pi_* Y_t), \end{aligned}$$

where $Y_t = JH_z(t, \cdot)$. Since

$$d(h_t \circ \pi) = \omega(\cdot, X_t),$$

therefore

$$X_t = \pi_* Y_t.$$

On the other hand, letting $\widetilde{z}(t) = e^{i\lambda t} z(t)$, (5.11) is rewritten as

$$-J\dot{\widetilde{z}}(t) = H_z\left(t, \widetilde{z}(t)\right).$$

By the uniqueness of initial value problems, we have

$$w(t) = \pi\widetilde{z}(t) = \pi[e^{i\lambda t} z(t)],$$

where $w(t)$ is defined by (5.10). Therefore $\pi(e^{i\lambda t} z(t))$ solves (5.10). The proof is finished.

Lemma 5.2. *Let (z^1, λ_1) and (z^2, λ_2) be two solutions of (5.11). Then $\pi\left(e^{i\lambda_1 t} z^1(t)\right) = \pi\left(e^{i\lambda_2 t} z^2(t)\right)$ implies $\lambda_1 = \lambda_2 \pmod{\pi}$.*

Proof. First, we claim that if $\widetilde{z}^1, \widetilde{z}^2$ solves

$$-J\dot{z} = H_z(t, z) \tag{5.12}$$

and $\pi\widetilde{z}^1 = \pi\widetilde{z}^2$, then $\exists\ \mu \in \mathbb{R}^1/2\pi\mathbb{Z}$, such that

$$\widetilde{z}^1(t) = e^{i\mu}\widetilde{z}^2(t).$$

Indeed, by definition, $\exists$ a function $\mu(t)$ such that $\widetilde{z}^1(t) = e^{i\mu(t)}\widetilde{z}^2(t)$. Substituting into equation (5.12), we have

$$e^{i\mu(t)}\dot{\widetilde{z}}^2(t) + ie^{i\mu(t)}\dot{\mu}(t)\widetilde{z}^2(t) = e^{i\mu(t)} JH_z(t, \widetilde{z}^2(t))$$

which implies either that $\exists\, t_0$ such that $\tilde{z}^2(t_0) = \theta$, so that $\tilde{z}^2(t) = \theta$, $\forall\, t$, and we can choose $\mu(t)$ to be constant, or that $\dot{\mu}(t) = 0\ \forall\, t$. And again we have $\mu(t) =$ const.

Next, we have

$$e^{i\mu} z^2(0) = z^1(0),$$
$$e^{i(\mu+\lambda_2)} z^2(2\pi) = e^{i\lambda_1} z^1(2\pi).$$

Since $z^i(2j\pi) \in \mathbb{R}^{n+1} \cap S^{2n+1}$ $(i = 1, 2,\ j = 0, 1)$, $e^{i\mu}$ and then $e^{i(\lambda_1 - \lambda_2)}$ must be real. Consequently,

$$\lambda_2 = \lambda_1 (\mathrm{mod}\ \pi).$$

Let us define an operator on $L^2([0, 2\pi], \mathbb{C}^{n+1})$,

$$A = -J\frac{d}{dt},$$

with domain

$$D(A) = \left\{z \in H^1([0, 2\pi], \mathbb{C}^{n+1}) \mid z(0), z(2\pi) \in \mathbb{R}^{n+1}\right\}.$$

Lemma 5.3. *The operator A is self adjoint, with spectrum $\sigma(A) = \frac{1}{2}\mathbb{Z}$; each eigenvalue $\frac{1}{2}k$ has multiplicity $(n + 1)$.*

Proof. Indeed, we have the following spectral decomposition:

$$L^2([0, 2\pi], \mathbb{C}^{n+1}) = \bigoplus_{k\in\mathbb{Z}} \mathrm{span}\{\cos\frac{1}{2}kte_j + i\sin\frac{1}{2}kte_{j+n+1} \mid j = 1, 2, \ldots, n+1\},$$

where $\{e_j + ie_{j+n+1} \mid j = 1, 2, \ldots, n + 1\}$ is the basis of $\mathbb{C}^{n+1}$.

Step 3. Reduction to a variational problem. According to the spectral decomposition, we decompose A into the positive, zero and negative parts: $A = A^+ + A^0 - A^-$, where A^+ and A^- are positive operators on their associated subspaces. Let us introduce a Hilbert space $E = D(|A|^{1/2})$, the domain of the square root of $|A|$, with norm

$$\|z\|_E = \left(|z|_{L^2}^2 + \|\,|A|^{1/2}z\|_{L^2}^2\right)^{1/2}.$$

In the following, if there is no confusion, we denote by $(\ ,\ )$ and by $\|\cdot\|$ the L^2 inner product and norm, respectively. Let us define

$$J(z) = \frac{1}{2}\left(\|(A^+)^{1/2}z\|^2 - \|(A^-)^{1/2}z\|^2\right) - \int_0^{2\pi} H(t, z(t))dt$$

on the manifold $S = \{z \in E \mid \|z\|_{L^2} = 1\}$.

Lemma 5.4. *Suppose that z_0 is a critical point of J on S with Lagrange multiplier λ_0. Then (z_0, λ_0) solves (5.11), and $J(z_0) = \lambda_0$.*

Proof. Letting (z_0, λ_0) be the critical point and the Lagrange multiplier, we have

$$\Big((A^+)^{1/2}z_0, (A^+)^{1/2}z\Big) - \Big((A^-)^{1/2}z_0, (A^-)^{1/2}z\Big) \\ - \int_0^{2\pi} \langle H_z(t, z_0) + 2\lambda_0 z_0, z\rangle dt = 0 \tag{5.13}$$

$\forall\, z \in E$, and it follows that

$$P_0(H_z(t, z_0) + \lambda z_0) = \theta,$$

where P_0 is the orthogonal projection onto the space associated with eigenvalue 0. Consequently,

$$(A^{\pm})^{1/2}z_0 = \pm(A^{\pm})^{1/2}(|A|^{-1})(H_z(t, z_0) + \lambda z_0),$$

and therefore

$$z_0 \in D(|A|) \subset D(A).$$

Then the weak solution equation becomes

$$Az_0 = H_z(t, z_0) + \lambda_0 z_0,$$

so that z_0 solves (5.11).

In particular, if we choose $z = z_0$ in (5.13), then

$$\Big(\|A^+)^{1/2}z_0\|^2 - \|(A^-)^{1/2}z_0\|^2\Big) - 2(\lambda_0 z_0, z_0) - \int_0^{2\pi} \langle H_z(t, z_0), z_0\rangle dt = 0.$$

Since $z_0(t) \in S^{2n+1}$ (Lemma 5.1) $\forall\, t \in [0, 2\pi]$, and $H(t, z)$ is positively 2-homogeneous in a neighborhood of the unit ball, we have

$$(H_z(t, z_0), z_0) = 2H(t, z_0).$$

Thus

$$\lambda_0 = \frac{1}{2}\Big(\|(A^+)^{1/2}z_0\|^2 - \|(A^-)^{1/2}z_0\|^2\Big) - \int_0^{2\pi} H(t, z_0)dt = J(z_0).$$

Remark 5.1. We take the working space $E = D(|A|^{1/2})$ with norm $\|z\|_E = \big(\|z\|^2_{L^2} + \|\,|A|^{1/2}z\|^2_{L^2}\big)^{1/2}$ rather than the space $H^{1/2}([0, 2\pi], \mathbb{C}^{n+1})$ since the trace operator is not well-defined on $H^{1/2}$, so that the boundary value condition cannot be formulated in $H^{1/2}$.

Finally, we observe that there is a natural symmetry for the functional. Namely, $J(-z) = J(z)$. Indeed, the function $H(t,z)$ is S^1-invariant, so that

$$H(t,-z) = H(t, e^{i\pi}z) = H(t,z).$$

Moreover the boundary value condition $D(A)$ is also invariant with respect to this group action. Consequently, J is well defined on the space $P = S/\mathbb{Z}_2$, where S is the unit L^2 sphere in E.

Returning to the original problem, we point out that we are not concerned with how many distinct critical values of J there are, *but how many distinct critical points there are associated with critical values in an interval with length* $< \pi$.

After the preparation, we shall give a proof of Theorem 5.3.

The Galerkin approximation will be applied. Let

$$E_k = \bigoplus_{|\ell|\le k} \operatorname{span}\left\{\cos\frac{1}{2}\ell t e_j + i\sin\frac{1}{2}\ell t e_{j+n+1} \mid j = 1,2,\dots,n+1\right\},$$

$$P_k = P \cap E_k,$$

and

$$J_k = J\big|_{P_k}.$$

Then $\dim E_k = (2k+1)(n+1)$, $\dim P_k = (2k+1)(n+1) - 1$.

Lemma 5.5. *J satisfies $(PS)^*$ with respect to $\{P_k \mid k = 1,2,\dots\}$.*

Proof. Suppose that $z_k \in P_k$ is a sequence satisfying

$$|J(z_k)| = \left|\frac{1}{2}(Az_k, z_k) - \int_0^{2\pi} H(t,z_k)dt\right| \le C_1 \tag{5.14}$$

and

$$\begin{aligned}(dJ_k(z_k), w)_E &= (Az_k, w) - \int_0^{2\pi} \langle H_z(t,z_k) + \lambda_k z_k, w\rangle dt \\ &= o(\|w\|_E) \quad \forall\, w \in E.\end{aligned} \tag{5.15}$$

We decompose z_k into $z_k^+ + z_k^0 + z_k^-$, according to the spectral decomposition, to positive, zero, negative eigenvalues, respectively.

First, setting $w = z_k^\pm$ in (5.15), we have

$$\|z_k^\pm\|_E^2 \le |\lambda_k| + o(\|z_k^\pm\|_E) + C_2$$

where C_2 is a bound for $|H_z(t,z)|$. Set $w = z_k$ in (5.15). Then by (5.14),

$$\begin{aligned}|\lambda_k| &= \left|(Az_k, z_k) - \int_0^{2\pi} \langle H_z(t,z_k), z_k\rangle dt\right| + o(\|z_k\|_E) \\ &\le 2C_1 + 2\pi M + o(\|z_k\|_E),\end{aligned}$$

where M is a bound for $|H|$ on the sphere S^{2n+1}. Since

$$\|z_k^0\|_E = \|z_k^0\|_{L^2} \le 1,$$

λ_k and $\|z_k\|$ are bounded.

Next, we prove that $\{z_k\}$ possesses a convergent subsequence in the E topology.

By the construction of the space E, we see that the injection

$$i: D(|A|) \to D(|A|^{1/2})$$

is compact. Let Π^*, $* = +, -, 0$ be the orthogonal projection onto the subspaces with positive, negative and zero eigenvalues. Again, from (5.15), we obtain

$$\left\| z_k^\pm - |A|^{-1}\Pi^\pm(H_z(t, z_k) + \lambda_k z_k)\right\|_E = o(1).$$

Thus $z_k^\pm$ are subconvergent, and, because $\Pi^0 E$ is finite dimensional, we conclude that z_k possesses a convergent subsequence. The proof is finished.

Let $P_k^+ = P_k \cap E^+$, where $E^+ = \Pi^+ E$, $\dim P_k^+ = k(n+1) - 1$. We find singular relative homology classes in $H_*(P_k, P_k \setminus P_k^+)$. From the special structure of P_k, there are nontrivial classes:

$$\theta \ne [z_1] < [z_2] < \cdots < [z_{k(n+1)}].$$

Lemma 5.6. *Set*

$$c_i^k = \inf_{\alpha\in[z_i]} \sup_{z\in|\alpha|} J_k(z), \quad i = 1, 2, \dots, k(n+1). \tag{5.16}$$

Then

(1) c_i^k *are critical values of* J_k,

(2) $c_1^k \le c_2^k \le \cdots \le c_{k(n+1)}^k$,

(3) $\frac{1}{4}(\ell - 1) - 2M\pi \le c_i^k \le \frac{1}{4}\ell$, *for* $(\ell - 1)(n+1) < i \le \ell(n+1)$, $\ell = 1, 2, \dots, k$.

where M is the maximum of $H(t,\cdot)$ on $[0, 2\pi] \times S^{2n+1}$.

Proof. (2) is trivial. We prove (3). On the one hand, $\forall\ \alpha \in [z_i]$, $|\alpha| \cap P_{\ell-1}^+ \ne \emptyset$, so that

$$\sup_{z\in|\alpha|} \langle Az, z\rangle \ge \frac{1}{2}(\ell - 1),$$

which, combined with $0 \le H(t,z) \le M$, yields the left hand inequality. On the other hand, $\exists\ \alpha \in [z_i]$ such that $|\alpha| \subset P_\ell^+$, provided by the special structure of P_k, and it follows that

$$\langle Az, z\rangle \le \frac{1}{2}\ell, \quad \forall\ z \in |\alpha|.$$

Hence the right hand inequality holds.

From (3), $[z_1],[z_2],\dots,[z_{k(n+1)}]$ are nontrivial in $(P_k,(J_k)_{-2M\pi})$, (1) follows from the minimax principle.

Now we take limits, letting

$$c_i = \lim_{k\to\infty} c_i^k.$$

Provided by the $(PS)^*$ condition, they are all critical values satisfying

$$-2M\pi \le c_1 \le \dots \le c_{\ell(n+1)} \le \frac{1}{4}\ell, \quad \ell = 1,2,\dots .$$

Moreover, according to the argument in Theorem 4.3 of Chapter II, they correspond to $\ell(n+1)$ distinct critical points of J.

Finally, we come to the proof of our theorem.

Proof of Theorem 5.3. As we have mentioned before, it remains to prove that there are at least $n+1$ distinct critical points associated with critical values in an interval with length $< \pi$.

Indeed, if the conclusion is false, the critical values

$$-2M\pi \le c_1 \le c_2 \le \dots \le c_{k(n+1)} \le \frac{1}{4}k, \quad k = 1,2,\dots ,$$

so obtained, satisfy

$$\#\{c_i \mid (\ell-1)/4 < c_i \le \ell/4\} \le n, \quad \forall\, \ell \ge 1.$$

Suppose $T = \#\{c_i \mid -2M\pi \le c_i \le 0\}$. T is finite, because

$$c_i^k \ge \frac{1}{4}\ell - 2M\pi \quad \text{for} \quad i \ge \ell(n+1),\ k \ge \ell.$$

It follows that

$$c_i \ge \frac{1}{4}\ell - 2M\pi.$$

Hence, we have the estimate $T \le 2M(n+1)$. Then

$$\begin{aligned} m(n+1) &\le \#\{c_i \mid -2M\pi \le c_i \le m/4\} \\ &\le T + m\cdot n, \quad \forall\, m, \end{aligned}$$

which is impossible for m large.

This contradiction shows that

$$\#\,(L \cap \varphi_1(L)) \ge n+1,$$

where $L = \mathbb{R}P^n$. The proof is finished.

Remark 5.2. The above method applies equally well when proving

Theorem 5.4. (AC_1) *is true for* (CP^n, ω), *i.e., there are at least* $n+1$ *fixed points for* $\varphi_{2\pi}$.

We are satisfied with pointing out the modifications:

(1) The boundary value condition in (5.11) is replaced by a periodic condition, i.e., $z(0) = z(2\pi)$.
(2) If two periodic solutions (z^1, λ_1) and (z^2, λ_2) on $\mathbb{C}^{n+1}$ correspond to one solution on $\mathbb{C}P^n$, then $\lambda_1 = \lambda_2 \pmod{2\pi}$.
(3) Let T_μ be the orthogonal representation on $H^{1/2}([0, 2\pi], \mathbb{C}^{n+1})$ $(= D(|A|^{1/2})$ for the periodic case) of the Hopf S^1 action $S_\mu: T_\mu z = S_\mu \cdot z \ \forall\, z \in E_k$, $k = 1, 2, 3, \dots$.

Then the functional J on S is S^1 invariant, so is well defined on $P = S/S^1$, and therefore $P_k = P \cap E_k$,

$$E_k = \bigoplus_{|\ell| \le k} \operatorname{span}\{\cos \ell t e_j \pm \sin \ell t e_{j+n+1} \mid j = 1, 2, \dots, n+1\},$$

is the $(2k+1)(n+1)-1$ dimensional complex projective space, $k = 1, 2, \dots$.

Remark 5.3. The Arnold conjectures were formulated in [Arn1–2]. The breakthrough is the pioneer work of Conley–Zehnder [CoZ1], in which Theorem 5.1 was proved. Soon after, a series of papers devoted to these problems appeared, cf. Floer [Flo1–4], Hofer [Hof3], Sikarov [Sik1]. The far-reaching contribution of Floer enabled him to solve (AC_1) and (AC_2) under the conditions $\pi_2(M) = 0$ and $\pi_2(M, L) = 0$, respectively. Unfortunately, we have no space in which to introduce his theory. For Theorem 5.2, cf. Szulkin [Szu1]. Theorem 5.3 is due to Givental and Chang–Jiang [ChJ1], and Theorem 5.4 to Fortune [For1].

CHAPTER V

Applications to Harmonic Maps and Minimal Surfaces

Geometric variational problems are of one of the most important parts of applications of infinite dimensional Morse theory. The closed geodesic problem, the minimal surface and the constant mean curvature problems, the harmonic map equation, the Yamabe problem and the Yang-Mills equation are not only interesting in themselves, but also for motivations in the development of infinite dimensional Morse theory.

Here, however, we are only concerned with a few examples showing how Morse theory is applied, because each topic deserves to be treated in specific books. Readers who are interested in these problems are referred to Klingenberg [Kli1], M. Atiyah and R. Bott [AtB1], H. Brézis and M. Coron [BrC2], R. Schoen [ScR1], C. Taubes [Tau1,2] and A. Bahri and M. Coron [BaC1].

1. Harmonic Maps and the Heat Flow

Let (M,g) and (N,h) be two smooth compact Riemannian manifolds with $m = \dim M$ and $n = \dim N$. For a map $u: M \to N$, $e(u) = \frac{1}{2}|\nabla u|^2$ is called the energy density. In local coordinates, it is written as $e(u) = \frac{1}{2}g^{ij}(x)h_{\alpha\beta}(u(x))u^\alpha_{,i}u^\beta_{,j}$, where $x = (x_1, x_2, \dots, x_m)$, $u = (u^1, u^2, \dots, u^n)$, $u^\alpha_{,i} = \frac{\partial u^\alpha}{\partial x_i}$, $\alpha, \beta = 1, 2, \dots, n$, $i, j = 1, 2, \dots, m$, $\big(h_{\alpha\beta}(u)\big)$ is the metric h on N, $\big(g^{ij}(x)\big)$ is the inverse of the metric g on M, and the convention summation is used. The energy of the map u is defined to be

$$E(u) = \int_M e(u)dV_g, \tag{1.1}$$

where V_g is the volume element over M. For a C^∞ map u, we introduce a C^∞ deformation φ_t $t \in (-\varepsilon, \varepsilon)$ for some $\varepsilon > 0$ with $\varepsilon_0 = a$ and consider the equation

$$\frac{d}{dt}E(\varphi_t)\big|_{t=0} = 0.$$

The equation is expressed as follows:

$$\int_M \langle\tau(u), v\rangle dV_g = 0, \qquad \forall\, v \in C^\infty\big(u^{-1}T(N)\big),$$

where $\tau(u) = \text{Trace}_g \nabla u$, or, in the local coordinates,

$$\tau(u)^\alpha(x) = \Delta_M u^\alpha(x) + g^{ij}(x)\ {}^N\Gamma^\alpha_{\beta\gamma}(u(x))u^\beta_i(x)u^\gamma_j(x),$$

where ${}^N\Gamma^\alpha_{\beta\gamma}(u)$ denotes the Christoffel symbol of the manifold N, Δ_M is the Laplace–Beltrami operator with respect to the metric g, and $C^\infty\left(u^{-1}T(N)\right)$ is the set of smooth sections of $u^{-1}T(N)$, i.e., the set of vector fields along u.

In the following we use the shorthand notations:

$$\Gamma(u)(\nabla u, \nabla u) = g^{ij}\ {}^N\Gamma^\alpha_{\beta\gamma}u^\beta_i u^\gamma_j$$

and

$$\Delta u = \tau(u);$$

then we get the Euler–Lagrange equation

$$\Delta u = 0. \tag{1.2}$$

The solution of (1.2) is called a harmonic map. It is easily seen that in the case $N = \mathbb{R}^1$, $M =$ a domain in $\mathbb{R}^n$, harmonic maps are harmonic functions.

In the case $\partial M \neq \emptyset$, harmonic maps are determined by the boundary value conditions. For instance, given $\psi \in C^\infty(\partial M, N)$, we find maps $u \in C^\infty(M, N)$ satisfying (1.2) and the boundary value condition

$$u\big|_{\partial M} = \psi.$$

If we want to study the existence (and multiplicity) problem for the harmonic maps from the critical point of view, then we should first find a suitable framework, Hilbert Riemannian manifold or Banach Finsler manifold, on which to develop our theory.

The Sobolev function spaces $W^\ell_p(\Omega)$ are extended to spaces of maps as follows: Provided by the Whitney–Nash embedding theorem, the target manifold N can be isometrically imbedded into $\mathbb{R}^k$ for some k, and then one naturally defines

$$W^\ell_p(M, N) = \left\{u \in W^\ell_p(M, \mathbb{R}^k) \mid u(x) \in N \quad \text{a.e. } x \in M\right\},$$

where $1 \leq p \leq \infty$, and ℓ is any integer. Since the energy functional is quadratic, it seems that $W^1_2(M, N)$ is a natural candidate. Unfortunately, $W^1_2(M, N)$ is not a Banach manifold for $m \geq 2$. Another reason for rejecting this choice comes from regularity. According to special nonlinearity, one can prove the smoothness of the solution only if we know it is continuous, but $W^1_2(M, N)$ is not sufficient to provide a continuous function.

In this sense, we are forced to take a second choice: $C^1(\bar{M}, N)$ or $W_p^1(M, N)$, $p > 2$. A severe problem occurs: The Palais–Smale condition is missing!

Carefully analyzing the role of the Palais–Smale condition in the proof of the deformation theorems, we find out that it is strongly tied to the gradient flow. We observe, however, that the heat equation for harmonic maps

$$\begin{cases} \partial_t f(t, \cdot) = \Delta f(t, \cdot) \\ f(0, \cdot) = \varphi(\cdot) \\ f(t, \cdot)\big|_{\partial M} = \psi(\cdot) \end{cases} \tag{1.3}$$

produces a flow $f(t, \cdot)$, which depends on the initial data φ and reduces the energy:

$$\begin{aligned} \frac{d}{dt} E(f(t, \cdot)) &= \int_M \langle \nabla f(t, \cdot), \nabla \partial_t f(t, \cdot) \rangle dV_g \\ &= -\int_M \langle \Delta f(t, \cdot), \partial_t f(t, \cdot) \rangle dV_g \\ &\quad + \int_{\partial M} g^{ij} h_{\alpha\beta} \partial_t f^\beta (t, \cdot) f^\alpha_{,i} \cdot n_j \, dS_g \\ &= -\int_M |\Delta f(t, \cdot)|^2 dV_g \leq 0. \end{aligned} \tag{1.4}$$

One expects to replace the gradient flow by the heat flow $f(t, \cdot)$ $(= f_\varphi(t, \cdot)$, to indicate the initial data).

This is possible if one can prove the following conclusions:

(1) the global existence of the heat flow, i.e., $f(t, \cdot)$ is defined on the whole half-axis $t \geq 0$;

(2) the limit of $f(t, \cdot)$, as $t \to \infty$, should in some sense be a harmonic map;

(3) the flow $f_\varphi(t, \cdot)$, as a map depending on t and the initial value φ, is continuous.

Before going into these statements, we introduce some notation:

$$C_\psi^{2+\gamma}(\overline{M}, N) = \{u \in C^{2+\gamma}(\overline{M}, N) \mid u\big|_{\partial M} = \psi\}, \quad 0 < \gamma < 1,$$

$Q_T = [0, T] \times M$, for $T > 0$; $C^{1+(\gamma/2), 2+\gamma}(Q_T, \mathbb{R}^k)$ equals the completion of $C^\infty(Q_T, \mathbb{R}^k)$ functions under the norm

$$\begin{aligned} \|f\|_{C^{1+\frac{\gamma}{2}, 2+\gamma}} = \|f\|_C + \sup_{\substack{0 \leq t, t' \leq T \\ x, y \in M}} &\left[\frac{|\partial_t f(t, x) - \partial_t f(t', x)|_C}{|t - t'|^{\gamma/2}} \right. \\ &\left. + \frac{|\partial_x^2 f(t, x) - \partial_x^2 f(t, y)|_C}{d(x, y)^\gamma} \right]; \end{aligned}$$

$W_p^{1,2}(Q_T, \mathbb{R}^k)$ has the same meaning, but under the norm

$$\|f\|_{W_p^{1,2}} = \|f\|_{L^p} + \|\partial_t f\|_{L^p} + \sum_{j=1}^{m} \|\partial_{x_j}^2 f\| L^p, \qquad 1 \leq p < \infty;$$

$$C^{1+(\gamma/2),2+\gamma}(Q_T, N) = \left\{ f \in C^{1+(\gamma/2),2+\gamma}(Q_T, \mathbb{R}^k) \mid f(t,x) \in N \right\}$$

has the same meaning for $W_p^{1,2}(Q_T, N)$.

The proof is fairly long. It is sufficient to point out the main steps and some crucial estimates.

Step 1. **Local Existence.** There exist $\varepsilon > 0$ and a unique $f \in C^{1+\frac{\gamma}{2},2+\gamma}(Q_\varepsilon, N)$ satisfying (1.4) in Q_ε, and then there is a maximal existence interval $[0, \omega)$.

The proof is standard. One uses the inverse function theorem on the map A:

$$f \mapsto \left(\partial_t f - \Delta f, f(0,\cdot), f(t,\cdot)\big|_{[0,T]\times\partial M}\right)$$

$$\begin{aligned} W_p^{1,2}(Q_T, \mathbb{R}^k) \to L^p(Q_T, \mathbb{R}^k) &\times B_p^{2(1-\frac{1}{p})}(M, \mathbb{R}^k) \\ &\times B_p^{1-\frac{1}{2p},2-\frac{1}{p}}([0,T] \times \partial M, \mathbb{R}^k). \end{aligned}$$

for $p > 4$, where $B_p^{2(1-1/p)}(M, \mathbb{R}^k)$ and $B^{1-(1/2p),2-(1/p)}([0,T] \times \partial M, \mathbb{R}^k)$ are Besov spaces, the trace spaces of $W_p^{1,2}$ on $\{0\} \times M$ and $[0,T] \times \partial M$, respectively.

Step 2. **Global Existence.** If the following a priori estimate holds— For solutions of (1.3), there exists a constant C_0 depending on φ, ψ, (M,g) and (N,h) such that

$$\sup_{[0,\omega)\times M} |\nabla f(t,x)| \leq C_0 \tag{1.5}$$

—then global existence follows.

Indeed, suppose $\omega < +\infty$. Combining (1.5) with the Schauder estimate for linear parabolic systems (cf. Ladyszenskaya, Solonhikov, Ural'ceva [LSU1]), we get immediately

$$\|f\|_{C^{1+\frac{\gamma}{2},2+\gamma}(Q_\omega)} \leq C_\gamma(\omega), \text{ a constant, for any } 0 < \gamma < 1.$$

Thus the solution of the evolution equation (1.3) may be extended beyond the maximal interval $[0, \omega)$. This is a contradiction. Consequently, $\omega = +\infty$.

In order to prove (1.5), we need a blow-up analysis and the following L^p estimates.

Step 3. **The L^p Estimates.**

Lemma 1.1. *Suppose that $1 < p, q < \infty$ and that $t \mapsto g(t,\cdot) \in L^p(M)$ for a.e. $t \in [0,T]$. Assume that*

$$\int_0^T \|g(t,\cdot)\|^q_{L^p(M)} dt < \infty,$$

and

$$\begin{aligned} \partial_t f - \Delta_M f &= g \quad \text{in } Q_T, \\ f(0,\cdot) &= 0 \quad \text{on } M, \\ f(t,\cdot)|_{\partial M} &= 0. \end{aligned}$$

Then we have a constant $C = C_{p,q}$ such that

$$\int_0^T \|f(t,\cdot)\|^q_{W^2_p(M)} dt \le C \int_0^T \|g(t,\cdot)\|^q_{L^p(M)} dt. \tag{1.6}$$

Proof. The linear equation is considered an evolution equation associated with the analytic semigroup, whose generator Δ_M is sectorial on the space $L^p(M)$ with domain

$$D(\Delta_M) = \{u \in W^2_p(M) | u|_{\partial M} = 0\}.$$

First we prove (1.6) in the case $p = 2$. With no loss of generality, we may assume $T = +\infty$, and we may define f and g as 0 for $t < 0$. Let $f\hat{}(\tau,\cdot)$ and $g\hat{}(\tau,\cdot)$ be the Fourier transforms of $f(t,\cdot)$ and $g(t,\cdot)$ with t respectively. We have

$$(i\tau - \Delta_M) f\hat{}(\tau,\cdot) = g\hat{}(\tau,\cdot),$$

i.e.,

$$\Delta_M f\hat{}(\tau,\cdot) = \Delta_M (i\tau - \Delta_M)^{-1} g\hat{}(\tau,\cdot).$$

The vector-valued Mihlin multiplier theorem (see N. Dunford, J.T. Schwartz [DS1]) provides the inequality

$$\int_0^T \|\Delta_M f(t,\cdot)\|^q_{L^2} dt \le C_q \int_0^T \|g(t,\cdot)\|^q_{L^2} dt,$$

and then by the elliptic theory, we have

$$\int_0^T \|f(t,\cdot)\|^q_{W^2_2} dt \le C_q \int_0^T \|g(t,\cdot)\|^q_{L^2} dt.$$

Next, in combining the L^q estimates for the parabolic equation with the above special case, the interpolation inequalities provide (1.6).

Lemma 1.2. *There exists a positive number $\varepsilon_0 > 0$ such that for a solution of the system* (E) *in a domain* $[0,T] \times D$, *where* $D = B_\rho(x_0) \cap \overline{M}$, *for some* $x_0 \in \overline{M}$, *and* $\rho > 0$, *if*

$$\sup_{t\in[t_0,t_1]} \int_D |\nabla f(t,\cdot)|^2 dV_g < \varepsilon_0$$

for some $t_0, t_1 \in (0,T)$, *then for any* $\rho' \in (0,\rho)$ *and* $(t_0', t_1') \subset (t_0, t_1)$, *we have some* $\alpha = 1 - \frac{4}{p} > 0$ *and a constant* C *depending on* ε_0, α, ρ', ρ, *and* t_0, t_1, t_0', t_1' *only such that*

$$\sup_{t\in[t_0',t_1']} \|f(t,\cdot)\|_{C^{1+\alpha}(D')} \le C\bigg[1 + \|\psi\|_{C^{2+\gamma}(\partial M\cap D)} + \left(\int_{[t_0,t_1]\times D} |\nabla f|^{2p} dt dV_g\right)^{1/p}\bigg]$$

for $p > 4$, *where* $D' = B_{\rho'}(x_0) \cap \overline{M}$.

Proof. Define a cutoff function $\varphi_1 \in C^\infty(Q_T)$, satisfying

$$0 \le \varphi_1 \le 1$$

and

$$\varphi_1(t,x) = \begin{cases} 1 & (t,x) \in [t_0', t_1'] \times D' \\ 0 & (t,x) \notin [t_0, t_1] \times D. \end{cases}$$

For $F = \varphi_1 \cdot f$,

$$\begin{aligned} \partial_t F - \Delta_M F &= \Gamma(f)(\nabla F, \nabla f) - \Gamma(f)(f\nabla\varphi_1, \nabla f) - 2\nabla f \cdot \nabla\varphi_1 + f(\partial_t - \Delta_M)\varphi_1, \\ F(t_0,\cdot) &= 0, \\ F(t,\cdot)\big|_{\partial M} &= \varphi_1 \cdot \psi \end{aligned}$$

According to the Sobolev embedding theorem (cf. Nikol'ski [Nik1]) and linear L^p theory, we have $\alpha = 1 - 4/p > 0$ such that

$$\begin{aligned} &\sup_{t\in[t_0',t_1']} \|f(t,\cdot)\|_{C^{1+\alpha}(D')} \\ &\quad\le \sup_{t\in[t_0,t_1]} \|F(t,\cdot)\|_{C^{1+\alpha}(D)} \\ &\quad\le C_\alpha \|F\|_{W_p^{1,2}([t_0,t_1]\times D)} \\ &\quad\le C\big[1 + \|\psi\|_{C^{2+\gamma}(\partial M\cap D)} + \|\nabla f\|_{L^p([t_0,t_1]\times D)} \\ &\qquad + \|\nabla F \cdot \nabla f\|_{L^p([t_0,t_1]\times D)}\big]. \end{aligned} \tag{1.7}$$

However, provided by the Sobolev imbedding theorem together with Lemma 1.1, letting $p_1 = 2p/(p+1)$, we have

$$
\begin{aligned}
&\int_{t_0}^{t_1} \|\nabla F\|^{2p}_{L^{2p}(D)} dt \\
&\le \int_{t_0}^{t_1} \|F\|^{2p}_{W^2_{p_1}(D)} dt \\
&\le C\big[1 + \|\psi\|^{2p}_{C^{2+\gamma}(\partial M \cap D)} + \int_{t_0}^{t_1} |\nabla f\|^{2p}_{L^{p_1}(D)} dt \\
&\quad + \int_{t_0}^{t_1} \|\nabla F \cdot \nabla f\|^{2p}_{L^{p_1}(D)} dt\big]
\end{aligned}
\tag{1.8}
$$

Applying the Hölder inequality,

$$
\begin{aligned}
\int_{t_0}^{t_1} \|\nabla F \cdot \nabla f\|^{2p}_{L^{p_1}(D)} dt &\le \int_{t_0}^{t_1} \|\nabla f\|^{2p}_{L^2(D)} \|\nabla F\|^{2p}_{L^{2p}(D)} dt \\
&\le \varepsilon_0^p \int_{t_0}^{t_1} \int_D |\nabla F(t,x)|^{2p} dV_g dt.
\end{aligned}
\tag{1.9}
$$

For sufficiently small $\varepsilon_0 > 0$, we put inequalities (1.8) and (1.9) together and obtain

$$
\begin{aligned}
\int_{t_0}^{t_1} \|\nabla F(t,\cdot)\|^{2p}_{L^{2p}(D)} dt \le C_{\varepsilon_0} \Big[1 &+ \|\psi\|^{2p}_{C^{2+\gamma}(\partial M \cap D)} \\
&+ \int_{t_0}^{t_1} \|\nabla f(t,\cdot)\|^{2p}_{L^{p_1}(D)} dt \Big].
\end{aligned}
\tag{1.10}
$$

Again by the Hölder inequality,

$$
\begin{aligned}
\|\nabla F \cdot \nabla f\|_{L^p([t_0,t_1]\times D)} &\le \|\nabla F\|_{L^{2p}} \cdot \|\nabla f\|_{L^{2p}} \\
&\le C_{\varepsilon_0} \big[1 + \|\psi\|_{C^{2+\gamma}(\partial M \cap D)} + \|\nabla f\|_{L^{2p}([t_0,t_1]\times D)}\big] \\
&\quad \times \|\nabla f\|_{L^{2p}([t_0,t_1]\times D)}
\end{aligned}
$$

Returning to (1.7), we have

$$
\begin{aligned}
\sup_{t\in[t_0',t_1']} \|f(t,\cdot)\|_{C^{1+\alpha}(D')} \le C\big[1 &+ \|\psi\|_{C^{2+\gamma}(\partial M \cap D)} \\
&+ \|\nabla f\|^2_{L^{2p}([t_0,t_1]\times D)}\big]
\end{aligned}
$$

Lemma 1.3. *Let $\omega > 0$ be finite or infinite. Assume that $\forall T < \omega$, $f \in W^{1,2}_p(Q_T, N)$, $p > 4$, is a solution of (1.3). If there is a relatively open set $D \subset M$ and a sequence of intervals $I_j \subset [0,\omega)$ with $\mathrm{mes}(I_j) \ge \delta > 0$ such that*

$$\sup_{t\in I_j} \int_D |\nabla f(t,\cdot)|^2 dVg < \varepsilon_0.$$

Then for any open subset $D' \subset\subset D$, for any sequence $\{t_j\}$ with $t_j \in I_j$ and $t_j \to \omega$, there is a subsequence $t_{j'}$ such that $f(t_{j'}, \cdot)$ is $C^1(\overline{D}', N)$ convergent to some $\widetilde{u} \in W_2^1(M, N)$.

Proof. Since

$$\int_M |\nabla f(t, \cdot)|^2 dVg \leq E(\varphi),$$

the family of maps $\{f(t_j, \cdot) | j = 1, 2, \dots\}$ is weakly compact in $W_2^1(M, \mathbb{R}^k)$, so that there is a subsequence $\{t_{j'}\}$ along which $f(t_{j'}, \cdot) \to \widetilde{u}$ weakly in $W_2^1(M, \mathbb{R}^k)$.

Starting from (1.10) with $p = 2$, we obtain a constant, which depends on ε_0, ψ and δ, dominating the norms $\|\nabla F\|_{L^4(I_j \times D)}$ $\forall\, j$. Applying (1.7), $\|F\|_{W_2^{1,2}(I_j \times D)}$ is also dominated. Then, the Sobolev embedding theorem implies the boundedness of $\|\nabla F\|_{L^{2p}(I_j \times D)}$ $\forall\, p > 4$. Thus, we have

$$\|f(t, \cdot)\|_{C^{1+\alpha}(D')} \leq \text{const.} \qquad \forall\, t \in I_j,$$

provided by Lemma 1.2. This implies a subsequence $\{t_{j'}\}$ such that $f(t_{j'}, \cdot)$ C^1-converges to $\widetilde{u}$.

Lemma 1.4. *Suppose that $f \in W_p^{1,2}(Q_T, N)$ $\forall\, T < \omega$ is a solution of (1.3), where $p > 4$. Then there is a sequence $t_{j'} \to \omega - 0$ and a finite number of points $\{x_1, \dots, x_\ell\} \subset M$ such that*

$$f(t_{j'}, \cdot) \to \widetilde{u}(\cdot) \quad \textit{in} \quad C^{1+\alpha'}(M \setminus \{x_1, \dots, x_\ell\}, N)$$

for some $\widetilde{u} \in W_2^1(M, N)$, and $0 < \alpha' < \alpha = 1 - 4/p$.

Proof. According to a covering theorem due to Besicovitch, there is an open covering of M consisting of disks $\{B_r(y_i) \mid i = 1, \dots, p\}$ such that

(a) $M \subset \bigcup_{i=1}^p B_{r/2}(y_i)$,

(b) $\forall\, x \in M$, there exist at most h disks $B_r(y_i)$ covering x, where h is independent of r.

Then

$$\sum_i \int_{B_r(y_i)} |\nabla f(t, \cdot)|^2 dVg \leq hE(f(t, \cdot)) \leq hE(\varphi).$$

Hence $\forall\, t$, $\exists$ at most $\ell = \left[\frac{2hE(\varphi)}{\varepsilon_0}\right] + 1$ disks $B_r(y_i)$, $i = 1, 2, \dots, \ell$, on which

$$\int_{B_r(y_i)} |\nabla f(t, \cdot)|^2 dVg \geq \frac{\varepsilon_0}{2}.$$

Fixing ℓ such disks, there is a sequence $t_j^* \uparrow \omega - 0$ such that

$$\int_{B_r(y_i)} |\nabla f(t_j^*, \cdot)|^2 dVg < \frac{\varepsilon_0}{2}, \qquad \forall\, i > \ell.$$

However, let $\varphi \in C_0^\infty(B_r(y_i))$ be a cutoff function, $0 \le \varphi \le 1$, $\varphi = 1$, on $B_{\frac{3r}{4}}(y_i)$, $|\nabla\varphi| \le \frac{8}{r}$. Multiplying the equation (1.3) by $h_{\alpha\beta}(f)\partial_t f^\alpha \varphi^2$, it follows from Young's inequality that

$$\begin{aligned}\int_{Q_{t'}^t} & h_{\alpha\beta}(f)\partial_t f^\alpha \partial_t f^\beta \varphi^2 dV_g ds + \int_{Q_{t'}^t} \frac{d}{dt} e(f)\varphi^2 dV_g ds \\ &\le C \int_{Q_{t'}^t} |\nabla f|\,|\partial_t f|\,|\nabla\varphi|\varphi dV_g ds \\ &\le \int_{Q_{t'}^t} h_{\alpha\beta}(f)\partial_t f^\alpha \partial_t f^\beta \varphi^2 dV_g ds + \frac{C_1}{r^2}\int_{Q_{t'}^t} |\nabla f|^2 dV_g ds,\end{aligned}$$

where $Q_{t'}^t = Q_t \setminus Q_{t'}$. Therefore,

$$\begin{aligned}\int_{B_{\frac{3}{4}r}(y_i)} |\nabla f(t,\cdot)|^2 dV_g &\le 2\int_M e\big(f(t',\cdot)\big)\varphi^2 dV_g + \frac{C_1}{r^2}\int_{t'}^t E\big(f(s,\cdot)\big)ds \\ &\le \int_{B_r(y_i)} |\nabla f(t',\cdot)|^2 dV_g + \frac{C_1(t-t')}{r^2}E(\varphi),\end{aligned}$$

which assures a uniform bound $\delta > 0$ such that

$$\sup_{|t-t_j^*|\le\delta} \int_{B_{\frac{3}{4}r}(y_i)} |\nabla f(t,\cdot)|^2 dV_g < \varepsilon_0 \qquad \forall\, i > \ell.$$

We apply Lemma 1.3 to these remaining disks. Then there is a sequence $t_{j'} \uparrow \omega - 0$ such that $f(t_{j'},\cdot)$ is $C^{1+\alpha'}$ convergent on $M \setminus \bigcup_{\ell i=1} B_{\frac{r}{2}}(y_i)$. Letting $r = 2^{-k}$, $k = 1, 2, \ldots$, by the diagonal process, there is a subsequence, still denoted by $\{t_{j'}\}$, so that $f(t_{j'},\cdot)$ $C^{1+\alpha'}$-converges on $M \setminus \{x_1, \ldots, x_\ell\}$, because the upper bound of the number of exceptional disks is independent of r.

Step 4. **On Asymptotic Behavior.** Now, we derive conclusion (2) from (1.5) and the a priori estimates.

Lemma 1.5. *There exist a harmonic map $\widetilde{u}$ and a sequence $t_j \uparrow +\infty$, such that $f(t_j,\cdot) \to \widetilde{u}(\cdot)$ in $C^{1+\alpha'}(\overline{M}, N)$ for $0 < \alpha' < \alpha$.*

Proof. We cover M by small balls $\bigcup_{i=1}^{p} B_{r/2}(x_i)$ such that C_0^2 $\mathrm{mes}(B_r(x_i)) < \varepsilon_0$. According to (1.5)

$$\sup_{t\in[k,k+1]} \|\nabla f(t,\cdot)\|_{L^2(B_r(x_i))} \le \varepsilon_0,$$

$k = 1, 2, \ldots$, $i = 1, \ldots, p$. It follows from Lemma 1.2 that

$$\sup_{t\in[1,\infty]} \|f(t,\cdot)\|_{C^{1+\alpha}(M)} \le C_1, \quad \text{a constant.}$$

From (1.4),

$$\int_0^\infty \int_M |\partial_t f(t,\cdot)|^2 dt dV_g \le E(\varphi),$$

so there is a sequence $t_j \nearrow +\infty$ such that

$$\partial_t f(t_j,\cdot) \to 0 \quad \text{in} \quad L^2(M, R^k).$$

On the one hand, according to Lemmas 1.3 and 1.4, with

$$\int_{B_r(x_i)} |\nabla f(t_j, x)|^2 dV_g < \varepsilon_0,$$

it follows that $f(t_j,\cdot) \to \widetilde{u}(\cdot)$ in $C^{1+\alpha'}\big(M \setminus \{x_1, \dots, x_\ell\}, N\big)$ with some $\{x_1, \dots, x_\ell\} \subset M$. On the other hand, $f(t_j,\cdot)$ is bounded in $C^{1+\alpha}(M,N)$, $\alpha' < \alpha$. We conclude that $f(t_j,\cdot) \to \widetilde{u}$ in $C^{1+\alpha'}(M,N)$. Thus

$$\int_M \big[g(\nabla\widetilde{u}, \nabla\varphi_1) + \Gamma(\widetilde{u})(\nabla\widetilde{u}, \nabla\widetilde{u})\varphi_1\big] dV_g = 0$$

$\forall\ \varphi_1 \in C_0^\infty(M,N)$. By applying the elliptic regularity theorem again, we conclude that $\widetilde{u} \in C^{2+\gamma}(M,N)$, and

$$\begin{cases} \Delta\widetilde{u} = 0, \\ \widetilde{u}|_{\partial M} = \psi. \end{cases}$$

Step 5. **Blow up Analysis.** Let

$$m = \inf\{E(u) \mid u \in W_2^1(M,N), \quad u\big|_{\partial M} = \psi\}$$

$$b = \begin{cases} \inf\{E(v) \mid v : S^2 \to N & \text{harmonic, and nonconstant}\} \\ +\infty & \text{if there is no such map.} \end{cases}$$

Lemma 1.6. *Suppose that* $\theta_T = \max_{(t,x)\in Q_T} |\nabla f(t,x)|$ *is not bounded. Then*

$$E(\varphi) \ge m + b.$$

Proof. We may find sequences $T_k \nearrow \omega$ and $a_k \in \overline{M}$ such that

$$|\nabla f(T_k, a_k)| = \max_{x\in\overline{M}} |\nabla f(T_k, x)| = \theta_{T_k},$$

$k = 1, 2, \dots$. From now on, we write θ_{T_k} simply as θ_k.

Neglecting subsequences, we may only consider the following two possibilities:

(1) $\theta_k \operatorname{dist}(a_k, \partial M) \to +\infty$,
(2) $\theta_k \operatorname{dist}(a_k, \partial M) \to \rho < +\infty$;

in both cases, we may assume $a_k \to a \in \overline{M}$.

Take a local chart U of a. Let

$$D_k = \left\{ y \in R^2 \mid a_k + \frac{y}{\theta_k} \in U \right\}$$

and

$$I_k = \left[-\theta_k^2 T_k, \theta_k^2(\omega - T_k)\right).$$

Define a function on $I_k \times D_k$ as follows:

$$v_k(\tau, y) = f\left(T_k + \frac{\tau}{\theta_k^2}, a_k + \frac{y}{\theta_k}\right),$$

$k = 1, 2, \ldots$. Then we see

$$\partial_\tau v_k = \Delta v_k, \tag{1.11}$$

and

$$\max_{(\tau,y)\in I_k\times D_k} |\nabla_y v_k(\tau, y)| \le 1, \qquad k = 1, 2, \ldots. \tag{1.12}$$

Let

$$h_k(\tau) = \int_{D_k} |\partial_\tau v_k(\tau, y)|^2 dy.$$

Then $h_k(\tau) \ge 0$, and $\forall\, e > 0$

$$\begin{aligned}\int_{-e}^{0} h_k(\tau)d\tau &\le \int_{T_k - e/\theta_k^2}^{T_k} dt \int_M |\partial_t f(t,x)|^2 dV_g \\ &= E\left(f\left(T_k - \frac{e}{\theta_k^2}, \cdot\right)\right) - E(f(T_k, \cdot)) \to 0\end{aligned}$$

as $k \to \infty$.

Thus, neglecting a subsequence, we may assume

$$h_k(\tau) \to 0 \quad \text{a.e.} \quad \tau \in [-e, 0],$$

i.e., for almost all $\tau \in [-e, 0]$,

$$\int_{D_k} |\partial_\tau v_k(\tau, y)|^2 dy \to 0. \tag{1.13}$$

In case (1), $a \in \mathring{M}$, and $D_k \to R^2$ in the sense that $\forall\, R > 0$, $\exists\, k_0 > 0$, the ball B_R centered at θ in R^2 is included in D_k for $k \ge k_0$.

On the one hand, by (1.13) ,

$$\partial_\tau v_k(\tau^*, y) \to 0, \quad L^2(B_R, R^k), \qquad \forall\, R > 0, \tag{1.14}$$

for almost all $\tau^* \in [-e, 0]$.

On the other hand, by Lemma 1.2, we have

$$\sup_{\tau\in[-e,0]} \|v_k(\tau,\cdot)\|_{C^{1+\alpha}(B_R)} \leq C\big[1 + (e4\pi R^2)^{1/p}\big]. \tag{1.15}$$

This implies a subsequence, where we do not change the subscripts, so that

$$v_k(\tau^*, y) \to \widetilde{v}(y), \quad C^{1+\alpha'}(R^2)$$

for some $\tau^* \in [-e, 0]$ (actually in a countable dense subset of $[-e, 0]$). We conclude that

$$\Delta\widetilde{v} = 0 \quad \text{in} \quad R^2.$$

According to the singularity removable theorem due to Sacks–Uhlenbeck (cf. [SaU1]), $\widetilde{v}$ is extendible to a harmonic map $\overline{v}: S^2 \to N$.

We are going to show that $\overline{v}$ is nonconstant. Indeed,

$$|\nabla_y v_k(0,\theta)| = \frac{1}{\theta_k}|\nabla_x f(T_k, a_k)| = 1,$$

since v_k satisfies (1.11) on $I_k \times D_k$ with the condition (1.12). The Schauder estimate applies to obtain an estimate:

$$\|v_k(\tau, y)\|_{C^{1+\frac{\gamma}{2},2+\gamma}([-e,0]\times(B_\delta(0)\cap D_k))} \leq C\big\{1 + [e\pi(2\delta)^2]^{1/p}\big\} \tag{1.16}$$

for some $\delta > 0$ small depending on U. The right hand side of the inequality is a constant independent of k. According to the embedding theorem (cf. [Nik1]),

$$\|\nabla_y v_k(\tau, y)\|_{C^{(1+\gamma)/2,1+\gamma}([-e,0]\times(B_\delta(\theta)\cap D_k)} \leq C_1,$$

where C_1 is a constant independent of k. Hence

$$|\nabla_y v_k(\tau^*, \theta) - \nabla_y v_k(0, \theta)| < C_1\tau^*.$$

We may choose $\tau^* > 0$ small enough so that

$$|\nabla v_k(\tau^*, \theta)| > \frac{1}{2}. \tag{1.17}$$

It proves that $\overline{v}$ is nonconstant.

Let $T_k' = T_k + \frac{\tau^*}{\theta_k^2}$, since $T_k' \to \omega$, $|\nabla f(T_k', \cdot)|$ blows up at at most finitely many points $\{x_1, \ldots, x_\ell\}$, which includes the limit set of $\{a_k\}$, according to Lemma 1.4.

We choose $\delta > 0$ small enough so that

$$\begin{aligned} E\big(f(T_k', \cdot)\big) &= \int_M |\nabla f(T_k', x)|^2 dV_g \\ &= \int_{M\setminus\cup_{j=1}^{\ell} B_\delta(x_j)} + \sum_{j=1}^{\ell}\int_{B_\delta(x_j)} |\nabla f(T_k', x)|^2 dV_g. \end{aligned}$$

Since

$$f(T'_k, \cdot) \to \widetilde{u}(\cdot) \qquad C^{1+\alpha'}\left(M \setminus \bigcup_{j=1}^{\ell} B_\delta(x_j), R^k\right),$$

and there exists at least one j_0 such that $a = x_{j_0}$, we have

$$\lim_{k\to\infty} \int_{M\setminus\cup_{j=1}^{\ell} B_\delta(x_j)} |\nabla f(T'_k, x)|^2 dV_g = \int_{M\setminus\cup_{j=1}^{\ell} B_\delta(x_j)} |\nabla \widetilde{u}(x)|^2 dV_g$$

and

$$\int_{B_\delta(x_{j_0})} |\nabla f(T'_k, x)|^2 dV_g \geq \int_{B_{1/2\theta_k\delta}(\theta)} |\nabla v_k(\tau^*, y)|^2 dy$$

for k large. First let $k \to \infty$; by definition

$$\lim_{k\to\infty} \int_{B_\delta(x_{j_0})} |\nabla f(T'_k, x)|^2 dV_g \geq b,$$

and then, because $\delta > 0$ is arbitrary,

$$\begin{aligned} E(\varphi) &\geq \lim_{k\to\infty} \int_M |\nabla f(T'_k, x)|^2 dV_g \\ &\geq \int_M |\nabla \widetilde{u}(x)|^2 dV_g + b \geq m + b. \end{aligned} \tag{1.18}$$

This is the desired conclusion.

In case (2), $a \in \partial M \cap U$. We choose a suitable coordinate (y_1, y_2) in R^2, such that the y_2-axis is parallel to the tangent at a of ∂M, and the y_1-axis points to the interior of U. Thus D_k tends to the half plane $R^2_+ = \{(y_1, y_2) \mid y_1 > -\rho\}$, and for each point on the boundary, $y_1 = -\rho$,

$$a_k + \frac{y}{\theta_k} \to a.$$

As in the proof of (1.15), now we have $\forall\, R > 0$,

$$\begin{aligned} \sup_{\tau\in[-e,0]} \|v_k(\tau, \cdot)\|_{C^{1+\alpha}(B_R\cap D_k)} \leq C\Bigg[1 &+ (e4\pi R^2)^{1/p} \\ &+ \left\|\psi\left(a_k + \frac{y}{\theta_k}\right)\right\|_{C^{2+\gamma}(\partial D_k \cap B_R)}\Bigg]. \end{aligned}$$

Since on the right hand side, there is a constant control independent of k, we find a function $\widetilde{v}^*$ on R^2_+ and a subsequence $v_k(\tau^*, \cdot)$ such that

$$v_k(\tau^*, y) \to \widetilde{v}^*(y) \qquad C^{1+\alpha'}(R^2_+),$$

and then

$$\Delta \widetilde{v}^* = 0 \quad \text{in } R_+^2,$$
$$\widetilde{v}^*\big|_{\partial R_+^2} = \psi(a).$$

On the one hand, similar to the proofs of (1.16) and (1.17), we see that $\widetilde{v}^*$ is nonconstant; and on the other hand, let us define a complex function

$$\eta(z) = h(\widetilde{v}_z^*, \widetilde{v}_z^*)$$

where h is the Riemannian metric on N, and

$$v_z^* = \frac{1}{2}(\partial_{y_1} - i\partial_{y_2})\widetilde{v}^*,$$
$$z = y_1 + iy_2.$$

Therefore,

$$\eta(z) = h(\widetilde{v}_{y_1}^*, \widetilde{v}_{y_1}^*) - h(\widetilde{v}_{y_2}^*, \widetilde{v}_{y_2}^*) - 2ih(\widetilde{v}_{y_1}^*, \widetilde{v}_{y_2}^*).$$

The harmonics of $\widetilde{v}^*$ implies the analyticity of the function η. The boundary condition on $\widetilde{v}$ implies that the function η can be analytically extended to the whole complex plane. From the condition

$$\eta(-\rho + iy_2) = 0,$$

we conclude that $\eta(z) \equiv 0$, and hence that $\widetilde{v}^*$ is a constant map. This is a contradiction, so Lemma 1.6 is proved.

In the following, we assume $\pi_2(N) = 0$. We shall expand the conclusion of Lemma 1.6 to the following:

$$E(\varphi) \geq m_{\mathcal{F}} + b,$$

where $\mathcal{F}$ is the homotopy class of φ, and

$$m_{\mathcal{F}} = \inf\{E(u) \mid u \in \mathcal{F}\}.$$

Only the inequality (1.18) should be fixed. It is known that $f(T_k', \cdot) \to \widetilde{u}(\cdot)$ in $C^{1+\alpha'}(M \setminus \bigcup_{j=1}^{\ell} B_\delta(x_j), R^k)$. We only want to show $\widetilde{u} \in \mathcal{F}$. Let $\delta > 0$ be small enough so that $B_\delta(x_i) \cap B_\delta(x_j) = \emptyset$, if $i \neq j$. Combine $\widetilde{u}(x_i)$ with the map $f(T_k', \cdot)\big|_{\partial B_\delta(x_i)}$ by the following map:

$$\widehat{f}_k(x) = \begin{cases} f(T_k', x) & \forall\, x \notin \bigcup_{i=1}^{\ell} B_\delta(x_i), \\ \exp_{\widetilde{u}(x_i)}\Big(\eta\Big(\frac{|x-x_i|}{\delta}\Big) \exp_{\widetilde{u}(x_i)}^{-1} f(T_k', x)\Big), & \forall\, x \in B_\delta(x_i), \end{cases}$$

where $\eta \in C^\infty(R^1)$ satisfies

$$\eta(r) = \begin{cases} 1 & r \geq 1 \\ 0 & r \leq \frac{1}{2}, \end{cases}$$

and exp is the exponential map. Since $\pi_2(N) = 0$, and

$$\widehat{f}_k\big|_{\partial B_\delta(x_i)} = f(T_k', \cdot)\big|_{\partial B_\delta(x_i)},$$

we see that $\widehat{f}_k$ remains in the same homotopy class $\mathcal{F}$. And from $\widehat{f}_k \to \widetilde{u}$ $C(M, N)$, we conclude $\widetilde{u} \in \mathcal{F}$.

Step 6. **Continuous dependence.** From the point of view of PDE, the continuous dependence

$$\varphi \longmapsto f_\varphi(t, x)$$

from $C_\psi^{2+\gamma}(\overline{M}, N)$ to $C_\psi^{1+\frac{\gamma}{2}, 2+\gamma}(Q_T, N)$ does hold.

The proof depends on the locally uniform boundedness of the heat flow $\varphi \longmapsto f_\varphi$, i.e, $\forall \varphi_0 \in E_c = \{u \in C_\psi^{2+\gamma}(\overline{M}, N) \mid E(u) \leq c\}$, where $c \leq m_{\mathcal{F}} + b$, $\exists\, \delta > 0$ and $C_1 > 0$ such that

$$\text{Sup}_{(t,x)\in[0,\infty))\times M} |\nabla f_\varphi(t, x)| \leq C_1, \ \forall \varphi \in B_\delta(\varphi_0), \tag{1.19}$$

where B_δ is the δ-ball in $C_\psi^{2+\gamma}(\overline{M}, N)$.

Indeed, if (1.19) does not hold, then $\exists\, \varphi_k \to \varphi_0$ in $C_\psi^{2+\gamma}(\overline{M}, N)$ and $\exists\, T_k$ with $T^* = \underline{\lim}_{T_k}$, satisfying

$$\|\nabla f_{\varphi_k}(T_k, \cdot)\|_{L^\infty(M)} \to \infty$$

and $\forall\, T < T^*$, $\exists\, C_2(T) < +\infty$ such that $\|\nabla f_{\varphi_k}(t, \cdot)\|_{L^\infty([0,T]\times M)} \leq C_2(T)$.

By local existence, we may assume $T_k \geq \varepsilon > 0$, and we shall prove $\|\nabla f_{\varphi_0}(T^* - 0, \cdot)\|_{L^\infty(M)} = \infty$, which contradicts Lemma 1.6 because $\varphi_0 \in E_c$.

For simplicity, we write $f_k = f_{\varphi_k}$, $k = 0, 1, 2 \ldots$. It is sufficient to prove that $\{f_k\}$ is a Cauchy sequence in $W_p^{1,2}(Q_T)$, $\forall T < T^*$. Since $\|\nabla f_k(t, \cdot)\|_{L^\infty(M)} \leq C_2(T) < \infty$, $\forall\, k\, \forall\, t \leq T < T^*$, for functions $p_k = d_k^2$, where $d_k(t, x) = \text{dist}\,(f_k(t, x), f_0(t, x))$, we have a constant $C_3(T) > 0$ satisfying

$$\partial_t p_k \leq \Delta p_k + C_3(T) p_k.$$

Thus, by the Maximum Principle,

$$\|d_k\|_{L^\infty(Q_T)} \leq C_4(T) \text{dist}\,(\varphi_k, \varphi_0).$$

Again, letting $f = f_k - f_0$ we write

$$\begin{aligned}\Delta(\Gamma(f)(\nabla f, \nabla f)) &= \Gamma(f_k)(\nabla f_k, \nabla f_k) - \Gamma(f_0)(\nabla f_0, \nabla f_0) \\ &= (\Gamma(f_k) - \Gamma(f_0))(\nabla f_k, \nabla f) + \Gamma(f_0)(\nabla f, \nabla f_0) \\ &\quad + \Gamma(f_0)(\nabla f_0, \nabla f).\end{aligned}$$

We have the following equations:

$$\begin{cases} \partial_t f = \Delta_M f + \Delta(\Gamma(f)(\nabla f, \nabla f)), \\ f(0, \cdot) = \varphi_k - \varphi_0, \\ f \,|_{[0,T]\times\partial M} = 0. \end{cases} \tag{1.20}$$

We apply the L^p estimates to (1.20), $p > 4$ and obtain

$$\|f\|_{W_p^{1,2}(Q,T)} \leq C_5(T)\|\varphi_k - \varphi_0\|_{C^{2+\gamma}(\overline{M},N)}. \tag{1.21}$$

This proves the conclusion.

Once (1.21) is established, $T^* = +\infty$, so (1.19) holds and then coninuous dependence follows, provided by a bootstrap iteration.

Nevertheless, this is not exactly what we need. As a family of maps $f(t, \cdot)$, $t \geq 0$, depending on φ, it is no longer a continuous flow under the strong topology on the Banach manifold $C_\psi^{2+\gamma}(\overline{M}, N)$. (The problem occurs at $t = 0$!). But it is continuous if we use a weaker topology, e.g., W_p^2, $1 - \frac{4}{p} > \gamma$. In the following, we shall employ the heat flow $f_\varphi(t, \cdot)$ as deformations, under a weaker topology $W_p^2(M, N)$, on the incomplete manifold $C_\psi^{2+\gamma}(\overline{M}, N)$.

For details, cf. Chang [Cha].

Step 7. **The First Deformation Lemma.** The critical set

$$K_c = \{u \in C_\psi^{2+\gamma}(\overline{M}, N) \mid \Delta u = 0\}$$

is compact in $C_\psi^{2+\gamma}$ as well as in the W_p^2-topology.

In extending critical point theory, we have the modified first deformation lemma: For a closed neighborhood U of K_c in the W_p^2 topology, $\exists\, \varepsilon > 0$ and a W_p^2 continuous deformation η: $[0, 1] \times E_{c+\varepsilon} \to E_{c+\varepsilon}$ satisfying

$$\begin{gathered}\eta(0, \cdot) = \mathrm{id}_{E_{c+\varepsilon}}, \\ \eta(1, E_{c+\varepsilon} \setminus U) \subset E_{c-\varepsilon}.\end{gathered}$$

The deformation is constructed by the solution of the following evolution equation:

$$\begin{cases} \partial_t f(t, \cdot) = \gamma\big(f(t, \cdot)\big)\Delta f(t, \cdot), \\ f(0, \cdot) = \varphi, \\ f(t, \cdot)\big|_{\partial M} = \psi, \end{cases} \tag{1.22}$$

where

$$\gamma(u) = \begin{cases} 1 & u \notin U_{\delta/4} \\ 0 & u \in U_{\delta/8} \end{cases}$$

and

$$U_\delta = \{u \in C^{2+\gamma}_\psi(\overline{M}, N) \mid \mathrm{dist}_{W^2_p}(u, K_c) \le \delta\} \subset U.$$

(1.19) is not a PDE, but after suitable reparametrization of (1.3), we are able to solve this equation. For details, cf. Theorem 7.1 in Chang [Cha10].

In summary, we have the following conclusion:

Theorem 1.1. *Let*

$$b = \inf\{E(v) \mid v\colon S^2 \to N,\ \ \text{nonconstant harmonic}\}$$

(if there is no nonconstant harmonic map from S^2 to N, then we define $b = +\infty$), and let $\mathcal{F}$ be a component of $C^{2+\gamma}_\psi(\overline{M}, N)$.

Assume that $\dim M = 2$, $\pi_2(N) = 0$, and that $\psi \in C^{2+\gamma}(\partial M, N)$, $\varphi \in \mathcal{F}$, with

$$E(\varphi) < m_{\mathcal{F}} + b,$$

where

$$m_{\mathcal{F}} = \inf\{E(u) \mid u \in \mathcal{F}\}.$$

Then we have

(1) *The heat flow, i.e., the solution of (1.3), globally exists.*
(2) *$\exists$ a harmonic map $\widetilde{u} \in \mathcal{F}$, and a sequence $t_j \uparrow +\infty$ such that*

$$f(t_j, \cdot) \to \widetilde{u}(\cdot) \quad \text{in} \quad C^1(\overline{M}, N).$$

(3) *If the infinitely dimensional manifold $C^{2,\gamma}_\psi(\overline{M}, N)$ is endowed with a weaker topology $W^2_p(M, N)$, $p > \frac{4}{1-\gamma}$, the flow*

$$(t, \varphi) \longmapsto f_\varphi(t, \cdot)$$

is continuous from $[0, \infty) \times \mathcal{F} \to \mathcal{F}$, where $f_\varphi(t, \cdot)$ denotes the flow with initial data φ.
(4) *The set*

$$K_c = \{u \in C^{2,\gamma}_\psi(\overline{M}, N) \mid \Delta u = 0, E(u) = c\}$$

is compact under the above topology, if $c < m_{\mathcal{F}} + b$.
(5) *Let $K = \bigcup_{c < m_{\mathcal{F}} + b} K_c$. Suppose that*

$$\mathrm{dist}_{W^2_p}(f_\varphi(t, \cdot), K) \ge \delta > 0 \qquad \forall\, t \in \mathbb{R}^1_+.$$

Then we have $\varepsilon = \varepsilon(\delta) > 0$ such that

$$\|\Delta f_\varphi(t,\cdot)\|_{L^2(M,N)} \geq \varepsilon.$$

(6) *For any closed neighborhood $U \subset C^{2,\gamma}_\psi(\overline{M}, N)$ of K_c, under the W^2_p-topology, where $c < m_{\mathcal{F}} + b$, $\exists\, \varepsilon > 0$, a closed neighborhood $V \subset U$, and a $W^2_p - \left(p > \frac{4}{1-\gamma}\right)$ strong deformation retract $\eta\colon [0,1] \times E_{c+\varepsilon} \to E_{c+\varepsilon}$, satisfying*

$$\eta(1, E_c \cap V) \subset E_c \cap U, \quad \text{and}$$
$$\eta(1, E_{c+\varepsilon} \setminus V) \subset E_{c-\varepsilon},$$

where $E_a = \{u \in \mathcal{F} \mid E(u) \leq a\}$ is the level set, $\forall\, a \in \mathbb{R}^1_+$.

Remark 1.1. The heat flow method was first used by J. Eells and Sampson [EeS1] in proving the existence of harmonic maps, where m is arbitrary and N has nonpositive sectional curvature. See also Hamilton [Ham1]. Without the restriction on curvatures, but with $m = 2$, see M. Struwe [Str4] and K.C. Chang [Cha10].

2. Morse Inequalities

In this section, we establish Morse inequalities for harmonic maps under the assumption that all harmonic maps are isolated. As shown in Chapter I, the crucial step in the proof is to prove the following deformation lemma:

Lemma 2.1. *Let $\mathcal{F}$ be a component of $C^{2+\gamma}_\psi(\overline{M}, N)$. Suppose that there is no harmonic map with energy in the interval $(c, d]$, where $d < m_{\mathcal{F}} + b$, and that there are at most finitely many harmonic maps on the level $E^{-1}(c)$. Assume that $\pi_2(N) = 0$. Then E_c is a strong deformation retract of E_d.*

In order to give the proof, first we must improve conclusion (2) of Section 1, under the condition that the set of smooth harmonic maps is isolated. Namely,

Lemma 2.2. *Let $E(\varphi) < m_{\mathcal{F}} + b$, and let*

$$c = \lim_{t \to +\infty} E\big(f_\varphi(t,\cdot)\big).$$

If K_c is isolated, then $f_\varphi(t,\cdot) \to \widetilde{u} \in K_c$ in the W^2_p-topology, $\forall\, p > \frac{4}{1-\gamma}$, as $t \to +\infty$.

Proof. According to Theorem 1.1, conclusion (2), combined with a bootstrap iteration, shows that $\exists\, \widetilde{u} \in K_c$ and $t_j \uparrow +\infty$ such that

$$f_\varphi(t_j,\cdot) \to \widetilde{u}, \quad C^{2,\gamma'}_\psi(\overline{M}, N), \qquad \forall\, \gamma' \in (0,\gamma).$$

If our conclusion were not correct, there would be a $\delta > 0$ such that the neighborhood $U_\delta = \{u \in C^{2,\gamma}_\psi(\overline{M}, N) \mid \text{dist}_{W^2_p}(u, \widetilde{u}) \le \delta\}$ contains the single element $\widetilde{u}$ in K_c, and a sequence $t'_j \uparrow +\infty$ such that $f_\varphi(t'_j, \cdot) \notin U_\delta$. Therefore $\exists\ (t^*_i, t^{**}_i)$ satisfying

(1) $t^*_i, t^{**}_i \to +\infty$,
(2) $f_\varphi(t^*_i, \cdot) \in \partial U_{2\delta}$, $f_\varphi(t^{**}_i, \cdot) \in \partial U_\delta$, and
(3) $f_\varphi(t, \cdot) \in U_{2\delta} \setminus U_\delta \quad \forall\, t \in (t^*_i, t^{**}_i)$.

On the one hand, we had

$$\delta \le \|f_\varphi(t^*_i, \cdot) - f_\varphi(t^{**}_i, \cdot)\|_{W^2_p} \le C_\delta |t^*_i - t^{**}_i|^{\gamma/2},$$

provided by the embedding theorem. On the other hand, according to Theorem 1.1, conclusion (5) states

$$\begin{aligned} E\big(f_\varphi(t^{**}_i, \cdot)\big) - E\big(f_\varphi(t^*_i, \cdot)\big) &= \int_{t^*_i}^{t^{**}_i} \int_M |\partial_t f(t, \cdot)|^2 dV_g dt \\ &= \int_{t^*_i}^{t^{**}_i} \int_M |\Delta f(t, \cdot)|^2 dV_g dt \\ &> \varepsilon(\delta) |t^{**}_i - t^*_i|. \end{aligned}$$

Since the left hand side of the inequality tends to zero as $i \to \infty$, this is a contradiction.

Now we return to the proof of Lemma 2.1. The basic idea is to reparametrize the heat flow $f_\varphi(t, \cdot)$.

Let $\tau = \rho(t)$, where

$$\rho(t) = (E(\varphi) - c)^{-1} \int_0^t \|\Delta f_\varphi(s, \cdot)\|^2_{L^2} ds,$$

if $E(\varphi) > c$, and let

$$g(\tau, \cdot) = f(t, \cdot).$$

Then we have the following relations:

(1) $\displaystyle \partial_\tau g(\tau, \cdot) = \frac{dt}{d\tau} \partial_t f(t, \cdot) = \frac{(E(\varphi) - c)}{\|\Delta g(\tau, \cdot)\|^2_{L^2}} \Delta g(\tau, \cdot),$

(2) $\displaystyle \frac{d}{d\tau} E(g(\tau, \cdot)) = -\int_M \langle \partial_\tau g(\tau, \cdot), \Delta g(\tau, \cdot)\rangle dV_g = -(E(\varphi) - c).$

Therefore

$$E\big(g(\tau, \cdot)\big) = (1 - \tau) E(\varphi) + \tau c, \qquad \forall\, \tau \in [0, 1].$$

(3) The functionρ: $[0, \infty) \to \mathbb{R}^1$is continuous and monotone increas-

ing which satisfies the following properties:

$$\begin{aligned}
&\rho(0) = 0,\\
&\rho(+\infty) = 1 \quad \text{if } f_\varphi(t,\cdot) \to \widetilde{u} \in K_c \text{ as } t \to +\infty,\\
&\rho(+\infty) > 1 \quad \text{if } \lim_{t\to+\infty} E(f_\varphi(t,\cdot)) < c.
\end{aligned}$$

Let us define a function $\eta: [0,1] \times E_d \to E_d$ as follows:

$$\eta(\tau,\varphi) = \begin{cases} g_\varphi(\tau,\cdot) & \text{if } (\tau,\varphi) \in [0,1] \times (E_d \setminus E_c),\\ \varphi & \text{if } (\tau,\varphi) \in [0,1] \times E_c. \end{cases}$$

In order to show that E_c is a deformation retract of E_d, only continuity at the following sets is needed:

(1) $\{1\} \times A$, where $A = \{\varphi \in E_d \setminus E_c \mid f_\varphi(\infty,\cdot) \in K_c\}$

(2) $[0,1] \times E^{-1}(c)$.

Verification for case (1). $\forall\, \varphi_0 \in A$, $\forall\, \varepsilon > 0$, we want to find $\delta > 0$ such that

$$\left.\begin{array}{c} \operatorname{dist}_{W_p^2}(\varphi,\varphi_0) < \delta \\ \tau > 1-\delta \end{array}\right\} \text{implies } \operatorname{dist}_{W_p^2}\big(g_\varphi(\tau,\cdot),\widetilde{u}\big) < \varepsilon,$$

where $\widetilde{u} = f_{\varphi_0}(\infty,\cdot)$.

Choose $\varepsilon_0 = \varepsilon_0(\delta_1)$ as in conclusion (5), i.e.,

$$\|\Delta f_\varphi(t,\cdot)\|_{L^2} \geq \varepsilon_0 \quad \text{if} \quad \operatorname{dist}_{W_p^2}(f_\varphi(t,\cdot),K) \geq \delta_1 \ \ \forall\, t,$$

and choose

$$0 < \delta_1 < \left(\frac{\varepsilon}{2C_\varepsilon}\right)^{2/\gamma} \frac{\varepsilon_0^2}{E(\varphi) - c}$$

such that

$$\operatorname{dist}_{W_p^2}\big(g_{\varphi_0}(1-\delta_1,\cdot),\widetilde{u}\big) < \frac{\varepsilon}{2}.$$

Again, we choose $\delta_2 > 0$ such that $\operatorname{dist}_{W_p^2}(\varphi,\varphi_0) < \delta_2$ implies

$$\operatorname{dist}_{W_p^2}\big(g_{\varphi_0}(1-\delta_1,\cdot), g_\varphi(1-\delta_1,\cdot)\big) < \frac{\varepsilon}{2}.$$

Therefore we have

$$\operatorname{dist}_{W_p^2}\big(g_\varphi(1-\delta_1,\cdot),\widetilde{u}\big) < \varepsilon \qquad \forall\, \varphi \in B_{\delta_2}(\varphi_0).$$

We want to prove

$$\operatorname{dist}_{W_p^2}(g_\varphi(\tau,\cdot),\widetilde{u}) < \varepsilon \qquad \forall\, (\tau,\varphi) \in (1-\delta_1, 1] \times B_{\delta_2}(\varphi_0).$$

If not, $\exists\, \tau'' > \tau' > 1-\delta_1$ and $\varphi_1 \in B_{\delta_2}(\varphi_0)$ such that

$$g_{\varphi_1}(\tau',\cdot) \in \partial B_{\varepsilon/2}(\widetilde{u}), \quad g_{\varphi_1}(\tau'',\cdot) \in \partial B_{\varepsilon}(\widetilde{u})$$

and

$$g_{\varphi_1}(\tau,\cdot) \in B_{\varepsilon}(\widetilde{u}) \setminus B_{\varepsilon/2}(\widetilde{u}) \quad \forall \tau \in (\tau',\tau'').$$

Then we have

$$\begin{aligned}\frac{\varepsilon}{2} &\le \mathrm{dist}_{W^2_p}\big(g_{\varphi_1}(\tau',\cdot), g_{\varphi_1}(\tau'',\cdot)\big)\\ &= \mathrm{dist}_{W^2_p}\big(f_{\varphi_1}(t',\cdot), f_{\varphi_1}(t'',\cdot)\big)\\ &\le C_\varepsilon |t'-t''|^{\gamma/2}.\end{aligned}$$

On the other hand,

$$\begin{aligned}\varepsilon_0^2|t'-t''| &\le \int_{t'}^{t''} \|\Delta f_{\varphi_1}(t,\cdot)\|^2_{L^2}dt\\ &= E\big(f_{\varphi_1}(t'',\cdot)\big) - E\big(f_{\varphi_1}(t',\cdot)\big)\\ &= E\big(g_{\varphi_1}(\tau'',\cdot)\big) - E\big(g_{\varphi_1}(\tau',\cdot)\big)\\ &= \big(E(\varphi)-c\big)|\tau''-\tau'|\\ &< \delta_1(E(\varphi)-c),\end{aligned}$$

which implies that

$$\delta_1 > \frac{\varepsilon_0^2}{E(\varphi)-c}|t'-t''| \ge \frac{\varepsilon_0^2}{(E(\varphi)-c)}\left(\frac{\varepsilon}{2C_\varepsilon}\right)^{2/\gamma}.$$

This is a contradiction.

Verification for case (2). $\forall\, \varphi_0 \in E^{-1}(c)$, $\forall\, \varepsilon > 0$, we want to find $\delta > 0$ such that $\mathrm{dist}(\varphi,\varphi_0) < \delta$ implies $\mathrm{dist}\big(\eta(\tau,\varphi),\varphi_0\big) < \varepsilon$.

Similar to the above argument, let us choose

$$0 < \delta_1 < \varepsilon_0^2\left(\frac{\varepsilon}{2C_\varepsilon}\right)^{2/\gamma}.$$

Find $0 < \delta < \varepsilon/2$ such that

$$E(\varphi) - c < \delta_1 \qquad \forall\, \varphi \in B_{\delta}(\varphi_0).$$

If our conclusion were not true, by the same procedure, we would have
(i) $\frac{\varepsilon}{2} \le C_\varepsilon|t'-t''|^{\gamma/2}$, and
(ii) $\varepsilon_0^2|t'-t''| \le (E(\varphi)-c)|\tau''-\tau'| < \delta_1$.
This is again a contradiction.

The continuity of η is proved, so that E_c is a strong deformation retract of E_d, $d < m_{\mathcal{F}} + b$.

Suppose that $\forall\, d < m_{\mathcal{F}}+b$ there are only isolated harmonic maps. Since $K \cap E_d$ is compact, they are finite. There are only isolated critical values (at most with limit $m_{\mathcal{F}} + b$)

$$m_{\mathcal{F}} = c_0 < c_1 < \cdots < c_i < \cdots < m_{\mathcal{F}} + b.$$

For each c_i, there are finitely many harmonic maps:

$$K_{c_i} = \{u_{ij} \mid j = 1, 2, \ldots, m_i\}.$$

$\forall\, d < m_{\mathcal{F}} + b$, let

$$M_q^d = \sum_{c_i < d} \sum_{j=1}^{m_i} \text{rank } C_q(u_{ij}; G)$$

be the qth Morse type number, $q = 0, 1, 2, \ldots$, for the manifold $E_d \cap \mathcal{F}$ and let

$$\beta_q^d = \text{rank } H_q(E_d \cap \mathcal{F}, G)$$

be the qth Betti number, $q = 0, 1, 2, \ldots$, for $E_d \cap \mathcal{F}$.

Comparing Corollary 4.1 and Theorem 4.3 of Chapter I with the above deformation, we obtain Morse inequalities for harmonic maps below a certain level.

Theorem 2.1. *Let $\mathcal{F}$ be a component of $C_\psi^{2,\gamma}(\overline{M}^2, N)$, and let $d < m_{\mathcal{F}} + b$. Assume that $\pi_2(N) = 0$, and that in the level set $E_d \cap \mathcal{F}$ there are only isolated harmonic maps. Then there exists a formal power series with nonnegative coefficients $Q^d(t)$ such that*

$$\sum_{q=0}^{\infty} M_q^d t^q = \sum_{q=0}^{\infty} \beta_q^d t^q + (1+t) Q^d(t).$$

3. Morse Decomposition

In this section, we will study the handle body decomposition of the level sets of the energy function, under the assumption that all harmonic maps in these level sets are nondegenerate.

Let u_0 be a harmonic map from M to N. Let $E = u_0^* TN$ be the pull back bundle over M. Let $\mathcal{O}$ be a neighborhood in $C^\infty(M, N)$ which contains the section $u_0(M)$. It is obvious that $\mathcal{O}$ is diffeomorphic to a neighborhood

$\mathcal{O}_E$ of the zero section of the tangent space $T_{u_0}(E)$. The diffeomorphism is realized by the exponential map

$$\begin{array}{ccc} \sigma \in \mathcal{O}_E & \xrightarrow{\exp_{u_0(x)}} & \mathcal{O} \\ \cap & & \cap \\ T_{u_0}(E) & & C^\infty_\psi(M,N). \end{array}$$

After this we do not concern ourselves with the tangent vector σ with its exponential map $\exp_{u_0(x)}\sigma(x)$. We shall restrict our studies to the neighborhood $\mathcal{O}_E$ of the vector space $T_{u_0}(E)$. The Taylor expansion of the energy functional at u_0 is as follows:

$$E(u) = E(u_0) + \frac{1}{2}d^2E(u_0)(\sigma,\sigma) + R(\sigma),$$

where $u(x) = \exp_{u_0(x)}\sigma(x)$, and the remainder $R(\sigma)$ satisfies

$$|R(\sigma)| = o\left(\int_M |\nabla\sigma|^2\right).$$

and

$$|dR(\sigma)| = o\left(\left(\int_M |\nabla\sigma|^2\right)^{1/2}\right).$$

As to the Hessian $d^2E(u_0)$, it is well-known (see Eells–Lemaire [EL1]) that, $\forall\, \sigma,\eta \in C^\infty(T_{u_0}(E))$,

$$d^2E(u_0)(\sigma,\eta) = \int_M \langle J_{u_0}\sigma,\eta\rangle dV_g,$$

where

$$J_{u_0}\sigma = -\Delta^{u_0}\sigma - \text{Trace}\, R^N(du_0,\sigma)du_0$$

is the Jacobi operator.

Noticing that J_{u_0} is a linear self-adjoint elliptic differential operator, with domain $W^2_2 \cap \overset{\circ}{W}{}^1_2(T_{u_0}(E))$, J_{u_0} can be extended to be a continuous bilinear form on the Hilbert space $\overset{\circ}{W}{}^1_2(T_{u_0}(E))$. And since

$$d^2E(u_0)(\sigma,\sigma) \geq \|\sigma\|^2_{W^1_2} - C_1(u_0)\|\sigma\|^2,$$

where $C_1(u_0)$ is a constant depending on u_0, the negative eigenspace of J_{u_0} must be finitely dimensional. The dimension of the negative eigenspace of J_{u_0} is called the Morse index of the harmonic map u_0 and is denoted by $\text{ind}(u_0)$. u_0 is called nondegenerate if J_{u_0} is invertible.

For the self adjoint operator J_{u_0}, it is well known that we have a spectral decomposition E_λ and two projections P_+ and P_-, which correspond the positive and negative eigenspaces respectively. For any $\sigma \in C^{2,\gamma}_{\psi}(u_0^*TN)$, we have

$$\sigma_\pm := P_\pm \sigma \in C^{2,\gamma}_{\psi}(u_0^*TN).$$

The two square roots

$$A_\pm := \left(P_\pm(\pm J_{u_0})P_\pm\right)^{1/2}$$

are well defined, and we have that

$$\|A_\pm \sigma\|_{L^2} \text{ is equivalent to } \|\sigma_\pm\|_{W_2^1}.$$

In the following, we shall denote $\|A_\pm\sigma\|_{L^2}$ by $|\sigma_\pm|$, and let $|\sigma|^2 = |\sigma_+|^2 + |\sigma_-|^2$. Thus, the energy function is written as follows:

$$E(u) = c + \frac{1}{2}\left(|\sigma_+|^2 - |\sigma_-|^2\right) + R(\sigma).$$

For any given $0 < \gamma < 1$, we choose $\tau > 0$, satisfying

$$\frac{\tau}{1-\tau} < \sqrt{\frac{1-\gamma}{1+\gamma}},$$

and $\delta > 0$ such that, for a W_p^2-ball B_δ with radius δ, centered at the zero section of $C^{2,\gamma}(u_0^*TN)$, we have

$$|R(\sigma)| < \frac{1}{2}\gamma|\sigma|^2 \tag{3.1}$$

and

$$|dR(\sigma)| < \tau|\sigma|. \tag{3.2}$$

$\forall\, \sigma \in U = B_\delta$. (In the following we always denote B_δ by U.) These imply that

$$\frac{1}{2}(1-\gamma)|\sigma_+|^2 - \frac{1}{2}(1+\gamma)|\sigma_-|^2 \le E(u) - \sigma \le \frac{1}{2}(1+\gamma)|\sigma_+|^2 - \frac{1}{2}(1-\gamma)|\sigma_-|^2. \tag{3.3}$$

Now we are going to construct a series of deformations, which deform the level set $E_{c+\varepsilon}$ (for suitable $\varepsilon > 0$) to $E_{c-\varepsilon}$ attached with cells:

(1) According to Lemma 2.1 we have a strong deformation retract η_1, which deforms $E_{c+\varepsilon}$ into E_c, for $\varepsilon > 0$ small, if $E^{-1}(c, c+\varepsilon) \cap K = \emptyset$.

(2) By conclusion (6) of Theorem 1.1, we have $\varepsilon > 0$ and a strong deformation retract η_2, which deforms E_c into $E_{c-\varepsilon} \cup (E_c \cap U)$ and satisfies $\eta_2(1, E_c \cap V) \subset E_c \cap U$, $\eta_2(1, E_c \setminus V) \subset E_{c-\varepsilon}$.

(3) Let us define two conical neighborhoods:

$$C_\gamma = \left\{\sigma \in U \mid |\sigma_+| \le \sqrt{\frac{1-\gamma}{1+\gamma}}|\sigma_-|\right\},$$

$$\widehat{C}_\gamma = \left\{\sigma \in U \mid |\sigma_+| \le \sqrt{\frac{1+\gamma}{1-\gamma}}|\sigma_-|\right\}.$$

Inequality (3.3) implies that

$$C_\gamma \subset E_c \cap U \subset \widehat{C}_\gamma.$$

Lemma 3.1. *There exists a strong deformation retract η_3, which deforms $E_{c-\varepsilon} \cup (E_c \cap U)$ into $E_{c-\varepsilon} \cup C_\gamma$.*

Proof. Noticing that $\forall\, \sigma \notin E_{c-\varepsilon} \cup C_\gamma$, with $\sigma \in U$, we have

$$|\sigma_-| \le \sqrt{\frac{1+\gamma}{2}}\delta.$$

Let $K = \sqrt{\frac{2}{1+\gamma}} - 1$ (> 0), and define a flow on U as follows:

$$\eta(t,\sigma) = (1-t)\sigma_+ + (1+tK)\sigma_-.$$

We have

(a) $\eta(0,\sigma) = \sigma$.

(b) $\eta(1,\sigma) = \sqrt{\frac{2}{1+\gamma}}\sigma_- \in U$ if $\sigma \notin C_\gamma$.

(c) Letting $\varphi(t) = E(\eta(t,\cdot))$, we have

$$\begin{aligned}\varphi'(t) &= -\frac{|\eta_+|^2}{1-t} - K|\eta_-|^2 + \langle dR(\eta(t,\cdot)), -\sigma_+ + K\sigma_-\rangle \\ &\le (1-\tau)\left[-\frac{|\eta_+|^2}{1-t} - K|\eta_-|^2 + \frac{\tau}{1-\tau}\left(\frac{1}{1-t} + K\right)|\eta_+|\,|\eta_-|\right] \\ &= (1-\tau)\left[-\frac{|\eta_+|}{1-t}\left(|\eta_+| - \frac{\tau}{1-\tau}|\eta_-|\right) - K|\eta_-|\left(|\eta_-| - \frac{\tau}{1-\tau}|\eta_+|\right)\right],\end{aligned}$$

where $\eta = (\eta_+, \eta_-)$. If $\eta \in (E_c \cap U) \setminus C_\gamma \subset \widehat{C}_\gamma \setminus C_\gamma$, then we have

$$|\eta_+| \ge \sqrt{\frac{1-\gamma}{1+\gamma}}|\eta_-| \ge \frac{\tau}{1-\tau}|\eta_-|$$

and

$$|\eta_-| \ge \sqrt{\frac{1-\gamma}{1+\gamma}}|\eta_+| \ge \frac{\tau}{1-\tau}|\eta_+|.$$

It follows that

$$\varphi'(t) < 0 \qquad \forall\, \eta \in \widehat{C}_\gamma \setminus C_\gamma. \tag{3.4}$$

Combining (a) and (b) with (c), we obtain

$$\eta(t,\sigma) \in (E_c \cap U) \quad \forall (t,\sigma) \in [0,1] \times ((E_c \backslash U) \backslash (C_\gamma \cup E_{c-\varepsilon})),$$

provided by the fact that $C_\gamma \subset E_c \cap U$.

From (a) and (b), we see that if $\sigma \notin E_{c-\varepsilon} \cup C_\gamma$, but $\sigma \in E_c \cap U$, then there is a unique $t^* \in (0,t)$ such that $\eta(t^*,\sigma) \in E^{-1}(c-\varepsilon) \cup \partial C_\gamma$. The uniqueness and the continuous dependence of t^* on σ are verified by the transversality $\eta \pitchfork E^{-1}(c-\varepsilon) \cup \partial C_\gamma$, which follows from inequality (3.4).

Let us define

$$\eta_3(t,\sigma) = \begin{cases} \eta(t^*t,\sigma) & \text{if} \quad \sigma \in (E_c \cap U) \setminus (E_{c-\varepsilon} \cup C_\gamma) \\ \sigma & \quad\quad \sigma \in E_{c-\varepsilon} \cup C_\gamma. \end{cases}$$

This is the deformation we need.

(4) Noticing that $\forall\, \sigma \in E_{c-\varepsilon} \cap C_\gamma$,

$$-\varepsilon \geq E(u) - c \geq \frac{1-\gamma}{2}|\sigma_+|^2 - \frac{1+\gamma}{2}|\sigma_-|^2,$$

we have

$$|\sigma_-| > \sqrt{\frac{2\varepsilon}{1+\gamma}}, \tag{3.5}$$

so that $E_{c-\varepsilon} \cap C_\gamma \subset S := \left\{\sigma \in C_\gamma \mid |\sigma_-| > \sqrt{\frac{2\varepsilon}{1+\gamma}}\right\}$.

On the other hand, $\forall\, \sigma \in S$,

$$|\sigma_-| \geq k_0|\sigma_+| + \delta_0,$$

where

$$k_0 = \frac{1}{2}\sqrt{\frac{1+\gamma}{1-\gamma}} \quad \text{and} \quad \delta_0 = \frac{1}{2}\sqrt{\frac{2\varepsilon}{1+\gamma}}.$$

Let us define

$$T_{k_0,\delta_0} = \left\{\sigma \in C_\gamma \mid |\sigma_-| \geq k_0|\sigma_+| + \delta_0\right\}.$$

In the following, we prove

Lemma 3.2. *There is a strong deformation retract η_4 which deforms $E_{c-\varepsilon} \cup C_\gamma$ into $E_{c-\varepsilon} \cup T_{k_0,\delta_0} \cup \{\theta_+\} \times B^k_{\delta_0}$, where $k = ind(u_0)$.*

Proof. We define

$$\eta_4(t,\sigma) = \begin{cases} \sigma & \sigma \in E_{c-\varepsilon} \cup T_{k_0,\delta_0}, \\ \sigma_- + \left[1 - t\left(1 - \frac{|\sigma_-|-\delta_0}{k_0|\sigma_+|}\right)\right]\sigma_+ & \sigma \in C_\gamma, \delta_0 \leq |\sigma_-| \leq k_0|\sigma_+| + \delta_0, \\ \sigma_- + (1-t)\sigma_+ & \sigma \in C_\gamma \cap \{|\sigma_-| \leq \delta_0\}. \end{cases}$$

(5) Choose $\varepsilon > 0$ small enough that

$$\varepsilon < \frac{\delta^2(1-\gamma)}{2}. \tag{3.6}$$

Define

$$0 < \mu < \sqrt{\frac{1 - \left(\gamma + \frac{2\varepsilon}{\delta^2}\right)}{1 + \left(\gamma + \frac{2\varepsilon}{\delta^2}\right)}}. \tag{3.7}$$

We consider the energy function on the conical section of the sphere ∂B_δ: $S_\mu = \{\sigma \in \partial B_\delta \mid |\sigma_+| < \mu|\sigma_-|\}$. Letting $\sigma \in S_\mu$, we have

$$\begin{aligned} E(u) - c &\leq \frac{1+\gamma}{2}|\sigma_+|^2 - \frac{1-\gamma}{2}|\sigma_-|^2 \\ &\leq \left(\frac{1+\gamma}{2}\mu^2 - \frac{1-\gamma}{2}\right)|\sigma_-|^2 \\ &= -\frac{1}{2}\left(\frac{1-\mu^2}{1+\mu^2} - \gamma\right)|\sigma_-|^2(1+\mu^2). \end{aligned}$$

Since

$$\delta^2 = |\sigma_+|^2 + |\sigma_-|^2 \leq (1+\mu^2)|\sigma_-|^2,$$

then

$$E(u) - c \leq -\frac{1}{2}\left(\frac{1-\mu^2}{1+\mu^2} - \gamma\right)\delta^2 < -\varepsilon,$$

i.e., $S_\mu \subset \overset{\circ}{E}_{c-\varepsilon}$.

Lemma 3.3. *The exit set of the flow*

$$\eta(t,\sigma) = e^{-k_1 t}\sigma_+ + e^{k_2 t}\sigma_-$$

on the ball B_δ, is the set S_μ, where $\frac{k_2}{k_1} = \mu^2$, $k_1, k_2 > 0$.

Proof. The flow η remains on the plane generated by the two vectors σ_+ and σ_-. Suppose that η meets ∂B_δ at time t_0, and let $\eta_+ = e^{-k_1 t_0}\sigma_+$, $\eta_- = e^{k_2 t_0}\sigma_-$.

Choosing suitable coordinates $(\eta_+, \eta_-) = \delta(\cos\theta, \sin\theta)$, we assume that the flow η leaves the ball B_δ. By comparing the tangents of the ball with the tangents of the flow, we see

$$-\frac{k_2}{k_1}\operatorname{tg}\theta > -\operatorname{ctg}\theta,$$

i.e.,

$$|\eta_+| < \mu|\eta_-|.$$

In other words, $(\eta_+, \eta_-) \in S_\mu$. □

Lemma 3.4. *There is a strong deformation retract η_5 which deforms the set $E_{c-\varepsilon} \cup T_{k_0,\delta_0} \cup \left(\{\theta_+\} \times B^k_{\varepsilon_0}\right)$ into $E_{c-\varepsilon} \cup \left(\{\theta_+\} \times B^k_\delta\right)$.*

Proof. We use the flow η defined in Lemma 3.3. Because $S_\mu \subset \overset{\circ}{E}_{c-\varepsilon}$, if $\sigma \notin E_{c-\varepsilon}$, then there must be a $t^* \in (0,\infty)$ such that $\eta(t^*,\sigma) \in E^{-1}(c-\varepsilon)$.

On the other hand, $\eta(t,\cdot)$ is transversal to the level set $E^{-1}(c-\varepsilon)$, provided by the fact that

$$\begin{aligned}
\frac{d}{dt}E(\eta(t,\sigma)) &= -\langle\eta_+, k_1\eta_+\rangle - \langle\eta_-, k_2\eta_-\rangle + \langle dR(\eta), -k_1\eta_+ + k_2\eta_-\rangle \\
&\leq -k_1|\eta_+|^2 - k_2|\eta_-|^2 + \tau\big(|\eta_+| + |\eta_-|\big)\big(k_1|\eta_+| + k_2|\eta_-|\big) \\
&= -(1-\tau)\left[k_1|\eta_+|^2 + k_2|\eta_-|^2 - \frac{\tau}{1-\tau}(k_1+k_2)|\eta_+|\,|\eta_-|\right] \\
&= -(1-\tau)k_1\left[|\eta_+|^2 + \mu^2|\eta_-|^2 - \frac{\tau}{1-\tau}(1+\mu^2)|\eta_+|\,|\eta_-|\right] \\
&< 0,
\end{aligned}$$

if we choose

$$\frac{\tau}{1-\tau} < \frac{2\mu}{1+\mu^2}. \tag{3.8}$$

Therefore, $\forall\ \sigma \in T_{k_0,\delta_0} \setminus E_{c-\varepsilon}$, $t^* = t^*(\sigma)$ is uniquely determined and continuous. We define our deformation retract as follows:

$$\eta_\pm(t,\sigma) = \begin{cases} \sigma & \text{if } \sigma \in E_{c-\varepsilon} \\ \eta(t^*(\sigma)t, \sigma) & \text{if } \sigma \in T_{k_0\delta_0} \setminus E_{c-\varepsilon} \\ e^{k_2 t^*(\delta_0\sigma_-/|\sigma_-|)}\sigma_- & \text{if } \sigma \in \{\theta_+\} \times B^k_{\delta_0}. \end{cases}$$

For any two strong deformation retracts

$$X_1 \xrightarrow{\varphi_1} X_2 \xrightarrow{\varphi_2} Y,$$

we define their composition as follows:

$$\varphi(t,x) = \begin{cases} \varphi_1(2t,x) & t \in \left[0, \frac{1}{2}\right] \\ \varphi_2(2t-1, \varphi_1(1,x)) & t \in \left[\frac{1}{2}, 1\right]. \end{cases}$$

This is again a strong deformation retract $\varphi\colon X_1 \to Y$, which is denoted by $\varphi = \varphi_2 \circ \varphi_1$.

Now we come to our main conclusion in this section.

Theorem 3.1. *Assume that $\pi_2(N) = 0$, and let $\mathcal{F}$ be a component of $C^{2,\gamma}_\psi(M^2, N)$. Suppose that on the level $E^{-1}(c) \cap \mathcal{F}$, $c < m_{\mathcal{F}} + b$, there are only nondegenerate harmonic maps $u_1, \ldots, u_\ell$, with Morse indices $m_1, \ldots, m_\ell$ respectively. Then the level set $E_{c-\varepsilon} \cap \mathcal{F}$, attached with ℓ handles, whose dimensions correspond to these indices, is a strong deformation retract of $E_{c+\varepsilon} \cap \mathcal{F}$, for suitable $\varepsilon > 0$.*

Proof. We choose $\gamma = \frac{1}{5}$, $\tau = \frac{1}{3}$, and $\mu = \frac{1}{2}$. Then we have $\delta > 0$ small enough such that (3.1) and (3.2) hold. Choose $\varepsilon > 0$ small enough such that $\varepsilon < \frac{\delta^2}{10}$ and that conclusion (6) holds. Inequalities (3.6), (3.7) and (3.8) are satisfied automatically. The strong deformation retract now is defined to be

$$\rho = \rho_5 \circ \rho_4 \circ \rho_3 \circ \rho_2 \circ \rho_1.$$

Combining Lemmas 3.1 and 3.2 with 3.4, we obtain our conclusion.

Corollary 3.1. *Suppose that u_0 is a nondegenerate harmonic map with $E(u_0) = c$ and $u_0 \in C^{2,\gamma}_{\psi}(\overline{M}, N)$, $\gamma > 0$. Assume that $c < m_{\mathcal{F}} + b$, if $u_0 \in \mathcal{F}$ and $\pi_2(N) = 0$. Then we have*

$$C_q(u_0; G) = \delta_{qk} G,$$

where $k = ind(u_0)$.

Sections 2 and 3 are adapted from Chang [Cha11].

4. Existence and Multiplicity for Harmonic Maps

We present here a few theorems about the existence and the multiplicity for harmonic maps. We follow the notations in previous sections.

Theorem 4.1. (Sacks–Uhlenbeck [SaU1], Lemaire [Lem1]). *If $\pi_2(N) = 0$, then for any homotopy class $\mathcal{F}$ of maps from M to N (with prescribed boundary value $\psi \in C^{2,\gamma}(\partial M, N)$ in the case $\partial M \neq \emptyset$), there exists a harmonic map.*

Proof. We choose any $d \in (m_{\mathcal{F}}, m_{\mathcal{F}} + b)$. Obviously,

$$\beta_0^d := \text{rank } H_0(E_d \cap F; G) \neq 0.$$

It follows $M_0^d \neq 0$. Consequently, there exists a minimum u of the energy function. Therefore $u \in C^{2,\gamma}_{\psi}(M, N)$ is a harmonic map in $\mathcal{F}$.

Theorem 4.2. (Brezis–Coron [BrC1], Jost [Jos1]). *Suppose that $N = S^2$, and that $\psi \in C^{2,\gamma}(\partial M, S^2)$ is not a constant. Then there exist at least two homotopically different harmonic maps.*

Proof. By the argument used in Theorem 4.1, we obtain a minimal energy harmonic map $\widetilde{u}$ among all homotopy classes $E(\widetilde{u}) = m$.

The second harmonic map will be obtained by constructing a map v homotopically different from $\widetilde{u}$ having energy

$$E(v) < m + b. \tag{4.1}$$

The construction of the map v is as follows: Choose a small disc D_0 on M, take an isometric copy D_1, and identify D_0 and D_1 along their boundary to obtain a 2-sphere S^2. Take a map $w: S^2 \to N = S^2$, which represents the generator of $\pi_2(N)$, and coincides with $\widetilde{u}$ on D_0, such that the map

$$v = \begin{cases} \widetilde{u} & M \setminus D_0 \\ w\big|_{D_1} & \text{on } D_0 \text{ and identify } D_0 \text{ with } D_1 \end{cases}$$

satisfies (4.1).

We shall construct w explicitly. Since $\psi \neq$ const., $\widetilde{u} \neq$ const., we can choose a point $x_0 \in M$, for which $\nabla\widetilde{u}(x_0) \neq \theta$. Rotating S^2, one can assume that $\widetilde{u}(x_0)$ is the south pole. Let $\pi\colon S^2 \to \mathbb{C}$ be the stereographic map from the north pole.

We choose local coordinates $z \in \mathbb{C}$ in a neighborhood of x_0, such that $z(x_0) = 0$, $|z| < \varepsilon$, $\varepsilon > 0$. Thus

$$|\pi \circ \widetilde{u}(z) - \nabla(\pi \circ \widetilde{u})(0)z| = O(|z|^2).$$

We denote $\nabla(\pi \circ \widetilde{u})(0)$ by a, which is a nonzero complex number, and write $z = re^{i\theta}$.

Letting

$$t(z) = (\pi \circ \widetilde{u})(\varepsilon e^{i\theta}) \left(\frac{r}{\varepsilon^2} - \frac{1-\varepsilon}{\varepsilon} \right) + a \left(\frac{1}{\varepsilon} - \frac{r}{\varepsilon^2} \right) \varepsilon e^{i\theta},$$

we have

$$t(\varepsilon e^{i\theta}) = \pi \circ \widetilde{u}(\varepsilon e^{i\theta}) \quad \text{and} \quad t\big((\varepsilon - \varepsilon^2)e^{i\theta}\big) = a\varepsilon e^{i\theta}.$$

Then we define a function $\varphi\colon \mathbb{C} \to \mathbb{C}$ as follows:

$$\varphi(z) = \begin{cases} az & |z| \le \varepsilon - \varepsilon^2 \\ t(z) & \varepsilon - \varepsilon^2 < |z| \le \varepsilon \\ |\pi \circ \widetilde{u}(\varepsilon e^{i\theta})|^2 \overline{(\pi \circ \widetilde{u}(\varepsilon^2 z^{-1})}^{-1} & |z| \ge \varepsilon. \end{cases}$$

If $\varepsilon > 0$ is small, φ is continuous and surjective. The map w is defined to be $\pi^{-1} \circ \varphi \circ \pi$.

Noticing that π is conformal, and that the energy is conformal invariant, we may compute the energy of w by the energy of $\pi^{-1}\varphi$.

Now

$$E(v) = \frac{1}{2} \int_{M \setminus B(x_0, \varepsilon)} |\nabla \widetilde{u}|^2 + \frac{1}{2} \int_{B(x_0, \varepsilon - \varepsilon^2)} |\nabla w|^2 + \frac{1}{2} \int_{B_\varepsilon \setminus B_{\varepsilon - \varepsilon^2}} |\nabla(\pi^{-1} t)|^2.$$

Since

$$\frac{1}{2} \int_{M \setminus B(x_0, \varepsilon)} |\nabla \widetilde{u}|^2 = E(\widetilde{u}) - O(\varepsilon^2) \qquad (a \neq 0)$$

$$\frac{1}{2} \int_{B(x_0, \varepsilon - \varepsilon^2)} |\nabla w|^2 \le \pi = b,$$

and

$$\int_{B_\varepsilon\setminus B_{\varepsilon-\varepsilon^2}} |\nabla(\pi^{-1}t)|^2$$
$$\leq C\int_{\varepsilon-\varepsilon^2}^{\varepsilon}\int_0^{2\pi}\left[|(\pi\circ\widetilde{u})(\varepsilon e^{i\theta})-a\varepsilon e^{i\theta}|^2\frac{1}{\varepsilon^4}+\left|(\pi\circ\widetilde{u})_\theta(\varepsilon e^{i\theta})\left(\frac{r}{\varepsilon^2}-\frac{1-\varepsilon}{\varepsilon}\right)\right.\right.$$
$$\left.\left.+ia\left(\frac{1}{\varepsilon}-\frac{r}{\varepsilon^2}\right)e^{i\theta}\right|^2\right]rdrd\theta$$
$$=O(\varepsilon^3),\quad \text{where}\quad C=\mathrm{Max}\|d\pi^{-1}\|^2.$$

Therefore, for $\varepsilon>0$ small enough, we have

$$E(v)<m+b.$$

The remaining part of the proof is the same as that in Theorem 4.1.

Theorem 4.3. *Given a Riemann surface M with boundary ∂M, if $\psi\in C^{2+\gamma}(\partial M,S^n)$, $\gamma>0$, $n\geq 3$, is not a constant map, then there exist at least two harmonic maps from M to S^n in the (nontrivial) homotopy class $\mathcal{F}\subset C_\psi^{2+\gamma}(M,S^n)$.*

Proof. The proof is similar to the above case. Since $\pi_2(S^n)=0$, $n\geq 3$, first, we apply Theorem 4.1 to obtain a minimum of the energy functional in $\mathcal{F}$, which is a harmonic map $u_{\mathcal{F}}$.

Then we shall prove that there exists $d<m_{\mathcal{F}}+b$ such that $\pi_{n-2}(E_d\cap\mathcal{F})\neq 0$, where $\mathcal{F}$ is regarded as a component of $C_\psi^{2+\gamma}(M,S^n)$; or, equivalently, we construct an essential map $\sigma\in C(S^{n-2},E_d\cap\mathcal{F})$ such that

$$\sup_{s\in S^{n-2}} E(\sigma(s))\leq d<m_{\mathcal{F}}+b.$$

The existence of such a map σ was constructed by Benci–Coron [BeC1], in a manner very similar to the construction of v in Theorem 4.2. We are satisfied with pointing out the main idea, and we omit the details. As in the proof of Theorem 4.2, $\psi\neq$ const., $\widetilde{u}\neq$ const. Again we choose a local chart U, outside which σ is defined to be $\widetilde{u}$, and inside which, we choose a small disk $B_\delta(z_0)$ on which $\sigma(s)(z)\colon S^{n-2}\times B_\delta(z_0)\to S^n$ is a homeomorphism. The map σ is connected smoothly. After a careful construction, this makes

$$E(\sigma(s))<m_{\mathcal{F}}+b.$$

By the fact that $\pi_2(S^n)=0$, we see that $\sigma(s)\in\mathcal{F}\ \forall\ s\in S^{n-2}$.

We point out first that σ is essential. Indeed,

$$\sigma(s)(z)=\widetilde{u}(z)\quad\text{for}\quad z\in M\setminus U$$

so, outside the ball $B_\delta(z_0)$, $\sigma(s)(z)$ is contractible. And inside $B_\delta(z_0)$, $\sigma: S^{n-2} \times B_\delta(z_0) \to S^n$ is a homeomorphism. Therefore, the image of σ is homotopic to a n-topological ball. Thus σ cannot shrink to a constant map, i.e., σ is essential.

Next, we apply the Hurewicz theorem, which implies the existence of an integer $0 < k \le n-2$ such that $H_k(E_d \cap \mathcal{F}, \mathbb{Z}) \neq 0$. The existence of the second harmonic map in $\mathcal{F}$ follows from Morse inequalities (Theorem 2.3).

Remark 4.1. Theorem 4.3 was obtained by Benci–Coron [BeC1] in case $M = D^2$, the 2-disk. See also W.Y. Ding [Din1] and K.C. Chang [Cha10]. For other results in this direction, readers are referred to Sacks–Uhlenbeck [SaU1,2], Schoen–Yau [ScY1] and Jost [Jos1].

Readers who need to know more about harmonic maps are referred to Eells–Lemaire [EeL1–2].

5. The Plateau Problem for Minimal Surfaces

The Problem. Given a Jordan curve Γ in $\mathbb{R}^n$, one asks for a surface S with minimal area spanning Γ. For technical reasons, we assume that Γ is defined by $\alpha: S^1 \to \mathbb{R}^n$, which is a C^2-diffeomorphism.

It is known from differential geometry that such a surface S, if it exists, has mean curvature 0 (see, for example, Ossermann [Oss1]).

We introduce isothermal coordinates on S, which parametrize S by a function $\varphi: D \to \mathbb{R}^n$, where

$$D = \{z = (x, y) \mid x^2 + y^2 < 1\}, \quad \text{with} \quad \partial D = S^1.$$

Thus, φ satisfies the following nonlinear differential system:

$$\begin{cases} \Delta\varphi = 0 & \text{in } D, \\ |\varphi_x|^2 - |\varphi_y|^2 = \langle \varphi_x, \varphi_y \rangle = 0 & \text{in } D, \\ \varphi\big|_{\partial D}: \partial D \to \Gamma & \text{is an oriented parametrization of } \Gamma, \end{cases} \tag{5.1}$$

where $\langle\ ,\ \rangle$ is the scalar product on $\mathbb{R}^n$, and $|\cdot| = \langle \cdot, \cdot \rangle^{1/2}$.

Although solutions of the differential system (5.1) no longer require S having minimal area, they are also called minimal surfaces.

We use complex notations,

$$z = x + iy,$$

and introduce the conformal group on the disc:

$$G = \left\{ g(z) = e^{i\varphi_0} \frac{a + z}{1 + \bar{a}z} \mid a \in \mathbb{C}, |a| < 1, \varphi_0 \in \mathbb{R} \right\}.$$

It is easily seen that if φ is a solution of (5.1), then $\varphi \circ g$, $\forall\, g \in G$, are all solutions of (5.1), i.e., the system is G-equivariant.

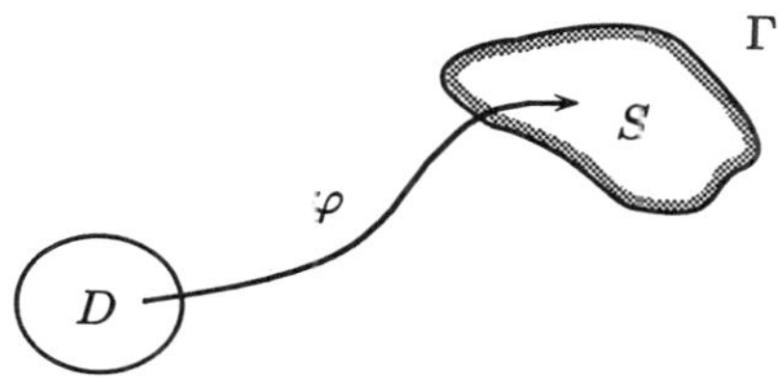

In order to avoid the nondetermination of solutions, we normalize the solutions as follows: Let P_1, P_2, P_3 be an oriented triple of distinct points on Γ, and assume

$$\varphi\left(e^{\frac{2ij\pi}{3}}\right) = P_j, \qquad j = 1, 2, 3.$$

We shall study this problem via critical point theory. Inspired by the early work of Courant [Cou1], the conformal condition (the second group of equations) can be solved by minimizing the Dirichlet integral. This suggests the following strategy: Define a parametrization set

$$\mathcal{M} = \left\{u \in C \cap H^{1/2}([0, 2\pi], \mathbb{R}^1) \mid u\left(\frac{2j}{3}\pi\right) = \frac{2j}{3}\pi, \quad j = 0, 1, 2, 3, \quad \text{and } u \text{ is nondecreasing}\right\}.$$

This is a closed convex subset of the Banach space $X = C \cap H^{1/2}([0, 2\pi], \mathbb{R}^1)$.

(1) $\forall\, u \in \mathcal{M}$, solve the Dirichlet problem:

$$\begin{cases} \Delta\varphi = 0 & \text{in } \overset{\circ}{D}, \\ \varphi\big|_{\partial D} = \alpha \circ u. \end{cases}$$

The unique solution $\varphi \in H^1(D, \mathbb{R}^n)$, depending on u continuously, is denoted by $\varphi = \mathcal{R}(\alpha \circ u)$. Then, we define a functional on $\mathcal{M}$:

$$J(u) = \frac{1}{2}\int_D |\nabla \mathcal{R}(\alpha \circ u)|^2 dx \wedge dy.$$

(2) We find the critical point of J with respect to the closed convex subset $\mathcal{M}$, and then

(a) verify that the generalized critical points solve the conformal condition, and

(b) the extended critical point theory provides the existence and multiplicity results.

Before going to the minimal surface problem, we study the Banach space $X = C \cap H^{1/2}([0, 2\pi], \mathbb{R}^n)$ and related Sobolev spaces.

Recall that the Sobolev space $H^1(D, \mathbb{R}^n)$ has the following norm:

$$\|x\|^2_{H^1(D)} = \|x\|^2_{L^2(D)} + \|\nabla x\|^2_{L^2(D)}.$$

The trace space of $H^1(D, \mathbb{R}^n)$ is the fractional order space $H^{1/2}(S^1, \mathbb{R}^n)$,

$$\|x\|^2_{1/2} = |x|^2_{L^2(S^1)} + \int\int_{S^1 \times S^1} \frac{|x(\xi) - x(\eta)|^2}{|\xi - \eta|^2} d\xi \, d\eta.$$

In particular, $\forall\, x \in H^{1/2}(S^1, \mathbb{R}^n)$, we have the Fourier series expansion

$$x(\theta) = \mathrm{Re} \sum_{m=-\infty}^{\infty} c_m e^{im\theta},$$

where

$$c_m = \frac{1}{2\pi} \int_0^{2\pi} x(\theta) e^{-im\theta} d\theta, \qquad m = 0, \pm 1, \pm 2, \ldots .$$

It is known (cf. Adams [Ada1]) that

$$\|x\|^2_{1/2} = \|x\|^2_{L^2} + |x|^2_{1/2} = \sum_{m=-\infty}^{\infty} (1 + |m|)|c_m|^2$$

is an equivalent norm, where $|x|_{1/2} = (\Sigma m |c_m|^2)^{1/2}$ is called a seminorm which induces a semi-inner product $(\ ,\ )_{1/2}$.

We need the following preparations:

(1) Let $u \mapsto \alpha \circ u$ be the Nemytski operator from the Banach linear submanifold $\widehat{X} = \{u \in C \cap H^{1/2}([0, 2\pi], \mathbb{R}^1) \mid u(0) = 0,\, u(2\pi) = 2\pi\}$ to X. Then α is a C^1-map and $J \in C^1(\widehat{X})$.

Claim. The Gateaux derivative of $u \mapsto \alpha \circ u$ at u_0 is $\alpha'(u_0)$, so we only want to verify that $u \mapsto \alpha'(u)$ is continuous in $\mathcal{L}(T_u(\widehat{X}, X))$ topology. Since the embedding

$$H^{1/2}([0, 2\pi]) \hookrightarrow L^p[0, 2\pi] \qquad \forall\, p > 1$$

is continuous, we have

$$\|\alpha' \circ u\|_{\mathcal{L}(T_u(\hat{X}), H^{1/2})} \le \left(\int_0^{2\pi} |\alpha' \circ u(\theta)|^r d\theta \right)^{1/r} \quad \text{for } \frac{1}{r} = 1 - \frac{2}{p}.$$

Therefore, the $\mathcal{L}(T_u(\widehat{X}), H^{1/2})$-continuity of $\alpha' \circ u$ with respect to u in $\widehat{X}$ follows from the Lebesgue dominance theorem.

It remains to verify the $\mathcal{L}(T_u(\widehat{X}), C)$-continuity of $\alpha' \circ u$ with respect to u in $\widehat{X}$. Since

$$|(\alpha' \circ u) \cdot v| \le |\alpha' \circ u|_{C([0,2\pi],\mathbb{R}^n)} \cdot |v|_C,$$

the C^1-continuity of α implies that

$$\|\alpha' \circ u - \alpha' \circ u_0\|_{\mathcal{L}(T_u(\hat{X}),C)} \to 0 \quad \text{as} \quad \|u - u_0\|_C \to 0.$$

This is just what we need.

(2) Suppose that $\varphi = R(\gamma)$, with $\gamma \in H^{1/2}(S^1, \mathbb{R}^n)$, i.e.,

$$\begin{cases} \Delta\varphi = 0 & \text{in } D \\ \varphi\big|_{\partial D} = \gamma. \end{cases}$$

According to the Poisson formulas, we have

$$\begin{aligned} \varphi(r,\theta) &= \operatorname{Re} \sum_{m=-\infty}^{\infty} c_m r^m e^{im\theta} \\ &= \operatorname{Re} \sum_{m=-\infty}^{\infty} c_m z^m, \qquad z = re^{i\theta}, \end{aligned}$$

where

$$\gamma(\theta) = \operatorname{Re} \sum_{m=-\infty}^{\infty} c_m e^{im\theta}.$$

(3) $\forall\, x, y \in H^{1/2}(S^1, \mathbb{R}^n)$, suppose that φ, ψ are corresponding solutions of the Dirichlet problem, with boundary data x and y respectively. Then we have

$$\begin{aligned} (x,y)_{1/2} &= \operatorname{Re} \sum_{m=-\infty}^{\infty} |m|\, c_m \bar{d}_m = \int_D \nabla\varphi \cdot \nabla\psi \, dx dy \\ &= \int_0^{2\pi} \langle \varphi_r, \psi \rangle d\theta. \end{aligned}$$

(4) $\forall\, x, y \in H^{1/2} \cap C(S^1, \mathbb{R}^n)$, we have $\langle x, y\rangle \in H^{1/2} \cap C(S^1, \mathbb{R}^1)$ and

$$|\langle x,y\rangle|_{1/2} \le \|x\|_C \cdot |y|_{1/2} + \|y\|_C \cdot |x|_{1/2}.$$

Claim. We only verify that

$$\begin{aligned} &\int\!\!\int_{S^1\times S^1} \frac{|\langle x,y\rangle(\xi) - \langle x,y\rangle(\eta)|^2}{|\xi-\eta|^2} d\xi\, d\eta \\ &= \int\!\!\int_{S^1\times S^1} \frac{|\langle x(\xi), y(\xi) - y(\eta)\rangle + \langle x(\xi) - x(\eta), y(\eta)\rangle|}{|\xi-\eta|^2} d\xi\, d\eta \\ &\le \|x\|_C \cdot |y|_{1/2} + \|y\|_C \cdot |x|_{1/2}. \end{aligned}$$

(5) If $\rho \in C^1(\mathbb{R}^2, \mathbb{R}^n)$ and $\sigma \in H^{1/2}(S^1, \mathbb{R}^2)$, then $\rho \circ \sigma \in H^{1/2}(S^1, \mathbb{R}^n)$, and $|\rho \circ \sigma|_{1/2} \leq \|\nabla \rho \circ \sigma\|_{L^\infty} |\sigma|_{1/2}$.

Claim.

$$\begin{aligned}\frac{|\rho \circ \sigma(\xi) - \rho \circ \sigma(\eta)|^2}{|\xi - \eta|^2} &\leq \frac{|\rho \circ \sigma(\xi) - \rho \circ \sigma(\eta)|^2}{|\sigma(\xi) - \sigma(\eta)|^2} \cdot \frac{|\sigma(\xi) - \sigma(\eta)|^2}{|\xi - \eta|^2} \\ &\leq \|\nabla \rho \circ \sigma\|_{L^\infty}^2 \frac{|\sigma(\xi) - \sigma(\eta)|^2}{|\xi - \eta|^2}.\end{aligned}$$

Integrating both sides of the inequality, we obtain the desired conclusion.

Next, we need to verify that critical points of J with respect to $\mathcal{M}$ satisfy the conformal condition. Letting $\varphi = \mathcal{R}(\alpha \circ u)$ and

$$F(z) = \partial_z \varphi = \varphi_x - i\varphi_y \colon D \to \mathbb{C}^n,$$

we have

$$\begin{aligned}|F(z)|^2 &= \langle \varphi_x - i\varphi_y, \varphi_x - i\varphi_y \rangle \\ &= |\varphi_x|^2 - |\varphi_y|^2 - 2i\langle \varphi_x, \varphi_y \rangle.\end{aligned}$$

Therefore,

$$|F(z)|^2 = 0 \quad \text{if and only if} \quad \varphi \text{ is conformal.}$$

However, we observe that

$$\partial_{\bar{z}} |F(z)|^2 = 2\langle \partial_{\bar{z}} F(z), F(z) \rangle = \langle \Delta \varphi(z), F(z) \rangle = 0;$$

therefore $|F(z)|^2$ is analytic in D.

In polar coordinates,

$$|F(z)|^2 = z^{-2}\left[r^2 |\varphi_r| - |\varphi_\theta|^2 - 2ir\langle \varphi_r, \varphi_\theta \rangle\right].$$

Lemma 5.1. $\forall\, u \in \mathcal{M}$, *letting* $\varphi = \mathcal{R}(\alpha \circ u)$, *we have*

$$\langle \varphi_r, \varphi_\theta \rangle \big|_{r=1} \in C^1(\partial D)^* \subset \mathcal{D}', \text{ the Schwartz distribution space.}$$

Proof. First, we assume $u \in C^\infty(\partial D)$. In this case

$$\langle \varphi_r, \varphi_\theta \rangle \big|_{r=1} \in C(\partial D).$$

Then

$$\langle H_u, \sigma \rangle = \int_0^{2\pi} \langle \varphi_\gamma, \varphi_\theta \rangle \cdot \sigma(\theta) d\theta, \qquad \forall\, \sigma \in C^1(\partial D),$$

defines a linear continuous functional on $C^1(\partial D)$.

In order to extend this functional continuously to $\mathcal{M}$, $\forall\, u, v \in C^\infty(\partial D)$, we let $\varphi = \mathcal{R}(\alpha \circ u)$, $\psi = \mathcal{R}(\alpha \circ v)$, and we make the following estimates:

$$\begin{aligned}
&\int_0^{2\pi} [\langle \varphi_r, \varphi_\theta\rangle - \langle \psi_r, \psi_\theta\rangle]\sigma \, d\theta \\
&= \int_0^{2\pi} [\langle (\varphi - \psi)_r, \varphi_\theta\rangle + \langle \psi_r, (\varphi - \psi)_\theta\rangle]\sigma \, d\theta \\
&= \int_D [\langle \nabla(\varphi - \psi), \nabla(\varphi_\theta \sigma)\rangle + \langle \nabla\psi, \nabla((\varphi - \psi)_\theta \sigma\rangle] dxdy\,,
\end{aligned}$$

where the function σ in the last integral is understood to be an extension of the same function defined on ∂D. The last integral is split into two terms:

$$\begin{aligned}
&\int_D [\langle \nabla(\varphi - \psi), \nabla\varphi_\theta\rangle\sigma + \langle \nabla\psi, \nabla(\varphi - \psi)_\theta\rangle\sigma] dxdy \\
&+ \int_D [\langle \nabla(\varphi - \psi), \varphi_\theta \nabla\sigma\rangle + \langle \nabla\psi, (\varphi - \psi)_\theta \nabla\sigma\rangle] dxdy.
\end{aligned}$$

Noticing that

$$\begin{aligned}
&\int_D \sigma\langle \nabla\psi, \nabla(\varphi - \psi)_\theta\rangle dxdy \\
&\qquad = -\int_D \langle \nabla(\varphi - \psi), \sigma_\theta \nabla\psi + \sigma\nabla\psi_\theta\rangle dxdy,
\end{aligned}$$

we have

$$\begin{aligned}
&\left|\int_D \sigma\langle \nabla(\varphi - \psi), \nabla(\varphi - \psi)_\theta\rangle dxdy\right| \\
&\qquad = \frac{1}{2}\left|\int_D \frac{\partial}{\partial\theta}|\nabla(\varphi - \psi)|^2 \cdot \sigma dxdy\right| \\
&\qquad = \frac{1}{2}\left|\int_D |\nabla(\varphi - \psi)|^2 \cdot \sigma_\theta dxdy\right| \\
&\qquad \le \frac{1}{2}\|\varphi - \psi\|^2_{H^1(D)} \cdot \|\sigma\|_{C^1(\partial D)}.
\end{aligned}$$

The remaining three terms are estimated by

$$\|\varphi - \psi\|_{H^1(D)}\big(2\|\psi\|_{H^1(D)} + \|\varphi\|_{H^1(D)}\big)\|\sigma\|_{C^1(\partial D)}.$$

In summary,

$$\begin{aligned}
&|\langle H_u - H_v, \sigma\rangle| \\
&\le C\Big[\|\varphi - \psi\|^2_{H^1(D)} + \|\varphi - \psi\|_{H^1(D)}\big(2\|\psi\|_{H^1(D)} + \|\varphi\|_{H^1(D)}\big)\Big] \\
&\|\sigma\|_{C^1(\partial D)} \\
&\le C\big(\|u - v\|_{H^{1/2}}, \|u\|_{H^{1/2}}, \|v\|_{H^{1/2}}\big)\|u - v\|_{H^{1/2}(\partial D)} \cdot \|\sigma\|_{C^1(\partial D)}.
\end{aligned}$$

Since $C^\infty(\partial D)$ is dense in $\mathcal{M}$ the domain of H can be extended to $\mathcal{M}$, such that $\forall\, u \in \mathcal{M}$, $H_u \in \big(C^1(\partial D)\big)^*$, the dual of $C^1(\partial D)$. The lemma is proved.

We turn now to finding out the derivatives of J. $\forall\, \sigma \in C^1(\partial D)$, with $|\sigma'(\theta)| < 1$, define

$$\rho_\varepsilon(\theta) = \theta + \varepsilon\sigma(\theta), \quad \text{for } |\varepsilon| < 1.$$

By definition, $\rho_0(\theta) = \theta$, and

$$\frac{d}{d\varepsilon}\rho_\varepsilon(\theta) = \sigma(\theta).$$

For any $u \in C^1(\partial D)$,

$$\begin{aligned}\frac{d}{d\varepsilon}J(u \circ \rho_\varepsilon) &= \int_0^{2\pi} \langle \varphi_r, \alpha_\theta \circ u\rangle u'(\theta) \cdot \sigma(\theta)d\theta \\ &= \int_0^{2\pi} \langle \varphi_r, \varphi_\theta\rangle \sigma\, d\theta \\ &= \langle H_u, \sigma\rangle.\end{aligned}$$

Generally speaking, however, $\forall\, u \in \mathcal{M}$, $u \circ \rho_\varepsilon^{-1}$ does not satisfy the three point condition, so we do not know if it is in $\mathcal{M}$.

In order to find the derivative of J with respect to $\mathcal{M}$, we need more work. Note that there is a conformal mapping $w_\varepsilon\colon D \to D$ satisfying

$$w_\varepsilon\colon \exp\left[iu \circ \rho_\varepsilon^{-1}\left(\frac{2j}{3}\pi\right)\right] \to \exp\left[\frac{2ij\pi}{3}\right], \qquad j = 0, 1, 2, 3.$$

If we define

$$\begin{aligned}\tau_\varepsilon(\theta) &= -i \ln w_\varepsilon(e^{i\theta}) \quad \text{and} \\ y_\varepsilon &= \tau_\varepsilon \circ u \circ \rho_\varepsilon^{-1},\end{aligned}$$

then

$$y_\varepsilon \in \mathcal{M}.$$

Lemma 5.2. *If $u \in \mathcal{M}$ is a critical point of J with respect to $\mathcal{M}$, then the distribution $H_u = 0$.*

Proof. Choose a sequence $u^k \in C^2(\partial D)$ such that $u^k \to u$ in $X = C \cap H^{1/2}$. Observing that J is invariant under τ_ε, we have

$$0 = \frac{d}{d\varepsilon}J(\tau_\varepsilon \circ u^k) = \frac{d}{d\varepsilon}J(y_\varepsilon^k \circ \rho_\varepsilon),$$

where $y_\varepsilon^k = \tau_\varepsilon \circ u^k \circ \rho_\varepsilon^{-1}$. Therefore

$$\left\langle J'(u^k), \frac{dy_\varepsilon^k}{d\varepsilon}\Big|_{\varepsilon=0} + \frac{du^k}{d\theta} \circ \sigma \right\rangle = 0.$$

Since $y_\varepsilon^k \in \mathcal{M}$, and u is a critical point of J with respect to $\mathcal{M}$, we have

$$\underline{\lim}\left\langle J'(u^k), \frac{dy^k}{d\varepsilon}\Big|_{\varepsilon=0} \right\rangle \geq 0,$$

which implies

$$\overline{\lim}\left\langle J'(u^k), \frac{du^k}{d\theta} \circ \sigma \right\rangle \leq 0.$$

Exploiting Green's formula, it follows that

$$\left\langle J'(u^k), \frac{du^k}{d\theta} \cdot \sigma \right\rangle = \langle H_{u^k}, \sigma \rangle.$$

Therefore

$$\langle H_u, \sigma \rangle \leq 0 \qquad \forall\, \sigma \in C^1(\partial D),$$

i.e., $H_u = 0$.

Theorem 5.1. *Suppose that $u \in \mathcal{M}$ is a critical point of J with respect to $\mathcal{M}$. Then $\varphi = \mathcal{R}(\alpha \circ u)$ is conformal, i.e.,*

$$|\varphi_x|^2 - |\varphi_y|^2 = \langle \varphi_x, \varphi_y \rangle = 0.$$

Proof. Observe the analytic function $z^2|F(z)|^2$: Since

$$\text{Im } z^2|F(z)|^2\big|_{\partial D} = H_u = 0,$$

this implies Re $z^2|F(z)|^2 = \text{const}$. The constant is determined by the value at $z = 0$. Hence,

$$z^2|F(z)|^2 = 0 \qquad \forall\, z \in D,$$

which implies

$$|F(z)|^2 = 0.$$

Therefore φ is conformal.

We turn now to verify the (PS) condition of J on $\mathcal{M}$. The following Courant–Lebesgue lemma is useful.

Lemma 5.3. (Courant–Lebesgue). *For each constant $M > 0$, the set*

$$J_{M/2} = \left\{ u \in \mathcal{M} \mid \int_D |\nabla R(\alpha \circ u)|^2 dx \wedge dy \le M \right\}$$

is compact in the C^0-toplogy.

Proof. 1. Let C_ρ be the circular arc centered at $(x_0, y_0) \in \partial D$ with radius $\rho > 0$. We want to show that $\forall\, \delta \in (0,1)$, $\exists\, \rho \in (\delta, \sqrt{\delta})$ such that

$$\int_{C_\rho} |\varphi_s|^2 ds \le \frac{2M}{\rho \ln(1/\delta)},$$

for each $\varphi = \mathcal{R}(\alpha \circ u)$, with $u \in J_M$, where φ_s denotes the directional derivative along the arc C_ρ.

Claim. Define an integral

$$I = \int_\delta^{\sqrt{\delta}} \int_{C_r} |\varphi_s|^2 ds dr.$$

Since

$$|\varphi_s|^2 \le |\nabla \varphi|^2,$$

we have

$$I \le \int_D |\nabla \varphi|^2 \le M.$$

Letting

$$p(r) = r \int_{C_r} |\varphi_s|^2 ds,$$

it follows that

$$\int_\delta^{\sqrt{\delta}} p(r) \frac{dr}{r} \le M.$$

Thus, $\exists\, \rho \in (\delta, \sqrt{\delta})$ such that

$$\frac{1}{2} \ln \frac{1}{\delta} \cdot p(\rho) \le M,$$

i.e.,

$$\int_{C_\rho} |\varphi_s|^2 ds \le \frac{2M}{\rho \ln(1/\delta)}.$$

2. We show that $\{\alpha \circ u\}$ is equicontinuous.

Claim. For the Jordan curve Γ, $\forall\, \varepsilon > 0$, $\exists\, d > 0$ such that for any two points $P, P' \in \Gamma$, $\Gamma \setminus \{P, P'\}$ possesses one component with diameter $< \varepsilon$, provided $0 < \mathrm{dist}(P, P') < d$.

According to (1), $\forall\ z_0 \in \partial D$, $\forall\ \delta > 0$, $\exists\ \rho \in (\delta, \sqrt{\delta})$ such that the arc length of the curve C_ρ has the following estimate:

$$\begin{aligned}\ell(\varphi(C_\rho))^2 &\leq \left(\int_{C_\rho} |\varphi_s| ds\right)^2 \\ &\leq \left(\int_{C_\rho} |\varphi_s|^2 ds\right) \cdot 2\pi\rho \\ &\leq \frac{4\pi M}{\ln(1/\delta)}.\end{aligned}$$

We choose $\delta > 0$ such that

$$\frac{4\pi M}{\ln(1/\delta)} < d^2$$

and that at least two of the following three inequalities hold:

$$\left|z - e^{\frac{2\pi ji}{3}}\right| \geq \sqrt{\delta} \qquad \forall\ z \in \partial D, \text{ and } j = 0, 1, 2.$$

One may assume $\varepsilon < \text{Min dist}_{j\neq k}(p_j, p_k)$. Therefore $\forall\ z_0 \in \partial D$, $\exists\ \rho \in (\delta, \sqrt{\delta})$ such that $\ell(\varphi(C_\rho)) < d$ and the arc C_ρ divides ∂D into a large arc A' and a smaller arc A'' which correspond to the two arcs $\overline{A}'$ and $\overline{A}''$ on Γ respectively. (The large arc A' is defined to be the arc which includes at least two of the three points $e^{\frac{2\pi ji}{3}}$, $j = 0, 1, 2$). Let P, P' be the intersections of $\varphi(C_\rho) \cap \Gamma$. Then $\text{dist}(P, P') < d$. Consequently, $\text{diam}(\overline{A}'') < \varepsilon$. Therefore, for any two points $z, z' \in \partial D$ satisfying $|z - z'| < \delta$, we must have

$$\text{dist}\big(\varphi(z), \varphi(z')\big) < \varepsilon.$$

3. Obviously, $\{\alpha \circ u \mid u \in J_{M/2}\}$ is bounded. According to the Arzela–Ascoli theorem, $\{\alpha \circ u \mid u \in J_{M/2}\}$ is compact in $C(S^1, \mathbb{R}^n)$. Since $\alpha\colon \mathcal{M} \to \alpha(\mathcal{M}) \subset C(S^1, \mathbb{R}^n)$ is a homeomorphism, $J_{M/2}$ is compact.

Lemma 5.4. *The functional J satisfies the (PS) condition on $\mathcal{M}$.*

Proof. Assume that $\{u_m\} \subset \mathcal{M}$ satisfies $J(u_m) \leq M$, and

$$\| - J'(u_m)\|_{u_m} \to 0.$$

We want to show that it is subconvergent.

According to the Courant–Lebesgue lemma, there exists a subsequence $\{u_m\}$ (without changing the subscripts), such that

$$\begin{aligned}&u_m \to u^*(C), \quad \text{and} \\ &u_m \rightharpoonup u^*(H^{1/2}).\end{aligned}$$

Since $\mathcal{M}$ is closed under the C-topology, it suffices to prove that $u_m \to u^*(H^{1/2})$. Indeed, according to (3), we have

$$\begin{aligned} |u_m - u^*|^2_{1/2} &\leq \int_D |\nabla(\mathcal{R}(\alpha \circ u_m) - \mathcal{R}(\alpha \circ u^*))|^2 dxdy \\ &= \int_D |\nabla(\varphi_m - \varphi^*)|^2 dxdy \\ &= \int_{\partial D} \left\langle \frac{\partial}{\partial r}(\varphi_m - \varphi^*), (\varphi_m - \varphi^*) \right\rangle d\theta, \end{aligned} \tag{5.2}$$

where $\varphi_m = \mathcal{R}(\alpha \circ u_m)$ and $\varphi^* = \mathcal{R}(\alpha \circ u^*)$. Integrating by parts, we have

$$\begin{aligned} (\varphi_m - \varphi^*)|_{\partial D} &= \alpha \circ u_m(\theta) - \alpha \circ u^*(\theta) \\ &= \alpha' \circ u_m(\theta)(u_m(\theta) - u^*(\theta)) - \int_{u^*(\theta)}^{u_m(\theta)} \int_{\theta'}^{u_m(\theta)} \alpha''(\bar{\theta}) d\bar{\theta} d\theta' . \end{aligned}$$

Introducing $\rho: \mathbb{R}^2 \to \mathbb{R}^n$, and $\sigma: S^1 \to \mathbb{R}^2$ by

$$\rho(\xi, \eta) = \int_\xi^\eta \int_{\theta'}^\eta \alpha''(\bar{\theta}) d\bar{\theta} d\theta'$$

and

$$\sigma(\theta) = (u^*(\theta), u_m(\theta)),$$

we obtain

$$|\rho \circ \sigma|_{1/2} \leq \|\nabla\rho \circ \sigma\|_{L^\infty} \cdot |\sigma|_{1/2}$$

from (5). By differentiation,

$$\begin{aligned} \frac{\partial \rho}{\partial \xi} &= -\int_\xi^\eta \alpha''(\bar{\theta}) d\bar{\theta}, \\ \frac{\partial \rho}{\partial \eta} &= \alpha''(\eta)(\eta - \xi). \end{aligned}$$

It follows that

$$\|\nabla\rho \circ \sigma\|_{L^\infty} \leq C_\alpha \|u_m - u^*\|_{L^\infty}$$

and

$$|\rho \cdot \sigma|_{1/2} \leq C_\alpha(|u_m|_{1/2} + |u^*|_{1/2}) \cdot \|u_m - u^*\|_{L^\infty},$$

where C_α is a constant depending on α. The condition $u_m \rightharpoonup u^*$ implies $\varphi_m \rightharpoonup \varphi^*$, and then

$$\int_{\partial D} \langle \varphi_r^*, \varphi_m - \varphi^* \rangle d\theta \to 0.$$

Again by the Courant–Lebesgue lemma and (3), we have

$$\begin{aligned}\left|\int_{\partial D}\langle(\varphi_m)_r, \rho(u_m(\theta), u^*(\theta))d\theta\right| \\ &= \left|\int\int_D \langle \nabla\varphi_m, \nabla\mathcal{R}(\rho\circ\sigma)\rangle dxdy\right| \\ &\le C\|\varphi_m\|_{H^1}\cdot\|\rho\cdot\sigma\|_{H^{1/2}} \to 0.\end{aligned}$$

Moreover, returning to (5.2),

$$\begin{aligned}|u_m - u^*|^2_{1/2} &= \int_0^{2\pi}\langle(\varphi_m)_r, \alpha'\circ u_m(\theta)\cdot(u_m(\theta) - u^*(\theta))\rangle d\theta + o(1) \\ &= \langle J'(u_m), u_m - u^*\rangle + o(1) = \langle -J'(u_m), u^* - u_m\rangle + o(1) \\ &\le |-J'(u_m)|_{u_m}\left(\|u_m\|_{H^{1/2}\cap C} + \|u^*\|_{H^{1/2}\cap C}\right) + o(1).\end{aligned}$$

However, there is a constant $C > 0$ such that

$$\begin{aligned}|u_m|_{1/2} &\le C|\alpha\circ u_m|_{1/2} \\ &\le C\int_D |\nabla\varphi_m|^2 dxdy \\ &= \frac{1}{2}CM;\end{aligned}$$

therefore, $u_m \to u^*(H^{1/2})$. □

Now we can apply the critical point theory on the closed convex set $\mathcal{M}$ to obtain the following results.

Theorem 5.2. (Existence). *Assume that $\alpha: S^1 \to \Gamma$ is a C^2 diffeomorphism. Then the Plateau problem is solvable.*

Proof. It follows directly from the results in Section 6.2 of Chapter I, because J is bounded from below and satisfies the (PS) condition on $\mathcal{M}$.

Theorem 5.3. (Morse–Tompkins–Shiffman). *By the same assumption as Theorem 5.2, if, further, we assume that system* (5.1) *possesses two distinct solutions which associate with two local minima of the Dirichlet integral J, then system* (5.1) *has the third solution.*

Proof. It is a direct consequence of the Morse relations for J on $\mathcal{M}$.

Remark 5.1. Theorem 5.2 is due to J. Douglas [Dou1] (1936). In his paper, Γ is only assumed to be a Jordan curve. Theorem 5.3 was obtained by M. Morse, C.B. Tompkins [MoT1] and M. Shiffman [Shi1] independently in 1939. The above proof, based on the modern critical point theory is due to M. Struwe [Str2] (1984).

Theorem 5.2 was extended by J. Jost [Jos2], to where the disk D is replaced by a compact oriented surface of type (p, k), and $\mathbb{R}^n$ is replaced by a complete Riemannian manifold (N, h) with nonpositive sectional curvatures. His method is taken from hyperbolic geometry. While Theorem 5.3 is also extended to that generality by K.C. Chang and J. Eells [ChE1], an a priori assumption excluding the change of topological type is made. Recently, J. Jost and M. Struwe [JoS1] removed the a priori assumption. They successfully developed a Morse theory for minimal surfaces of varying topological type, but one in which the target space $N = \mathbb{R}^3$ is assumed.

Another related result is due to M. Ji and G.Y. Wang [JiW1]. They proved the Morse–Tomplins–Shiffman type result in the case $M = D$, the unit disk, and N equal to a compact Riemannian manifold which admits no minimal sphere.

We also mention the work of T. Tromba [Tro2–3], in which degree theory is used to study disk-type minimal surfaces, and of Struwe [Str3] for annulus type.

APPENDIX

Witten's Proof of Morse Inequalities

0. Introduction

In his paper "Supersymmetry and Morse Theory," E. Witten [Wit1] presented an analytic proof of Morse inequalities. It is the purpose of this appendix to introduce his proof.

According to de Rham–Hodge Theory, the Betti numbers of a differential manifold M are related to the dimensions of harmonic forms. In the first section, we shall briefly review Hodge theory.

The idea of Witten's proof is to introduce a perturbed elliptic complex for a given Morse function f as follows:

$$\begin{array}{ccccccc}
\cdots \to & A^{p-1}(M) & \xrightarrow{d_{p-1}} & A^{P}(M) & \xrightarrow{d_p} & A^{p+1}(M) & \to \cdots \\
 & e^{-tf}\downarrow & & e^{-tf}\downarrow & & e^{-tf}\downarrow & \\
\cdots \to & A^{p-1}(M) & \xrightarrow{d_t^{p-1}} & A^{p}(M) & \xrightarrow{d_t^{p}} & A^{p}(M) & \to \cdots
\end{array}$$

with

$$d_t^p = e^{-tf} d_p e^{tf}, \qquad p = 0, 1, 2, \dots, n-1,$$

and to compute the perturbed Laplacian

$$\begin{aligned}
\Delta_t^p &= d_t^{p^*} d_t^p + d_t^{p-1} d_t^{p-1^*} \\
&= \Delta + t[\cdots] + t^2[\cdots].
\end{aligned}$$

We range the eigenvalues of Δ_t^p as follows

$$0 \le \lambda_1^p(t) \le \lambda_2^p(t) \le \cdots \le \lambda_k^p(t) \le \cdots .$$

The Hodge theory implies that there are β_p eigenvalues equal to 0, where β_p is the pth Betti number, $p = 0, 1, \dots, n-1$.

In local coordinates,

$$\Delta_t^p = \Delta^p + t^2 |\nabla f|^2 + t \sum_{i,j} \frac{\partial^2 f}{\partial x_i \partial x_j} \left[dx^i \wedge i_{dx^j} \right].$$

Assume that x^* is a critical point of f. We approximate Δ_t^p in a neighborhood of x^* and obtain the approximate perturbed Laplacian:

$$\Delta_{t,x^*}^p = \sum_{i=1}^n \left(-\frac{\partial^2}{(\partial x^i)^2} + t^2 \mu_i^2 (x^i)^2 + t\mu_i \left[dx^i \wedge, i_{dx^i} \right] \right),$$

where $\{\mu_i\}$ are the eigenvalues of the Hessian $d^2 f(x^*)$.

If we put all these Δ_{t,x_j^*} together (in the product space) for all critical points $\{x_j^*\}$ as a new operator, and range all eigenvalues as follows:

$$0 \le te_1^p \le te_2^p \le \cdots,$$

the number of zero-eigenvalues is then proved to be the number of critical points with Morse indices p. The simple version of Morse inequalities

$$\beta_p \le m_p := \#\{x_j^* \mid \operatorname{ind}(f, x_j^*) = p\},$$

then will be proved if we have the following asymptotics:

$$\lim_{t \to \infty} \frac{\lambda_k^p(t)}{t} = e_k^p, \qquad k = 1, 2, \dots.$$

A revised elliptic complex is used to prove the final version of the Morse inequalities.

The material of this appendix is based on [Wit1], F. Annik [Ann1], G. Henniart [Hen1] and B. Helffer [Hel1].

1. A Review of Hodge Theory

Let (M, g) be a compact, connected, C^∞, n-Riemannian manifold without boundary.

TM denotes the tangent bundle;
T_xM denotes the tangent space at $x \in M$;
T^*M denotes the cotangent bundle;
T_x^*M denotes the cotangent space at $x \in M$;
$\Lambda^p T^*M$ denotes the anti-symmetric tensor product of T^*M.

The section of $\Lambda^p T^*M$ is called a p-form over M.

$C^\infty(\Lambda^p T^*M)$, the set of all C^∞ p-forms, is denoted by $A^p(M)$. $\forall\, \omega \in A^p(M)$, in the local coordinates $(x_1, \dots, x_n)$, is expressed as follows:

$$\omega = \sum_{1 \le i_1 < \cdots < i_p \le n} a_{i_1, \dots, i_p} dx^{i_1} \wedge \cdots \wedge dx^{i_p}.$$

We write $g_{ij} = g\left(\frac{\partial}{\partial x_i}, \frac{\partial}{\partial x_j} \right)$, $g^{ij} = g(dx^i, dx^j)$, $i \le i, j \le n$. Then $(g^{ij}) = (g_{ij})^{-1}$ is positive definite.

We may also extend g to p-forms:

$$g\left(dx^{i_1}\wedge\cdots\wedge dx^{i_p}, dx^{j_1}\wedge\cdots\wedge dx^{j_p}\right)=\sum_{k_1,\dots,k_p}\varepsilon_{k_1\cdots k_p}g^{i_1j_{k_1}}\cdots g^{i_pj_{k_p}},$$

where $(k_1,\dots,k_p)$ runs over $(1,\dots,p)$, and

$$\varepsilon_{k_1\cdots k_p}=\pm1 \quad\text{if}\quad \begin{pmatrix}1,\dots,p\\ k_1,\dots,k_p\end{pmatrix}\ \begin{matrix}\text{is even}\\ \text{is odd.}\end{matrix}$$

The differential operator d: $A^p(M)\to A^{p+1}(M)$ is defined to be

(1) a linear operator,

(2) $d\left(a_{i_1\cdots i_p}(x)dx^{i_1}\wedge\cdots\wedge dx^{i_p}\right)=\sum_{j=1}^n\frac{\partial a_{i_1\cdots i_p}}{\partial x_j}dx^j\wedge dx^{i_1}\wedge\cdots\wedge dx^{i_p}$.

From the definition, it is easily seen that

(i) $d^2=0$.

(ii) $d(\omega\wedge\theta)=d\omega\wedge\theta+(-1)^p\omega\wedge d\theta$, $\forall\,\omega\in A^p(M)$, $\forall\,\theta\in A^q(M)$.

The Hodge star operator $*$: $A^p(M)\to A^{n-p}(M)$ is defined as follows:

(1) $*(a(x)\omega+b(x)\theta)=a(x)*\omega+b(x)*\theta$,

(2) $*\left(dx^{i_1}\wedge\cdots\wedge dx^{i_p}\right)=\sum_{k_1\cdots k_p}\eta_{j_{k_1}\cdots j_{k_p}i_{p+1}\cdots i_n}g^{i_1j_{k_1}}\cdots g^{i_pj_{k_p}}dx^{i_{p+1}}\wedge\cdots\wedge dx^{i_n}$, where

$$1\le j_1<\cdots<j_p\le n,$$
$$1\le i_{p+1}<\cdots<i_n\le n,$$

$\{j_1,\dots,j_p,i_{p+1}\cdots i_n\}$ is a permutation of $\{1,\dots n\}$, $\{k_1,\dots,k_p\}$ is a permutation of $\{1,\dots,p\}$, $\eta_{\ell_1\cdots\ell_n}=|g|^{1/2}\varepsilon_{\ell_1,\dots,\ell_n}$, $|g|=\det(g_{ij})$, and

$$\varepsilon_{\ell_1\cdots\ell_n}=\pm1 \quad\text{if}\quad \begin{pmatrix}1,2,\dots,n\\ \ell_1,\ell_2,\dots,\ell_n\end{pmatrix}\ \begin{matrix}\text{is even}\\ \text{is odd.}\end{matrix}$$

Then we have

(i) $*1=\eta$, $*\eta=1$ where $\eta=|g|^{1/2}dx^1\wedge\cdots\wedge dx^n$,

(ii) $**\omega=(-1)^{p(n-p)}\omega$ $\forall\,\omega\in A^p(M)$,

(iii) $g(\omega,\theta)\eta=\omega\wedge(*\theta)$ $\forall\,\omega,\theta\in A^p(M)$.

Claim. We only want to verify this identity for $\omega=dx^{i_1}\wedge\cdots\wedge dx^{i_p}$, $\theta=dx^{j_1}\wedge\cdots\wedge dx^{j_p}$, with $1\le i_1<\cdots<i_p\le n$, $1\le j_1<\cdots<j_p\le n$.

LHS $=\sum\varepsilon_{k_1\cdots k_p}g^{i_1j_{k_1}}\cdots g^{i_pj_{k_p}}\cdot\eta$,

RHS $=$

$$dx^{i_1}\wedge\cdots\wedge dx^{i_p}\wedge\left(\sum_{k_1\cdots k_p}\eta_{t_{k_1}\cdots t_{k_p}i_{p+1}\cdots i_n}g^{j_1t_{k_1}}\cdots g^{j_pt_{k_p}}dx^{i_{p+1}}\wedge\cdots\wedge dx^{i_n}\right)$$

$$=|g|^{1/2}\sum_{k_1\cdots k_p}\varepsilon_{t_{k_1}\cdots t_{k_p}i_{p+1}\cdots i_n}g^{j_1t_{k_1}}\cdots g^{j_pt_{k_p}}dx^{i_1}\wedge\cdots\wedge dx^{i_n}.$$

Since $\{i_1, \dots, i_n\}$ must be a permutation of $\{1, \dots, n\}$, and $\{t_1 \cdots t_p, i_{p+1} \cdots i_n\}$ is a permutation of $\{1, \dots, n\}$ with $t_1 < t_2 < \cdots < t_p$, we have $i_1 = t_1, \dots, i_p = t_p$. Therefore,

$$\begin{aligned}\text{RHS} &= |g|^{1/2}\varepsilon_{i_1\cdots i_n} \sum_{k_1\cdots k_p} \varepsilon_{k_1\cdots k_p} g^{i_1 j_{k_1}} \cdots g^{i_p j_{k_p}} dx^{i_1} \wedge \cdots \wedge dx^{i_n} \\ &= |g|^{1/2} \sum_{k_1\cdots k_p} \varepsilon_{k_1\cdots k_p} g^{i_1 j_{k_1}} \cdots g^{i_p j_{k_p}} dx^1 \wedge \cdots \wedge dx^n \\ &= \text{LHS}.\end{aligned}$$

The scalar product on $A^p(M)$ is defined by

$$\begin{aligned}\langle \omega, \theta \rangle &= \int_M g(\omega, \theta)\eta \\ &= \int_M \omega \wedge (*\theta).\end{aligned}$$

It is real, symmetric, bilinear and positive definite. The completion of $A^p(M)$ with respect to $\langle\ ,\ \rangle$ is denoted by $A^p_{L^2}(M)$. It is a Hilbert space.

The codifferential operator $d^*\colon A^p(M) \to A^{p-1}(M)$ is defined to be the adjoint operator of d with respect to $\langle\ ,\ \rangle$, i.e.,

$$\langle d^*\omega, \rho \rangle = \langle \omega, d\rho \rangle$$

$\forall\, \omega \in A^p(M)$, $\forall\, \rho \in A^{p-1}(M)$.

Note. The scalar products on both sides are different!

Basic properties of d^*.

(i) $d^* = (-1)^{n(p-1)+1} * d*$.

Claim.

$$\begin{aligned}\langle d^*\omega, \rho \rangle &= \langle d\rho, \omega \rangle \\ &= (-1)^{p(n-p)} \langle d\rho, * * \omega \rangle \\ &= \int_M d\rho \wedge (*\omega) \\ &= (-1)^p \int_M \rho \wedge (d * \omega) \\ &= (-1)^{p+(p-1)(n-p+1)} \int_M \rho \wedge (* * d * \omega) \\ &= (-1)^{(p-1)(n-p)+1} \langle \rho, *d * \omega \rangle\end{aligned}$$

(ii) $d^*d^* = 0$.

The Laplacian. $\Delta^p: A^p(M) \to A^p(M)$ is defined to be $d^*d + dd^*$. A p-form ω satisfying $\Delta^p\omega = 0$ is called a p-harmonic form. Denote $H^p(M) = \ker(\Delta^p)$.

Example. $p = 0$. $\forall\, f \in C^\infty(M)$,

$$\Delta^0 f = d^*df = -|g|^{-1/2}\sum_{i,j}\frac{\partial}{\partial x_j}\left(|g|^{1/2}g^{ij}\frac{\partial}{\partial x_i}f\right).$$

This is the Laplace-Beltrami operator on (M,g). We have

(i) Let $D(\Delta^p)$, be the space of W_2^2-Sobolev sections of the vector bundle Λ^pT^*M. Then Δ^p is positive and self-adjoint.

Claim. $\forall\, \theta,\omega \in A^p(M)$, we have

$$\begin{aligned}\langle \Delta^p\omega,\theta\rangle &= \langle (d^*d+dd^*)\omega,\theta\rangle \\ &= \langle d\omega, d\theta\rangle + \langle d^*\omega, d^*\theta\rangle \\ &= \langle \omega, (d^*d+dd^*)\theta\rangle \\ &= \langle \omega, \Delta^p\theta\rangle.\end{aligned}$$

Friedrich's extension provides the self adjointness. The positiveness is obvious.

(ii) Δ^p is an elliptic operator. See (vi) in the following paragraph.

(iii) Δ^p possesses only discrete spectrum, i.e., it has only eigenvalues $\sigma(\Delta^p) = \{\lambda_1^p < \lambda_2^p < \cdots\}$, with $\lambda_1^p \ge 0$, $\lambda_k^p \to +\infty$ as $k \to \infty$, and each eigenvalue has only finite multiplicity.

This follows from Riesz-Schauder theory.

Exterior and interior product. $\forall\, \omega \in A^1(M)$,

$$\omega\wedge: dx^{i_1}\wedge\cdots\wedge dx^{i_p} \mapsto \omega\wedge dx^{i_1}\wedge\cdots\wedge dx^{i_p},\ A^p(M) \to A^{p+1}(M),$$

$$\begin{aligned} i_\omega: dx^{i_1}\wedge\cdots\wedge dx^{i_p} \mapsto \sum_{j=1}^p (-1)^{j+1} g(\omega, dx^{i_j})dx^{i_1}\wedge \\ \cdots\wedge \widehat{dx^{i_j}}\wedge\cdots\wedge dx^{i_p}, \\ A^p(M) \to A^{p-1}(M),\end{aligned}$$

are called the exterior and interior product with respect to ω respectively. These products are extended to $A^p(M)$ linearly.

(i) One has $\forall\, \theta \in A^{p-1}(M)$, $\forall\, \psi \in A^p(M)$,

$$\langle \omega\wedge\theta,\psi\rangle = \langle \theta, i_\omega\psi\rangle.$$

(ii) $\forall\, f \in A^0(M)$, $\forall\, \theta \in A^p(M)$,

$$d^*(f\theta) = fd^*\theta - i_{df}\theta.$$

Claim. $\forall\, \psi \in A^{p-1}(M)$,

$$\begin{aligned}\langle \psi, d^*(f\theta)\rangle &= \langle d\psi, f\theta\rangle \\ &= \langle f d\psi, \theta\rangle \\ &= \langle d(f\psi) - df \wedge \psi, \theta\rangle \\ &= \langle f\psi, d^*\theta\rangle - \langle \psi, i_{df}\theta\rangle \\ &= \langle \psi, f d^*\theta - i_{df}\theta\rangle.\end{aligned}$$

(iii) $\forall\, \omega_1, \omega_2 \in A^1(M)$, $\forall\, \theta \in A^p(M)$,

$$i_{\omega_1}(\omega_2 \wedge \theta) = g(\omega_1, \omega_2)\theta - \omega_2 \wedge i_{\omega_1}\theta.$$

Claim. We may verify this for $\omega_1 = dx^1$ and $\omega_2 = dx^1$ or dx^2 in suitable coordinates.

(iv) The principal symbol of the differential operator d is $\sigma_L d = i\xi\wedge$, where $\xi = \sum_{j=1}^n \xi_j dx^j$, $(\xi_1, \dots, \xi_n) \in T^*M$. Therefore

$$d\left(a_{i_1\cdots i_p} dx^{i_1} \wedge \cdots \wedge dx^{i_p}\right) = \sum \frac{\partial a_{i_1\cdots i_p}}{\partial x^j} dx^j \wedge dx^{i_1} \wedge \cdots \wedge dx^{i_p}.$$

Therefore, $\sigma_L d\omega = i \sum \xi_j dx^j \wedge \omega$, $\forall\, \omega \in A^p(M)$.

Note. For d, the symbol $\sigma d =$ the principal symbol $\sigma_L d$.

(v) $\sigma_L d^* = i \cdot i_\xi$, where $\xi = \sum \xi_j dx^j$.

Claim. Letting $\widetilde{\theta}$, $\widetilde{\omega}$ denote the Fourier transforms (in local coordinates) for $\theta \in A^p(M)$ and $\psi \in A^{p+1}(M)$ respectively,

$$\begin{aligned}\langle \theta, d^*\psi\rangle &= \langle d\theta, \psi\rangle \\ &= \langle \sigma_L d \cdot \widetilde{\theta}, \widetilde{\psi}\rangle = i\langle \xi \wedge \widetilde{\theta}, \widetilde{\psi}\rangle \\ &= i\langle \widetilde{\theta}, i_\xi \widetilde{\psi}\rangle.\end{aligned}$$

Therefore $\langle \widetilde{\theta}, \sigma d^* \cdot \widetilde{\psi}\rangle = i\langle \widetilde{\theta}, i_\xi \widetilde{\psi}\rangle$.

(vi) $\sigma_L \Delta = -|\xi|^2$.

Claim.

$$\begin{aligned}\sigma_L(d^*d + dd^*) &= \sigma_L d^* \sigma_L d + \sigma_L d \sigma_L d^* \\ &= -\left(\xi \wedge i_\xi + i_\xi \cdot \xi\wedge\right) \\ &= -|\xi|^2.\end{aligned}$$

By choosing ξ along an axis, say $\xi = (\xi_1, \dots, \xi_n) = |\xi| e_1$,

$$e_1 \wedge i_{e_1} + i_{e_1} \cdot e_1\wedge = 1.$$

Elliptic complex. Let M^n be a Riemannian manifold, and let $E = \{E_i\}_0^m$ be a family of vector bundles over M. Let $d = \{d_i\}_0^{n-1}$,

$$d_i\colon C^\infty(E_i) \to C^\infty(E_{i+1}), \qquad i = 0, 1, \dots, n-1,$$

be a family of pseudo differential operators (ψDO) of order r, satisfying

(1) $d_{i+1}d_i = 0$,

(2) $\forall\, x \in M$, $\forall\, \xi \in T_x^*M \setminus \{\theta\}$, the sequence

$$0 \longrightarrow E_0 \xrightarrow{\sigma_L d_0(x,\xi)} E_1 \xrightarrow{\sigma_L d_1(x,\xi)} \cdots \xrightarrow{\sigma_L d_{n-1}(x,\xi)} E_n \longrightarrow 0$$

is exact, where $\sigma_L d(x,\xi)$ is the principal symbol of the $\psi DO\, d$. We say that (E, d) is an elliptic complex.

Example (de Rham). We define (E, d) as follows:

$$E = \left\{\Lambda^p T^*M\right\}_{p=0}^{n}, \qquad d = \left\{d_p\right\}_{p=0}^{n-1},$$

where d_p is the differential operator. This is an elliptic complex.

Claim. We only want to verify the exactness of the sequence

$$0 \longrightarrow \Lambda^0 T^*M \xrightarrow{\sigma_L d_0} \Lambda^1 T^*M \xrightarrow{\sigma_L d_1} \cdots \xrightarrow{\sigma_L d_{n-1}} \Lambda^n T^*M \to 0.$$

Since $\forall\, \omega \in A^p(M)$,

$$\sigma_L d(x,\xi)\omega_x = i\xi \wedge \omega_x\,,$$

where $\xi = \sum_{j=1}^n \xi_i dx_j$, and it is easy to see that

$$\ker \sigma_L d(x,\xi) = \operatorname{Im} \sigma_L d(x,\xi).$$

(Choose ξ along an axis, say $\xi = e_1$,

$$e_1 \wedge (e_{i_1} \wedge \cdots \wedge e_{i_p}) = \begin{cases} 0 & \text{if } i_1 = 1 \\ e_1 \wedge e_{i_1} \wedge \cdots \wedge e_{i_p} & \text{if } i_1 > 1 \end{cases}$$

Therefore,

$$\begin{aligned}\ker \sigma_L d(x,\xi) &= \operatorname{Span}\left\{e_{i_1} \wedge \cdots \wedge e_{i_p} \mid 1 = i_1 < \cdots < i_p \leq n\right\} \\ &= \operatorname{Im} e_1 \wedge = \operatorname{Im} \sigma_L d(x,\xi).)\end{aligned}$$

Let (E, d) be an elliptic complex, define

$$D_i\colon C^\infty(E_i) \to C^\infty(E_i)$$

as follows:

$$D_i = d_{i-1}d_{i-1}^* + d_i^* d_i, \qquad i = 0, 1, \dots, n-1.$$

We have

(i) D_i is symmetric (and it has a self-adjoint extension), and positive. The proof is quite similar to those for Δ^p.

(ii) The ψDO D_i is elliptic, i.e.,

$$\sigma_L D_i = \sigma_L d_{i-1} \cdot \sigma_L d_{i-1}^* + \sigma_L d_i^* \cdot \sigma_L d_i \quad \text{is invertible.}$$

Claim. Assume that for $\theta \in E_i$, $(\sigma_L D_i)\theta = 0$, then

$$\begin{aligned} &g\left([(\sigma_L d_{i-1} \cdot (\sigma_L d_{i-1}^*) + (\sigma_L d_i^*) \cdot (\sigma_L d_i)]\theta, \theta\right) = 0, \\ &\quad \Rightarrow g\big((\sigma_L d_{i-1}^*)\theta, (\sigma_L d_{i-1}^*)\theta\big) + g\big((\sigma_L d_i)\theta, (\sigma_L d_i)\theta\big) = 0, \\ &\quad \Rightarrow (\sigma_L d_{i-1}^*)\theta = (\sigma_L d_i)\theta = 0. \end{aligned}$$

By the exactness of the sequence, $(\sigma_L d_i) \cdot \theta = 0 \Rightarrow \exists\, \psi \in E_{i-1}$ such that $\theta = \sigma_L d_{i-1}\psi$, therefore

$$\begin{aligned} 0 = (\sigma_L d_{i-1}^*)\theta &= (\sigma_L d_{i-1}^*)(\sigma_L d_{i-1})\psi, \\ &\Rightarrow g\big((\sigma_L d_{i-1})\psi, (\sigma_L d_{i-1})\psi\big) = 0, \\ &\Rightarrow \theta = \sigma_L d_{i-1}\psi = 0. \end{aligned}$$

In the following, we use the same notations d_i (and d_i^*), representing the differential operators with domains $C^\infty(E_i)$ as well as their closed extensions in $L^2(E_i)$, $i = 1, 2, \ldots, n-1$.

Hodge Theorem. *Let (E, d) be an elliptic complex, and let*

$$D_i = d_{i-1}d_{i-1}^* + d_i^* d_i, \qquad i = 0, 1, 2, \ldots, n-1.$$

Then we have

(i) $L^2(E_i) = N(D_i) \oplus R(d_{i-1}) \oplus R(d_i^*)$;

(ii) $N(d_i) = R(d_{i-1}) \oplus N(D_i)$;

(iii) $N(d_{i-1}^*) = R(d_i^*) \oplus N(D_i)$;

(iv) $H^i(E, d) = N(d_i)/R(d_{i-1})$ defined *to be the cohomology group of the elliptic complex. Then for each $i = 0, 1, \ldots, n-1$, the following isomorphism holds:*

$$H^i(E, d) \cong N(D_i),$$

where we denote $N(D) = ker(D)$, $R(D) = Im(D)$ for each linear operator.

Proof. D_i has a self-adjoint extension, which is denoted by the same notation. We have

$$L^2(E_i) = N(D_i) \oplus R(D_i).$$

(Because D_i is elliptic, D_i has closed range.)

By definition,

$$R(D_i) \subset R(d_i^*) + R(d_{i-1}),$$

however,

$$d_i d_{i-1} = 0$$

implies that

$$\begin{aligned} &\langle d_i^* \omega, d\theta_{i-1} \rangle = 0, \qquad \forall\, \omega \in C^\infty(E_{i+1}), \quad \forall\, \theta \in C^\infty(E_{i-1}), \\ &\Rightarrow R(d_i^*) \perp R(d_{i-1}) \Rightarrow R(D_i) \subset R(d_i^*) \oplus R(d_{i-1}). \end{aligned}$$

On the other hand,

$$\begin{aligned} &R(d_i^*) \subset N(d_i)^\perp, R(d_{i-1}) \subset N(d_{i-1}^*)^\perp \\ &\Rightarrow R(d_i^*) \oplus R(d_{i-1}) \subset N(d_i)^\perp + N(d_{i-1}^*)^\perp \subset N(D_i)^\perp = R(D_i). \end{aligned}$$

The last inclusion follows from

$$N(D_i) \subset N(d_i) \cap N(d_{i-1}^*).$$

We obtain the first conclusion: $R(D_i) = R(d_i^*) \oplus R(d_{i-1})$, and

$$L^2(E_i) = N(D_i) \oplus R(d_i^*) \oplus R(d_{i-1}).$$

For (ii), since $N(d_i) \subset R(d_i^*)^\perp$, we have

$$N(d_i) \subset N(D_i) \oplus R(d_{i-1}).$$

Conversely, $N(D_i) \subset N(d_i)$ is known, and

$$R(d_{i-1}) \subset N(d_i) \quad \text{follows from} \quad d_i d_{i-1} = 0.$$

(ii) follows. (iii) is obtained in a similar manner. (iv) is a direct consequence of (ii).

Corollary. *For the de Rham complex,*

$$H^i(M) = N(d_i)/R(d_{i-1})$$

is defined to be the ith cohomology group of M, which is isomorphic to $N(\Delta^i)$, $i = 0, 1, \dots, n-1$.

The Betti numbers

$$\begin{aligned} \beta_i &= \dim H_i(M) \\ &= \dim H^i(M) \\ &= \dim N(\Delta^i), \end{aligned}$$

$i = 0, 1, \ldots, n-1$.

2. The Witten Complex

Let $f: M^n \to \mathbb{R}^1$ be a C^∞-function. $x_0 \in M$ is called a critical point of f if

$$df(x_0) = \theta.$$

Let K be the set of critical points of f.

A function f is called nondegenerate if $d^2 f(x)$ is invertible for each $x \in K$.

For a given nondegenerate function f, we define a new complex (E, d_t) as follows:

$$E = \{A^p(M) \mid p = 0, 1, \ldots, n\}.$$

$\forall\, t \geq 0$, let

$$d_t^p = e^{-tf} d_p e^{tf}, \qquad p = 0, 1, \ldots, n-1,$$

and let

$$d_t = \{d_t^p \mid p = 0, 1, \ldots, n-1\},$$

i.e.,

$$\begin{array}{ccccccccc} 0 & \to & A^0(M) & \to \cdots \to & A^p(M) & \xrightarrow{d_p} & A^{p+1}(M) & \to \cdots \to & 0 \\ & & e^{-tf}\downarrow & & e^{-tf}\downarrow & & e^{-tf}\downarrow & & \\ 0 & \to & A^0(M) & \to \cdots \to & A^p(M) & \xrightarrow{d_t^p} & A^{p+1}(M) & \to \cdots \to & 0. \end{array}$$

It is easily verified that (E, d_t) is an elliptic complex, $\forall\, t \geq 0$.

Claim.

(1) $d_t^p d_t^{p-1} = e^{-tf} d_p d_{p-1} e^{tf} = 0$,
(2) $\sigma_L(d_t^p) = \sigma_L(d_p) = i\xi\Lambda$,

so that the sequence

$$0 \to \Lambda^0 T^* M \to \cdots \xrightarrow{\sigma_L(d_t^{p-1})} \Lambda^p T^* M \xrightarrow{\sigma_L(d_t^p)} \cdots \to 0$$

is exact.

Similarly, we define $\langle d_t^{p^*} \omega, \theta \rangle = \langle \omega, d_t^p \theta \rangle \ \forall\, \omega \in A^{p+1}, \forall\, \theta \in A^p$; therefore

$$d_t^{p^*} \omega = e^{tf} d_p^* e^{-tf}.$$

Then define

$$\Delta_t^p = d_t^{p^*} d_t^p + d_t^{p-1} d_t^{p-1^*}.$$

By the Hodge theorem for elliptic complexes,

$$\begin{aligned}\ker \Delta_t^p &\cong \ker d_t^p / \operatorname{Im} d_t^{p-1} \\ &\cong \ker d_p / \operatorname{Im} d_{p-1} \\ &\cong \ker \Delta^p, \qquad \Rightarrow \beta_p = \dim \ker \Delta_t^p.\end{aligned}$$

Claim. The second isomorphism holds, because $\alpha\colon \omega \to e^{-tf}\omega$ satisfies

(1) $\alpha\big|_{\ker d_p}\colon \ker d_p \to \ker d_t^p$ is an isomorphism,

(2) $\alpha \operatorname{Im} d_{p-1} \cong \operatorname{Im} d_t^{p-1}$.

Next, we compute Δ_t.

(1)

$$\begin{aligned}d_t\omega &= e^{-tf} d(e^{tf}\omega) \\ &= e^{-tf}\left(te^{tf} df \wedge \omega + e^{tf} d\omega\right) \\ &= tdf \wedge \omega + d\omega.\end{aligned}$$

(2)

$$\begin{aligned}d_t^*\omega &= e^{tf} d^*(e^{-tf}\omega) \\ &= e^{tf}\left(e^{-tf} d^*\omega - i_{de^{-tf}}\,\omega\right) \\ &= d^*\omega - e^{tf} i_{-te^{-tf}df}\,\omega \\ &= d^*\omega + t i_{df}\,\omega.\end{aligned}$$

(3)

$$\begin{aligned}\Delta_t\omega &= d_t d_t^*\omega + d_t^* d_t\omega \\ &= t\,df \wedge d_t^*\omega + d(d_t^*\omega) + d^*(d_t\omega) + t\, i_{df}(d_t\omega) \\ &= t\,df \wedge \left(d^*\omega + t\, i_{df}\omega\right) + d\left(d^*\omega + t\, i_{df}\omega\right) \\ &\quad + d^*(t\,df \wedge \omega + d\omega) + t\, i_{df}(t\,df \wedge \omega + d\omega) \\ &= \Delta\omega + t\left[df \wedge d^*\omega + d(i_{df}\omega) + d^*(df \wedge \omega) + i_{df} d\omega\right] \\ &\quad + t^2\left[df \wedge i_{df}\omega + i_{df}(df \wedge \omega)\right] \\ &= \Delta\omega + t^2 g(df, df)\omega + t\, P_{df}\omega\,,\end{aligned}$$

where

$$P_{df}\omega = i_{df} d\omega + d(i_{df}\omega) + d^*(df \wedge \omega) + df \wedge d^*\omega.$$

Let us express P_{df} explicitly in local coordinates. First we observe that

$$\begin{aligned}P_{df}(\varphi\omega) &= \varphi i_{df} d\omega + i_{df}(d\varphi \wedge \omega) + \varphi d(i_{df}\omega) + d\varphi \wedge (i_{df}\omega) \\ &\quad + \varphi d^*(df \wedge \omega) - i_{d\varphi}(df \wedge \omega) + \varphi\, df \wedge d^*\omega - df \wedge i_{d\varphi}\omega \\ &= \varphi\, P_{df}\omega + g(df, d\varphi)\omega - g(d\varphi, df)\omega \\ &= \varphi\, P_{df}\omega.\end{aligned}$$

Next, we assume that

$$K = \{x_j^* \mid j = 1, 2, \dots, s\}.$$

We may find coordinate charts $\{(U_j, \varphi_j) \mid j = 1, 2, \dots, s\}$ such that $x_j^* \in U_j$, $U_i \cap U_j = \emptyset$ if $i \neq j$, $\varphi_j : U_j \to \mathbb{R}^n$, with $\varphi_j(x_j^*) = \theta$, and assign a special metric g_j on U_j such that

$$g\left(\varphi_j^*\left(\frac{\partial}{\partial y^k}\right), \varphi_j^*\left(\frac{\partial}{\partial y^\ell}\right)\right) = \delta_{k\ell},$$

$j = 1, 2, \dots, s$, $k, \ell = 1, 2, \dots, n$, where $y = \varphi_j(x)$.

In this case, g_j on U_j is flat, so

$$\begin{aligned}
P_{df}\, dx^I &= d(i_{df}\, dx^I) + d^*(df \wedge dx^I) \\
&= d\left(\sum_k \frac{\partial f}{\partial x^k} i_{dx^k}\, dx^I\right) + d^*\left(\sum_\ell \frac{\partial f}{\partial x^\ell} dx^\ell \wedge dx^I\right) \\
&= \sum_{k,\ell} \frac{\partial^2 f}{\partial x^k \partial x^\ell} dx^\ell \wedge i_{dx^k}\, dx^I + \sum_k \frac{\partial f}{\partial x^k} d(i_{df^k}\, dx^I) \\
&\quad + \sum_\ell \frac{\partial f}{\partial x^\ell} d^*(dx^\ell \wedge dx^I) - \sum_{k,\ell} \frac{\partial^2 f}{\partial x^\ell \partial x^k} i_{dx^k}(dx^\ell \wedge dx^I) \\
&= \sum_{k,\ell} \frac{\partial^2 f}{\partial x^k \partial x^\ell} (dx^\ell \wedge i_{dx^k} - i_{dx^k}\, dx^\ell \wedge) dx^I
\end{aligned}$$

where we use the notation

$$dx^I = dx^{i_1} \wedge \cdots \wedge dx^{i_p}, \qquad I = \{i_1, \dots, i_p\}.$$

Let us introduce the commutator:

$$[dx^\ell \wedge, i_{dx^k}] = dx^\ell \wedge i_{dx^k} - i_{dx^k}\, dx^\ell \wedge .$$

We obtain

$$P_{df}\omega = \sum_{k,\ell} \frac{\partial^2 f}{\partial x^k \partial x^\ell} [dx^\ell \wedge, i_{dx^k}]\omega, \qquad \forall\, \omega \in A^p(M).$$

In summary, for a suitable metric g_j, in a neighborhood of a critical point x_j^* of f, we have

$$\Delta_t^p \omega = \Delta^p \omega + t^2 \sum_k \left(\frac{\partial f}{\partial x^k}\right)^2 \omega + t \sum_{k,\ell} \frac{\partial^2 f}{\partial x^k \partial x^\ell} [dx^\ell \wedge, i_{dx^k}]\omega.$$

It is important to note that neither the Betti numbers β_p, $p = 0, 1, \dots, n$, nor the Morse type numbers m_p, $p = 0, 1, \dots, n$, are influenced by the changing of the Riemannian metric g, so we could choose a suitable g to simplify our computations.

First, by the Morse lemma, we find neighborhoods U_j of critical points x_j^* of f, $j = 1, \dots, s$, as well as local charts φ_j, such that $U_j \cap U_{j'} = \emptyset$, if $j \neq j'$, $\varphi_j(x_j^*) = \theta$,

$$f(x) - f(x_j^*) = \frac{1}{2} \sum_{k=1}^{n} \mu_k^j y_k^2, \qquad y = \varphi_j(x) \text{ for } x \in U_j \,,$$

where

$$d^2 f(x_j^*) = \operatorname{diag}(\mu_1^j, \dots, \mu_n^j).$$

Second, let V_j be an open neighborhood of U_j, with $V_j \cap V_{j'} = \emptyset$ if $j \neq j'$, $j = 1, \dots, s$, and let

$$V_0 = M \setminus \bigcup_{j=1}^{s} \overline{U}_j.$$

Then $\{V_j\}_0^s$ is an open covering of M. We have a C^∞-partition of unity $\{\eta_j\}_0^s$: $1 = \sum_{j=0}^{s} \eta_j$, where $\operatorname{supp} \eta_j \subset V_j$ and $\eta_j = 1$ on U_j, $j = 1, \dots, s$. Define

$$\widetilde{g} = \eta_0 g + \sum_{j=1}^{s} \eta_j g_j.$$

This is the metric we need.

Provided by the new metric $\widetilde{g}$ on M, Δ_t^p equals the following operator in U_j:

$$\Delta_{t,x_j^*}^p = \sum_{k=1}^{n} \left(- \left(\frac{\partial}{\partial x_j} \right)^2 + t^2 \mu_k^{j2} x_k^2 + t \mu_k^j [dx^k \wedge, i_{dx^k}] \right) .$$

It is an operator of separable variables.

Notice that

$$H_t = - \left(\frac{d}{dx} \right)^2 + t^2 \mu^2 x^2$$

is the Hermite operator in mathematical physics (harmonic oscillation). It has eigenvalues

$$t\,|\mu|\,(1 + 2N),$$

with eigenfunctions

$$\varphi_N(x) = H_N\big(\sqrt{t|\mu|}\, x\big) e^{-\frac{t|\mu|x^2}{2}},$$

where H_N are the hermitian polynomials, $N = 0, 1, 2, \dots$.

Denote

$$H_t^{k,j} = -\left(\frac{\partial}{\partial x^k}\right)^2 + t^2\,\mu_k^{j2} x_k^2,$$

and

$$K^k = \left[dx^k \wedge, i_{dx^k}\right],$$

$k = 1, 2, \ldots, n$. We have

$$\Delta^p_{t,x_j^*} = \sum_{k=1}^n \left(H_t^{k,j} + t\,\mu_k^j K^k\right).$$

Since

$$K^k dx^I = \varepsilon_k^I dx^I,$$

where

$$\varepsilon_k^I = \begin{cases} 1 & \text{if } k \in I \\ -1 & \text{if } k \notin I, \end{cases}$$

obviously K^k is a scalar operator on $A^p(\mathbb{R}^n)$. Thus the operator Δ^p_{t,x_j^*} is self-adjoint, with eigenvalues

$$t \sum_{k=1}^n \left[(1 + 2\,N_k^j)|\mu_k^j| + \varepsilon_k^{I^j}\,\mu_k^j\right]$$

and eigenvectors (orthonormal)

$$\varphi^t_{N^j,I^j} = t^{n/4} \prod_{k=1}^n H_{N_k^j}\left(\sqrt{t|\mu_k^j|}\,x_k\right) \exp\left(-\frac{t}{2}\sum_{k=1}^n |\mu_k^j| x_k^2\right) dx^{I^j},$$

where $N^j = (N_1^j, \ldots, N_n^j)$ runs over $\mathbb{N}^n$, and $I^j = (i_1^j, \ldots, i_p^j)$ runs over $\mathbb{N}^p$ with $i_1^j < \cdots < i_p^j$, and $j = 1, 2, \ldots, s$.

We define the direct sum space

$$H = \bigoplus_{j=1}^s A^p_{L^2}(\mathbb{R}^n), \qquad (s\text{-copies}),$$

and a self-adjoint operator A_t^p

$$A_t^p(\omega_1, \ldots, \omega_s) = \left(\Delta^p_{t,x_1^*}\omega_1, \ldots, \Delta^p_{t,x_s^*}\omega_s\right).$$

We range the eigenvalues of A_t^p as follows:

$$0 \le te_1^p \le te_2^p \le \cdots \le te_k^p \le \cdots.$$

Theorem. $\dim \ker(A_t^p) = m_p := \#\{x_j^* \in K \mid \operatorname{ind}(f, x_j^*) = p\}$.

Proof. By definition, $\Delta^p_{t,x_j^*}\varphi^t_{N^j,I^j} = 0$ iff

$$\sum_{k=1}^{n}\left[(1+2N_k^j)|\mu_k^j| + \varepsilon_k^{I^j}\mu_k^j\right] = 0$$

$$\Leftrightarrow \begin{cases} N_k^j = 0 \\ |\mu_k^j| + \mu_k^j\varepsilon_k^{I^j} = 0, \quad k = 1, \dots, n \end{cases}$$

$$\Leftrightarrow \begin{cases} N_k^j = 0 \\ \mu_k^j > 0 & \text{if } k \notin I^j \\ \mu_k^j < 0 & \text{if } k \in I^j \end{cases} \Rightarrow \operatorname{ind}(f, x_j^*) = p.$$

Therefore, each x_j^* with Morse index p has a one-dimensional contribution to the null space, but if $\operatorname{ind}(f, x_j^*) \neq p$, there is no contribution and therefore

$$\dim \ker(A_t^p) = m_p.$$

3. The Weak Morse Inequalities

We shall prove the following inequalities:

$$m_p \geq \beta_p, \qquad p = 0, 1, \dots, n.$$

If we compare with the two operators Δ_t^p and A_t^p, we see that

$$m_p = \dim \ker(A_t^p) \quad \text{and} \quad \beta_p = \dim \ker(\Delta_t^p).$$

We range the eigenvalues of the operator Δ_t^p as follows:

$$0 \leq \lambda_1^p(t) \leq \cdots \leq \lambda_k^p(t) \leq \cdots.$$

The weak Morse inequalities hold, if we can prove

$$\lim_{t \to +\infty} \frac{\lambda_k^p(t)}{t} = e_k^p. \tag{(*)}$$

First, let us pull back the eigenvector in H

$$\varphi_\alpha^t = \left\{\varphi^t_{N^j,I^j}\right\}_{j=1}^{s}, \qquad \alpha = \{N^j, I^j \mid 1 \leq j \leq s\},$$

onto the differential manifold M.

We have charts (U_j, φ_j), where U_j is a neighborhood of x_j^*, φ_j is a coordinate such that the Morse lemma holds, and, on (U_j, φ_j), the metric $\widetilde{g}$ is Euclidean, $j = 1, \dots, s$.

Define a cut-off function $\rho \in C^\infty(\mathbb{R}^n)$, such that $0 \le \rho \le 1$, and

$$\rho(y) = \begin{cases} 1 & |y| \le 1 \\ 0 & |y| \ge 2. \end{cases}$$

And define

$$\psi_\alpha^t = \sum_{j=1}^{s} \rho\big(t^{2/5}\varphi_j(x)\big)(\varphi_\alpha^t)^j \circ \varphi_j(x),$$

where $(\varphi_\alpha^t)^j$ is the j-component of φ_α^t, which is a vector in $A^p(M)$. For $t > 0$ large, the support of ψ_α^t is concentrated in $\bigcup_{j=1}^s U_j$.

These vectors are considered to be "approximate eigenvectors" for the operator Δ_t^p.

In order to prove $(*)$, the following Rayleigh–Ritz principle is needed.

Theorem [Rayleigh–Ritz]. *Assume that A is a self-adjoint operator bounded below on a Hilbert space H. If A only possesses discrete spectrum, consisting of eigenvalues with finite multiplicities, $\lambda_1 \le \lambda_2 \le \cdots \le \lambda_n \le \cdots$, then*

$$\lambda_n = \sup_{\varphi_1,\dots,\varphi_{n-1}\in H} \; \inf_{\substack{x\in D(A)\\ x\in \mathrm{span}\{\varphi_1,\dots,\varphi_{n-1}\}^\perp}} \frac{\langle Ax, x\rangle}{\|x\|^2}.$$

Proof. According to the spectral decomposition theorem,

$$\langle Ax, x\rangle = \sum \lambda_i x_i^2,$$

where $x_i = \langle x, e_i\rangle$, and e_i is the ortho-normal eigenvector corresponding to λ_i, $i = 1, 2, \dots$. Therefore,

$$\begin{aligned} \lambda_n &= \inf_{\substack{x\in D(A)\\ x\in \mathrm{span}\{e_1,\dots,e_{n-1}\}^\perp}} \frac{\langle Ax, x\rangle}{\|x\|^2} \\ &\le \sup_{\varphi_1,\dots,\varphi_{n-1}\in H} \; \inf_{\substack{x\in D(A)\\ x\in \mathrm{span}\{\varphi_1,\dots,\varphi_{n-1}\}^\perp}} \frac{\langle Ax, x\rangle}{\|x\|^2}. \end{aligned}$$

On the other hand, $\forall\ \{\varphi_1, \dots, \varphi_{n-1}\} \ne \{e_1, \dots, e_{n-1}\}$, we may choose $x_0 = \sum_{j=1}^{n-1} x_j e_j$ such that $x_0 \perp \{\varphi_1, \dots, \varphi_{n-1}\}$ and

$$\lambda_n \ge \frac{\langle Ax_0, x_0\rangle}{\|x_0\|^2}.$$

This proves the equality in the theorem.

Let us make some computations:

(i) $\langle \psi_\alpha^t, \psi_\beta^t\rangle = \delta_{\alpha\beta} + O(\exp(-at^{1/5}))$ as $t \to +\infty$, where $a > 0$ is a constant.

Claim. Noticing that $\langle\varphi_\alpha^t, \varphi_\beta^t\rangle = \delta_{\alpha\beta}$ in the space H, we have

$$\begin{aligned}\langle\psi_\alpha^t, \psi_\beta^t\rangle &= \sum_{j=1}^s \int_{\mathbb{R}^n} \rho^2(t^{2/5}y)(\varphi_\alpha^t)^j(y)(\varphi_\beta^t)^j(y)dy \\ &= \langle\varphi_\alpha^t, \varphi_\beta^t\rangle - \sum_{j=1}^s \int_{\mathbb{R}^n} \left[1-\rho^2(t^{2/5}y)\right](\varphi_\alpha^t)^j(y)(\varphi_\beta^t)^j(y)dy \\ &= \delta_{\alpha\beta} + I.\end{aligned}$$

Since $0 \le \rho \le 1$,

$$\rho(y) = 1 \quad \text{for} \quad |y| \le 1.$$

Letting

$$z = \sqrt{t}\, y,$$

we have

$$\begin{aligned} I &\le \int_{|z|\ge t^{1/10}} \sum_{j=1}^s (\varphi_\alpha^1)^j(z)(\varphi_\beta^1)^j(z)dz, \\ &\le \int_{|z|\ge t^{1/10}} P(z)e^{-|z|^2}dx = O(\exp(-at^{1/5})),\end{aligned}$$

where the explicit expansions for Hermitian functions are used, $P(z)$ is a polynomial of z, and a is any positive number less than 1.

Before computing $\langle\psi_\alpha^t, \Delta_t^p\psi_\beta^t\rangle$, we need

(ii) $\forall\, h \in C^\infty(M)$,

$$\left[h, [h, \Delta^p]\right] = -2(\nabla h)^2.$$

Claim. $\forall\, \omega \in A^p(M)$,

$$\begin{aligned}\left[h, [h, \Delta^p]\right]\omega &= (h^2\Delta^p - 2h\Delta^p h + \Delta^p h^2)\omega \\ &= h^2 dd^*\omega - 2hdd^*(h\omega) + dd^*(h^2\omega) \\ &\quad + h^2 d^*d\omega - 2hd^*d(h\omega) + d^*d(h^2\omega) \\ &= h^2dd^*\omega - 2h^2dd^*\omega + h^2dd^*\omega \\ &\quad - 2h\,dh \wedge d^*\omega + 2h\,dh\wedge d^*\omega \\ &\quad + 2hdi_{dh}\omega - 2hdi_{dh}\omega - 2dh \wedge i_{dh}\omega \\ &\quad + h^2d^*d\omega - 2h^2d^*d\omega + h^2d^*d\omega \\ &\quad + 2hi_{dh}d\omega - 2hi_{dh}d\omega \\ &\quad - 2hd^*(dh\wedge\omega) + 2hd^*(dh\wedge\omega) - 2i_{dh}(dh\wedge\omega) \\ &= -2(dh\wedge i_{dh}\omega + i_{dh}(dh\wedge\omega)) \\ &= -2(\nabla h)^2\omega.\end{aligned}$$

(iii) $\langle \psi_\alpha^t, \Delta_t^p \psi_\beta^t \rangle = \frac{t}{2}(e_\alpha^p + e_\beta^p)\langle \psi_\alpha^t, \psi_\beta^t \rangle + O(\exp(-at^{1/5}))$ as $t \to +\infty$, where e_α^p and e_β^p are the eigenvalues associated with φ_α^1 and φ_β^1 respectively, and $a > 0$ is a constant.

Claim.

$$
\begin{aligned}
&\langle \psi_\alpha^t, \Delta_t^p \psi_\beta^t \rangle - \frac{t}{2}(e_\alpha^p + e_\beta^p)\langle \psi_\alpha^t, \psi_\beta^t \rangle \\
&= \sum_{j=1}^{s} \Big[\langle \rho(t^{2/5}y)(\varphi_\alpha^t)^j, \Delta_{t,x_j^*}^p \rho(t^{2/5}y)(\varphi_\beta^t)^j \rangle_{A^p(\mathbb{R}^n)} \\
&\qquad - \frac{1}{2}\langle \Delta_{t,x_j^*}^p (\varphi_\alpha^t)^j, \rho^2 \cdot (\varphi_\beta^t)^j \rangle_{A^p(\mathbb{R}^n)} - \frac{1}{2}\langle \rho^2 \cdot (\varphi_\alpha^t)^j, \Delta_{t,x_j^*}^p (\varphi_\beta^t)^j \rangle_{A^p(\mathbb{R}^n)} \Big] \\
&= \frac{1}{2}\sum_{j=1}^{s} \langle (\varphi_\alpha^t)^j, \left(2\rho\Delta_{t,x_j^*}^p \rho - \Delta_{t,x_j^*}^p \rho^2 - \rho^2 \Delta_{t,x_j^*}^p \right) (\varphi_\beta^t)^j \rangle_{A^p(\mathbb{R}^n)} \\
&= -\frac{1}{2}\sum_{j=1}^{s} \langle (\varphi_\alpha^t)^j, [\rho, [\rho, \Delta_{t,x_j^*}^p]] (\varphi_\beta^t)^j \rangle_{A^p(\mathbb{R}^n)} \\
&= -\frac{1}{2}\sum_{j=1}^{s} \langle (\varphi_\alpha^t)^j, [\rho, [\rho, \Delta^p]] (\varphi_\beta^t)^j \rangle_{A^p(\mathbb{R}^n)} \\
&= \sum_{j=1}^{s} \langle (\varphi_\alpha^t)^j, \left(\nabla\rho(t^{2/5}y)\right)^2 (\varphi_\beta^t)^j \rangle_{A^p(\mathbb{R}^n)},
\end{aligned}
$$

because $\Delta_{t,x_j^*}^p = \Delta^p +$ terms without differentials, which commute with ρ.

Again, we see

$$
\left(\nabla\rho(t^{2/5}y)\right)^2 = t^{4/5}|(\nabla\rho)(t^{2/5}y)|^2,
$$

which is equal to zero outside $|t^{2/5}y| \le 2$, and therefore

$$
\begin{aligned}
&\langle (\varphi_\alpha^t)^j, \left(\nabla\rho(t^{2/5}y)\right)^2 (\varphi_\beta^t)^j \rangle_{A^p(\mathbb{R}^n)} \\
&\qquad = \int_{|z| \ge 2t^{1/10}} t^{4/5} P(z) e^{-|z|^2} dz = O(\exp(-at^{1/5})),
\end{aligned}
$$

where $P(z)$ is another polynomial of z.

Now, we turn to the first half of our conclusion:

$$
\overline{\lim_{t\to+\infty}} \frac{\lambda_k^p(t)}{t} \le e_k^p.
$$

Proof. We range $\{\psi_k^t \mid k = 1, 2, \ldots\}$ in such a way that φ_k^t corresponds to the eigenvalue e_k^p, $k = 1, 2, \ldots$. By the Gram–Schmidt procedure we obtain

$$
\widetilde{\psi}_k^t = \left(\psi_k^t - \sum_{j=1}^{k-1} c_{jk}^t \psi_j^t \right) \left\| \psi_k^t - \sum_{j=1}^{k-1} c_{jk}^t \psi_j^k \right\|^{-1},
$$

where

$$\sum_{j=1}^{k-1} c_{jk}^t \langle \psi_j^t, \psi_i^t \rangle = \langle \psi_k^t, \psi_i^t \rangle, \qquad i = 1, \dots, k-1.$$

Therefore,

$$c_{jk}^t = O(\exp(-at^{1/5})) \quad \text{as} \quad t \to +\infty.$$

It follows that

$$\langle \widetilde{\psi}_j^t, \Delta_t^p \widetilde{\psi}_k^t \rangle = \frac{t}{2}(e_j^p + e_k^p)\delta_{jk} + O(\exp(-at^{1/5})),$$

$j, k = 1, 2, \dots$, and that

$$\{\widetilde{\psi}_k^t \mid k = 1, 2, \dots\} \quad \text{is an orthonormal basis.}$$

By the Rayleigh-Ritz principle,

$$\begin{aligned}
\frac{\lambda_k^p(t)}{t} &= \sup_{\psi_1, \dots, \psi_{k-1} \in A^p(M)} \inf_{\substack{\psi \in D(\Delta_t^p) \\ \|\psi\|=1, \psi \in \operatorname{Span}\{\psi_1, \dots, \psi_{k-1}\}^\perp}} \langle \psi, \frac{1}{t}\Delta_t^p \psi \rangle \\
&\le \sup_{\psi_1, \dots, \psi_{k-1} \in A^p(M)} \inf_{\substack{\psi \in V \\ \|\psi\|=1, \psi \in \operatorname{span}\{\psi_1, \dots, \psi_{k-1}\}^\perp}} \langle P_V\psi, \frac{1}{t}\Delta_t^p P_V\psi \rangle
\end{aligned}$$

where $V = \operatorname{span}\{\widetilde{\psi}_1^t, \dots, \widetilde{\psi}_k^t\}$, and P_V is the orthogonal projection on V. Therefore

$$\begin{aligned}
\frac{\lambda_k^p(t)}{t} &\le \sup_{\psi_1, \dots, \psi_{k-1} \in A^p(M)} \inf_{\substack{\psi \in V \\ \|\psi\|=1, \psi \in \operatorname{span}\{P_V\psi_1, \dots, P_V\psi_{k-1}\}^\perp}} \langle P_V\psi, \frac{1}{t}\Delta_t^p P_V\psi \rangle \\
&= \sup_{\psi_1, \dots, \psi_{k-1} \in V} \inf_{\substack{\psi \in V \\ \|\psi\|=1, \psi \in \operatorname{span}\{\psi_1, \dots, \psi_{k-1}\}^\perp}} \langle \psi, \frac{1}{t}\Delta_t^p \psi \rangle \\
&= \sup_{\substack{\varphi \in V \\ \|\varphi\|=1}} \langle \varphi, \frac{1}{t}\Delta_t^p \varphi \rangle \\
&\le e_k^p + O\big(\exp(-at^{1/5})\big) \quad \text{as} \quad t \to +\infty.
\end{aligned}$$

This proves

$$\overline{\lim_{t \to +\infty}} \frac{\lambda_k^p(t)}{t} \le e_k^p.$$

The rest of this section is devoted to proving the second half of our conclusion, i.e.,

$$\underline{\lim}_{t \to +\infty} \frac{\lambda_k^p(t)}{t} \ge e_k^p.$$

On the manifold M, we define a cut-off function

$$J_j^t(x) = \begin{cases} 0 & x \notin U_j \\ \rho(t^{2/5}\varphi_j(x)) & x \in U_j \end{cases} \quad t \geq 1,\ j = 1, 2, \dots, s$$

and let

$$(J_0^t)^2 = 1 - \sum_{j=1}^{s} (J_j^t)^2.$$

Then we have

(iv) $\Delta_t^p = \sum_{j=0}^{s} J_j^t \Delta_t^p J_j^t - \sum_{j=0}^{s} (\nabla J_j^t)^2$.

Claim. Substituting $h = J_j^t$ in (ii), we obtain

$$\begin{aligned} &(J_j^t)^2 \Delta_t^p - 2J_j^t \Delta_t^p J_j^t + \Delta_t^p (J_j^t)^2 \\ &= \left[J_j^t, [J_j^t, \Delta_t^p]\right] \\ &= \left[J_j^t, [J_j^t, \Delta^p]\right] \\ &= -2(\nabla J_j^t)^2. \end{aligned}$$

Since

$$\sum_{j=0}^{s} (J_j^t)^2 = 1,$$

it follows that

$$\Delta_t^p = \sum_{j=0}^{s} J_j^t \Delta_t^p J_j^t - \sum_{j=0}^{s} (\nabla J_j^t)^2.$$

Lemma. *Suppose that* $e_k^p < r < e_{k+1}^p$. *Then for large* $t > 0$, *there is a finite rank operator* $F_k(t)\colon A_{L^2}^p(M) \to A_{L^2}^p(M)$ *with* $\dim \operatorname{Im} F_k(t) \leq k$, *such that*

$$\Delta_t^p \geq \text{rt Id} + F_k(t).$$

Proof. Since

$$\Delta_t^p = J_0^t \Delta_t^p J_0^t + \sum_{j=1}^{s} J_j^t \Delta_t^p J_j^t - \sum_{j=0}^{s} (\nabla J_J^t)^2,$$

the operator of the second term acts as the same as the operator A_t^p together with a cut-off function. Let P_k be the orthogonal projection onto the subspace spanned by the first k eigenvectors, corresponding to the eigenvalues $e_1^p, \dots, e_k^p$. Then the operator

$$\widehat{F}_k(t) = \sum_{j=1}^{s} J_j^t \widetilde{P}_k \Delta_t^p \widetilde{P}_k J_j^t$$

($\widetilde{P}_k$ stands for the pull back of P_k on $A^p(M)$) is of finite rank, with dim Im $\widehat{F}_k(t) \leq k$. We have $\forall\, \psi \in A^p(M)$,

(i)

$$\begin{aligned}\langle J_0^t \Delta_t^p J_0^t \psi, \psi\rangle &= \langle \Delta_t^p J_0^t \psi, J_0^t \psi\rangle \\ &= \langle \Delta^p J_0^t \psi, J_0^t \psi\rangle + t^2 |\nabla f|^2 \|J_0^t \psi\|^2 + t\langle P_{df} J_0^t \psi, J_0^t \psi\rangle \\ &= T_1 + T_2 + T_3,\end{aligned}$$

where $T_1 \geq 0$.

As for T_2, $\exists\, \varepsilon_0 > 0$ such that

$$|\nabla f|^2 \geq \varepsilon_0, \quad \text{for} \quad x \in V_0 = M \setminus \bigcup_{j=1}^{s} U_j.$$

Since $|\nabla f(x)|^2 = \sum_{k=1}^n (\mu_k^j y^k)^2$ for $x \in U_j$, $y = \varphi_j(x)$, we know

$$\|J_0^t \psi\|^2 \cdot |\nabla f(x)|^2 = O(t^{-4/5}) \|J_0^t \psi\|^2, \quad \text{for } x \in U_j,$$

and therefore

$$T_2 = t^2 |\nabla f|^2 \|\nabla J_0^t \psi\|^2 \geq \varepsilon_1 t^{6/5} \|J_0^t \psi\|^2, \quad \text{for some } \varepsilon_1 > 0.$$

As for T_3, P_{df} is a bounded operator, which commutes with the multiplications of a function, and therefore

$$T_3 \geq -Mt \|J_0^t \psi\|^2, \quad \text{for some constant} \quad M > 0.$$

In summary, $\langle J_0^t \Delta_t^p J_0^t \psi, \psi\rangle \geq t e_{k+1}^p \|J_0^t \psi\|^2$ for t large,

(ii)
$$\sum_{j=1}^{s} \langle J_j^t \Delta_t^p J_j^t \psi, \psi\rangle = \langle A_t^p \psi_t, \psi_t\rangle,$$

where $\psi_t \in H$ equals the element $\{\rho(t^{2/5} y) \psi(\varphi_j^{-1}(y))\}_{j=1}^{s}$. And, according to the orthogonal decomposition,

$$\begin{aligned}\langle A_t^p \psi_t, \psi_t\rangle &= \langle A_t^p (I - P_k) \psi_t, (I - P_k) \psi_t\rangle + \langle \widehat{F}_k(t) \psi, \psi\rangle \\ &\geq t e_{k+1}^p \|\psi_t\|^2 + \langle F_k(t) \psi, \psi\rangle \\ &= t e_{k+1}^p \sum_{j=1}^{s} \langle (J_j^t)^2 \psi, \psi\rangle + \langle F_k(t) \psi, \psi\rangle,\end{aligned}$$

where $F_k(t) = \sum_{j=1}^{s} J_j^t \widetilde{P}_k (\Delta_t^p - t e_{k+1}^p) \widetilde{P}_k J_j^t$.

(iii) We know that

$$\begin{aligned}
(\nabla J_j^t(x))^2 &= 0 \quad \text{if} \quad x \notin U_j \\
(\nabla J_j^t(x))^2 &= \sum_{k=1}^{n} \left(\left(\frac{\partial}{\partial x_k} \right) \rho(t^{2/5}\varphi_i(x)) \right)^2 \\
&= O(t^{4/5}) \sum_{k=1}^{n} \left(\left(\frac{\partial \rho}{\partial x_k} \right) (t^{2/5}\varphi_j(x)) \right)^2 \\
&= O(t^{4/5}) \quad \text{if} \quad x \in U_j,
\end{aligned}$$

$j = 1, 2, \dots, s$. And

$$J_0^t(x) = \left(1 - \sum_{j=1}^{s} (J_j^t(x))^2 \right)^{1/2},$$

so that

$$\begin{aligned}
(\nabla J_0^t(x))^2 &= 0 \quad \text{if} \quad x \in V_0 = M \setminus \bigcup_{j=1}^{s} \overline{U}_j, \\
(\nabla J_0^t(x))^2 &= \sum_{k=1}^{n} \left(\frac{\partial J_0^t}{\partial x_k} \right)^2 \\
&= t^{4/5} \sum_{k=1}^{n} \left[\left(\frac{\partial \rho}{\partial x_k} \right) (t^{2/5}\varphi_j(x)) \cdot \rho(t^{2/5}\varphi_j(x)) \right]^2 \\
&\qquad / [1 - \rho(t^{2/5}\varphi_j(x))^2] \\
&= O(t^{4/5}) \quad \text{if} \quad x \in U_j.
\end{aligned}$$

Then, finally, we obtain $\forall\, \psi \in A^p(M)$,

$$\begin{aligned}
\langle \Delta_t^p \psi, \psi \rangle &\geq t e_{k+1}^p \langle (J_0^t)^2 \psi, \psi \rangle + t e_{k+1}^p \sum_{j=1}^{s} \langle (J_j^t)^2 \psi, \psi \rangle \\
&\quad + \langle F_k(t)\psi, \psi \rangle + O(t^{4/5}) \|\psi\|^2 \\
&= t(e_{k+1}^p + O(t^{-1/5})) \|\psi\|^2 + \langle F_k(t)\psi, \psi \rangle.
\end{aligned}$$

If $e_k^p < r < e_{k+1}^p$, then for large $t > 0$, we have

$$\Delta_t^p \geq t \cdot r \cdot \mathrm{Id} + F_k(t).$$

Now we are going to prove $\underline{\lim}_{t \to +\infty} \frac{\lambda_k^p(t)}{t} \geq e_k^p$. The proof is divided into two cases.

(1) $e_{k-1} < e_k$. We choose $\varepsilon > 0$ such that

$$e_{k-1}^p < e_k^p - \varepsilon < e_k^p.$$

Then we have $F_{k-1}(t)$ (a bounded operator with rank $\leq k-1$) such that

$$\Delta_t^p \geq t(e_k^p - \varepsilon)\mathrm{Id} + F_{k-1}(t) \quad \text{for} \quad t > 0 \text{ large.}$$

According to the Rayleigh–Ritz principle,

$$\begin{aligned} \frac{\lambda_k^p(t)}{t} &= \sup_{\psi_1,\dots,\psi_{k-1}} \inf_{\substack{\psi \in D(\Delta_t^p) \\ \|\psi\|=1, \psi \in \mathrm{span}\{\psi_1,\dots,\psi_{k-1}\}^\perp}} \langle \psi, \frac{1}{t}\Delta_t^p \psi \rangle \\ &\geq e_k^p - \varepsilon \quad \text{for} \quad t > 0 \text{ large,} \end{aligned}$$

provided we take $\psi_1, \dots, \psi_{k-1}$ as a basis of the subspace Im $F_{k-1}(t)$. Since $\varepsilon > 0$ is arbitrary, we have

$$\underline{\lim}_{t \to +\infty} \frac{\lambda_k^p(t)}{t} \geq e_k^p.$$

(2) $e_{k-1}^p = e_k^p$. We may assume that $e_k^p > 0$, and then $\exists\, d > 1$ such that $e_{k-d}^p < e_{k-d+1}^p = \cdots = e_k^p$. According to case (1), we have

$$\underline{\lim}_{t \to +\infty} \frac{\lambda_k^p(t)}{t} \geq \underline{\lim}_{t \to +\infty} \frac{\lambda_{k-d+1}^p(t)}{t} \geq e_{k-d+1}^p = e_k^p.$$

This proves our conclusion.

Theorem. *Suppose that M is a compact, connected, orientable C^∞-manifold. Then there exists a Riemannian metric $\widetilde{g}$ such that*

$$\lim_{t \to +\infty} \frac{\lambda_k^p(t)}{t} = e_k^p.$$

4. Morse Inequalities

We have defined β_p, m_p, $p = 0, 1, \dots, n$ in Sections 1 and 2. Now we are going to prove the following inequalities:

$$\begin{aligned} m_0 &\geq \beta_0, \\ m_1 - m_0 &\geq \beta_1 - \beta_0 \\ &\cdots \\ m_n - m_{n-1} + \cdots + (-1)^n m_0 &\geq \beta_n - \beta_{n-1} + \cdots + (-1)^n \beta_0, \end{aligned}$$

or, in a compact form, letting

$$P^M(t) = \sum \beta_p t^p,$$
$$M^f(t) = \sum m_p t^p,$$

we have

$$M^f(t) = P^M(t) + (1+t)Q(t),$$

where $Q(t)$ is a formal power series with nonnegative coefficients.

Let $0 < \varepsilon < \operatorname{Min}\{e^p_{m_p+1} \mid p = 0, 1, \dots, n\}$. Fixing t large enough, we define a new cohomology complex as follows:

$$X^p = X^p_t = \Big\{\omega \in A^P(M) \mid \text{it is an eigenvector of } \Delta^p_t,$$
$$\text{with eigenvalue } \lambda^p_m(t) \text{ such that } \frac{\lambda^p_m(t)}{t} < \varepsilon\Big\}.$$

According to the theorem in Section 3, we see that

$$\dim X^p = m_p, \qquad p = 0, 1, \dots, n,$$

and we have

(i) $d^p_t: X^p \to X^{p+1}$, $d^{p-1}_t: X^p \to X^{p-1}$.

Claim. $\forall\, \omega \in X^p$, we have $\Delta^p_t\omega = \lambda^p_m(t)\omega$ with $\lambda^p_m(t) < \varepsilon t$. Therefore

$$\begin{aligned}
\Delta^{p+1}_t d^p_t\omega &= \left(d^{p+1^*}_t d^{p+1}_t + d^p_t d^{p^*}_t\right) d^p_t\omega \\
&= d^p_t d^{p^*}_t d^p_t\omega \\
&= d^p_t\left(d^{p^*}_t d^p_t + d^{p-1}_t d^{p-1^*}_t\right)\omega \\
&= d^p_t \Delta^p_t\omega = \lambda^p_m(t) d^p_t\omega.
\end{aligned}$$

This implies that $d^p_t\omega \in X^{p+1}$.

Similarly, one proves $d^{p-1^*}_t\omega \in X^{p-1}$, so we obtain a smaller cohomology complex,

$$0 \longrightarrow X^0 \xrightarrow{d^0_t} X^1 \xrightarrow{d^1_t} \cdots \xrightarrow{d^{n-1}_t} X^n \longrightarrow 0.$$

(ii) $\dim N(d^p_t)/R(d^{p-1}_t) = \beta_p$.

Warning. This is different from the property stated in Section 2 because the complex is different.

Claim. We see easily that (1) $N(\Delta^p_t) \subset X^p \cap N(d^p_t)$. (2) $\forall\, \omega \in X^p \cap N(d^p_t) \cap N(\Delta^p_t)^\perp$, we have

$$\Delta^p_t\omega = \lambda^p_m(t)\omega \quad \text{where} \quad \lambda^p_m(t) \neq 0,$$

and

$$\Delta_t^p\omega = \left(d_t^{p^*}d_t^p + d_t^{p-1}d_t^{p-1^*}\right)\omega = d_t^{p-1}d_t^{p-1^*}\omega.$$

Since $d_t^{p-1^*}\omega \in X^{p-1}$, we see

$$\omega = \frac{d_t^{p-1}d_t^{p-1^*}\omega}{\lambda_m^p(t)} \in R(d_t^{p-1}),$$

i.e., those p-forms in $X^p \cap N(d_t^p)$, which have contributions in $N(d_t^p)/R(d_t^{p-1})$, are just Δ_t^p harmonic forms. Therefore,

$$N(d_t^p)/R(d_t^{p-1}) \cong N(\Delta_t^p)$$

in the smaller cohomology complex.

Theorem. *Suppose that M is a compact, connected, orientable C^∞ manifold and that $f: M \to \mathbb{R}^1$ is a nondegenerate C^∞ function. Then the Morse inequalities hold.*

Proof. We start with the following cohomology complex:

$$0 \longrightarrow X^0 \xrightarrow{d_t^0} X^1 \xrightarrow{d_t^1} \cdots \xrightarrow{d_t^{n-1}} X^n \longrightarrow 0 \text{ for large } t.$$

We have shown that

(i) $\dim X^p = m_p$, and

(ii) $\dim N(d_t^p)/R(d_t^{p-1}) = \beta_p$.

Since

$$\dim X^p = \dim N(d_t^p) + \dim R(d_t^p),$$

and

$$\dim N(d_t^p) = \dim R(d_t^{p-1}) + \beta_p,$$

we obtain

$$m_p = \beta_p + \dim R(d_t^p) + \dim R(d_t^{p-1}),$$

$p = 1, \ldots, n$, where we assume $d_t^n = 0$. It follows that

$$\sum_{p=0}^{m}(-1)^{m-p}(m_p - \beta_p) = \dim R(d_t^m) \geq 0,$$

for $m = 0, 1, 2, \ldots, n$. And for the last one, it is an equality:

$$\sum_{p=0}^{n}(-1)^{n-p}m_p = \sum_{p=0}^{n}(-1)^{n-p}\beta_p.$$

References

[Ama1] Amann, H., Fixed point equations and nonlinear eigenvalue problems in ordered Banach spaces, *SIAM Rev.* **18** (1976), 620–709.

[Ama2] Amann, H., Saddle points and multiple solutions of differential equations, *Math. Z.* **169** (1979), 122–166.

[Ama3] Amann, H., A note on degree theory for gradient mappings, *Proc. Amer. Math. Soc.* **85** (1982), 591–597.

[AmH1] Amann, H. and Hess, P., A multiplicity result for a class elliptic boundary value problems, *Proc. Roy. Soc. Edinburgh,* Section A **84** (1979), 145–151.

[AmZ1] Amann, H. and Zehnder, E., Nontrivial solutions for a class of nonresonance problems and applications to nonlinear differential equations, *Annali Scuola Norm. Sup. Pisa* **7** (1980), 539–603.

[AmZ2] Amann, H. and Zehnder, E., Periodic solutions of asymptotically linear Hamiltonian systems, *Manuscripta Math.* **32** (1980), 149–189.

[Amb1] Ambrosetti, A., Differential equations with multiple solutions and nonlinear functional analysis, *Equadiff.*1982; *LN in Math.***1017** (1983), 1–22.

[Amb2] Ambrosetti, A., Elliptic equations with jumping nonlinearities, *J. Math. Phys. Sci.* (Madras), (1984), 1–10.

[AmC1] Ambrosetti, A. and Coti-Zelati, V., Morse theory and periodic solutions of Hamiltonian systems, *ISAS Trieste* YEAR?

[ACE1] Ambrosetti, A., Coti-Zelati, V., and Ekeland, I., Symmetry breaking in Hamiltonian systems, *JDE* **67** (1987), 165–184.

[AmP1] Ambrosetti, A. and Prodi, G., On the inversion of some differentiable mappings with singularities between Banach spaces, *Ann. Mat. Pura Appl.* **93** (1973), 231–247.

[AmR1] Ambrosetti, A. and Rabinowitz, P.H., Dual variational methods in critical point theory and applications, *J. Funct. Anal.* **14** (1973), 349–381.

[Ann1] Annik, Franc, Démonstration analytique des inégalités de Morse, *Univ. Libre de Bruxelles,* 1984–85.

[Arn1] Arnold, V.I., Problems in present day mathematics, XX: Fixed points of symplectic diffeomorphisms, *Mathematical Developments Arising from Hilbert Problems* (F.E. Browder, ed.) Proc. Symp. Pure Math., Vol. 28, AMS Providence 1976, p. 66.

[Arn2] Arnold, V.I., *Mathematical Methods of Classical Mechanics*, Springer, 1978.

[AtB1] Atiyah, M. and Bott, R., On the Yang–Mills equations over Riemann surfaces, *Phil. Trans. R. Soc. London A* **308** (1982), 523–615.

[Bah1] Bahri, A., Une méthode perturbative en théorie de Morse, Thèse, d'Etat, Univ. Paris 6, (1981).

[BaB1] Bahri, A. and Berestycki, H., A perturbation method in critical point theory and applications, *TAMS* **267** (1981), 1–32.

[BaB2] Bahri, A. and Berestycki, H., Forced vibrations of superquadratic Hamiltonian systems, *Acta Math.* **152** 3–4, (1984), 143–197.

[BaB3] Bahri, A. and Berestycki, H., Existence of forced oscillations for some nonlinear differential equations, *Comm. Pure Appl. Math.* **37** (1984), 403–442.

[BaC1] Bahri, A. and Coron, J.M., On a nonlinear elliptic equation involving the critical Sobolev exponent: the effect of the topology of the domain, *Comm. Pure Appl. Math.* **41**(1988), 253–294.

[BaL1] Bahri, A. and Lions, P.L., Morse index of some Min-Max critical points, I. Applications to multiplicity results, *Comm. Pure and Appl.* **XLI8** (1988), 1027–1038.

[BaL2] Bahri, A. and Lions, P.L., Remarques sur la théorie variationnelle des points critiques et applications, *CR Acad. Sci. Paris* **301** (1985), 145–147.

[BaR1] Bahri, A. and Rabinowitz, P., A minimax method for a class of Hamiltonian systems with singular potentials, *J. Funct. Anal.* **82** (1989), 412–428.

[BaR2] Bahri, A. and Rabinowitz, P., Periodic solutions of Hamiltonian systems of 3-body type, CMS Report #90-8, (1989), Univ. Wisconsin.

[BaC1] Bartsch, T. and Clapp, M., Bifurcation theory for symmetric potential operators, preprint.

[BCP1] Bartsch, T., Clapp, M., and Puppe, D., A mountain pass theorem for actions of compact Lie groups, preprint.

[Ben1] Benci, V., A new approach to Morse theory, *Recent Advances in Hamiltonian systems* (G. Dell' Antonli, ed.), World Sci. Publ.

[Ben2] Benci, V., Some applications of the generalized Morse-Conley index, *Conf. Semin. Mat. Univ. Bari* **218** (1987), 1–32.

[Ben3] Benci, V., A geometrical index for the group S^1 and some applications to the periodic solutions of ordinary differential equations, *Comm. Pure Appl. Math.* **34** (1981), 393–432.

[Ben4] Benci, V., On the critical point theory for indefinite functionals in the presence of symmetries, *TMAS* **274** (1982), 533–572.

[BeC1] Benci, V. and Coron, J.M., The Dirichlet problem for harmonic maps from the disk into the Euclidean n-sphere, *Analyse Nonlinéaire* **2** (1985), 119–141.

[BeP1] Benci, V. and Pacella, F., Morse theory for symmetric functionals in the sphere and application to a bifurcation problem, *Nonlinear Anal.* **9** (1985), 763–773.

[BeR1] Benci, V. and Rabinowitz, P.H., Critical point theorems for indefinite functionals, *Inv. Math.* **52** (1979), 241–273.

[BeL1] Berestycki, H. and Lions, P.L., Sharp existence results for a class of semilinear elliptic problems, *Bol. Brasil. Mat.* **12** (1981), 9–20.

[Ber1] Berger, M.S., *Nonlinearity and Functional Analysis,* Acad. Press, 1977.

[BeP1] Berger, M.S. and Podolak, E., On the solutions of a nonlinear Dirichlet problem, *Indiana Univ. Math. J.* **24** (1975), 837–846.

[Bir1] Birkhoff, G.D., Dynamical systems with two degrees of freedom, *TAMS* **18** (1917), 199–300.

[BoF1] Bonic, R. and Frampton, J., Smooth functions on Banach manifolds, *J. Math. and Mech.* **15** (1966), 877–898.

[Bot1] Bott, R., Nondegenerate critical manifolds, *Ann. of Math.* **60** (1954), 248–261.

[Bot2] Bott, R., Lectures on Morse theory, old and new, *Bull. AMS* **7** (1982), 331–358.

[Bre1] Brezis, H., On a characterization of flow invariant sets, *Comm. Pure Appl. Math.* **23** (1970), 261–263.

[BrC1] Brezis, H. and Coron, J.M., Large solutions for harmonic maps in two dimensions, *Comm. Math. Phys.* **92** (1983), 203–215.

[BrC2] Brezis, H. and Coron, J.M., Multiple solutions of H-systems and Rellich's conjecture, *Comm. Pure Appl. Math.* **37** (1984), 147–187.

[BrB1] Brown, K.J. and Budin, H., On the existence of positive solutions for a class of semilinear elliptic BVP, *SIAM J. Math. Anal.* **10** (1979), 875–883.

[Cam1] Cambini, A., Sul lemma di Morse, *Boll. Un. Mat. Ital.* **10** (1974), 713–723.

[CaL1] Castro, A. and Lazer, A.C., Critical point theory and the number of solutions of a nonlinear Dirichlet problem, *Ann. Mat. Pura Appl.* **70** (1979), 113–137.

[Cha1] Chang, K.C., Solutions of asymptotically linear operator equations via Morse theory, *Comm. Pure Appl. Math.* **34** (1981), 693–712.

[Cha2] Chang, K.C., Morse theory on Banach spaces and its applications, *Chinese Ann. Math. Ser.* **b4** (1983), 381–399.

[Cha3] Chang, K.C., A variant mountain pass lemma, *Sci. Sinica, Ser A* **26** (1983), 1241–1255.

[Cha4] Chang, K.C., Variational method and the sub- and super- solutions, *Sci. Sinica, Ser A* **26** (1983), 1256–1265.

[Cha5] Chang, K.C., Applications of homology theory to some problems in differential equations, *Nonlinear Functional Analysis* (F.E. Browder, ed.), Proc. Symp. Pure Math., AMS, 1986,

[Cha6] Chang, K.C., On a bifurcation theorem due to Rabinowitz, *J. Syst. Sci. Math. Sci.* **4** (1984), 191–195.

[Cha7] Chang, K.C., On the mountain pass lemma, *Equadiff.* **6**, *LN Math.* **1192** (1986), 203–207.

[Cha8] Chang, K.C., An extension on the minimax principle, *Proc. Symp. DD3* (Chern, S.S., etc. ed.), Science Press, Beijing, 1986.

[Cha9] Chang, K.C., On the periodic nonlinearity and the multiplicity of solutions, *Nonlinear Analysis TMA* **13** (1989), 527–537.

[Cha10] Chang, K.C., Heat flow and boundary value problem for harmonic maps, *Analyse Nonlinéaire* **6** (1989), 363–396.

[Cha11] Chang, K.C., Morse theory for harmonic maps, *Variational Methods, Proc. of a Conf. Paris, June 1988*, Berestycki, Coron, Ekeland, eds, Birkhäuser (1990), 431–446.

[Cha12] Chang, K.C., Infinite dimensional Morse theory and its applications, *Univ. de Montreal* **97** (1985).

[Cha13] Chang, K.C., Critical point theory and its applications, *Shanghai Sci. Techn. Press* (1986), (in Chinese).

[Cha14] Chang, K.C., Variational methods for nondifferentiable functionals and its applications to partial differential equations, *J. Math. Anal. Appl.* **80** (1981), 102–129.

[ChE1] Chang, K.C. and Eells, J., Unstable minimal surface coboundaries, *Acta Math. Sinica, New ser.* **2** (1986), 233–247.

[ChJ] Chang, K.C. and Jiang, M.Y., The Lagrange intersections for $(\mathbb{C}P^n, \mathbb{R}P^n)$, *Manuscripta Math.* **68** (1990), 89–100.

[ChL1] Chang, K.C. and Liu, J.Q., Morse theory under general boundary conditions, *J. System Sci. & Math. Sci.* **4**(1991), 78–83.

[ChL2] Chang, K.C. and Liu, J.Q., A strong resonance problem, *Chinese Ann. Math.* 11, B.2. (1990), 191–210.

[CLZ1] Chang, K.C., Long, Y., and Zehnder, E., Forced oscillations for the triple pendulum, *Analysis et cetera*, Rabinowitz, Zehnder, eds, Academic Press, 1990.

[CWL1] Chang, K.C., Wu, S.P., and Li, S., Multiple periodic solutions for an asymptotically linear wave equation, *Indiana Math. J.* **31** (1982), 721–731.

[Cla1] Clark, D.C., A variant of Ljusternik–Schnirelman theory, *Indiana Math. J.* **22** (1972), 65–74.

[Cof1] Coffman, C.V., Ljusternik–Schnirelman theory, Nodal properties and Morse index, *Nonlinear diffusion equations and their equilibrium states* 1, Springer, (1988), 245–266.

[Cof2] Coffman, C.V., Ljusternik–Schnirelman theory: complementary principles and the Morse index, *Nonlinear Analysis, TAM*, to appear.

[Con1] Conley, C.C., Isolated invariant sets and the Morse index, *CBMS Regional Conference Series* 38, AMS, 1978.

[CoZ1] Conley, C.C. and Zehnder, E., The Birkhoff–Lewis fixed point theorem and a conjecture of V. Arnold, *Invent. Math.* **73** (1983), 33–49.

[CoZ2] Conley, C.C. and Zehnder, E., Morse type index theory for flows and periodic solutions for Hamiltonian equations, *Comm. Pure Appl. Math.* **37** (1984), 207–253.

[CoW1] Costa, D. and Willem, W., Points critique multiples de fonctionnelles invariantes, *CR Acad. Sci. Paris* **298** (1984), 381–384.

[Cot1] Coti-Zelati, V., Morse theory and periodic solutions of Hamiltonian systems, Ph.D. thesis, Trieste, 1987.

[Cou1] Courant, R., *Dirichlet Principle, Conformal Mapping and Minimal Surfaces,* Wiley, Interscience, 1950.

[Dan1] Dancer, E.N., Degenerate critical points, homotopy indices and Morse inequalities, *J. Reine Angew. Math.* **350** (1984), 1–22.

[Die1] Dieck, T. Tom., *Transformation Groups,* Walter de Gruyter, 1987.

[Din1] Ding, W.Y., Ljusternik–Schnirelman theory for harmonic maps, *Acta Math. Sinica* **2** (1986), 105–122.

[DiL1] Ding, Y.H. and Liu, J.Q., Periodic solutions of asymptotically linear Hamiltonian systems, *J. Sys. Sci. & Math. Sci.* **9** (1989), 30–39.

[DoL1] Dong, G.C. and Li, S., On the infinitely many solutions of the Dirichlet problems for some nonlinear elliptic equations, *Scient. Sinica* (1982).

[Dou1] Douglas, J., Some new results in the problem of Plateau, *J. Math. Phys.* **15** (1936), 55–64.

[Du1] Dumford, N. and Schwartz, J. T., *Linear Operators, vol. 2,* Wiley: Interscience, 1962.

[DuL1] Duvaut, G. and Lions, J.L., *Les inéquations en mećanique et en physique,* Dunod, 1972.

[EeL1] Eells, J. and Lemaire, L., A report on harmonic maps, *Bull. London Math. Soc.* **16** (1978), 1–68.

[EeL2] Eells, J. and Lemaire, L., Another report on harmonic maps, *Bull. London Math. Soc.* **20** (1988), 385–524.

[EeS1] Eells, J. and Sampson, J.H., Harmonic mappings of Riemannian manifolds, *JAMS* **86** (1964), 109–160.

[Eke1] Ekeland, I., Une théorie de Morse pour les systèmes hamiltoniens convexes, *Analyse Nonlinéaire* 1, (1984), 19–78.

[Eke2] Ekeland, I., Periodic solutions of Hamiltonian equation and a theorem of P. Rabinowitz, *J. Diff. Equa.* **34** (1979), 523–534.

[EkL1] Ekeland, I. and Lasry, J.M., On the number of closed trajectories for a Hamiltonian flow on a convex energy surface, *Ann. of Math.* **112** (1980), 283–319.

[Fad1] Fadell, E.R., Cohomological method in non-free G-space with applications to general Borsuk–Ulam theorems and critical point theorems for invariant functionals, *Nonlinear Funct. Anal. and Its Appl.* 1–45 (Singh, S.P., ed.), Reidel Publ. Co. (1986).

[FHR1] Fadell, E.R., Husseini, S.Y. and Rabinowitz, P.H., Borsuk–Ulam theorems for arbitrary S^1 actions and applications, *TAMS* **274** (1982), 345–360.

[FaR1] Fadell, E.R. and Rabinowitz, P.H., Bifurcation for odd potential operators and an alternative topological index, *J. Funct. Anal.* **26** (1977), 48–67.

[FaR2] Fadell, E.R. and Rabinowitz, P.H., Generalized cohomological index theories for Lie group actions with an application to bifurcation questions for Hamiltonian systems, *Invent. Math.* **45** (1978), 139–174.

[Fel1] Felmer, P.L., Periodic solutions of spatially periodic Hamiltonian systems, CMS Report #90-3, Univ. of Wisconsin, (1990).

[Flo1] Floer, A., A refinement of the Conley index and an application to the stability of hyperbolic invariant sets, *Erg. Theo. and Dynamic Sys.* **7** (1987), 93–103.

[Flo2] Floer, A., Proof of the Arnold conjecture for surfaces and generalization to certain Kahler manifolds, *Duke Math. J.* **53** (1986), 1–32.

[Flo3] Floer, A., A Morse theory for Lagrangian intersections, *J. Diff. Geom.* **28** (1988), 513–547.

[Flo4] Floer, A., A cuplength estimate for lagrangian intersections, *Comm. Pure Appl. Math*, XLII (1989), 335-357.

[FlZ1] Floer, A. and Zehnder, E., The equivariant Conley index and bifurcations of periodic solutions of Hamiltonian systems, *Erg. and Dynam. Syst.* (1987).

[FoM1] Fonda, A. and Mawhin, J., Multiple periodic solutions of conservative systems with periodic nonlinearity, preprint, Univ. Catholique de Louvain, 1988.

[For1] Fortune, C., A symplectic fixed point theorem for CP^n, *Invent. Math.* **81** (1985), 29–46.

[FoW1] Fournier, G. and Willem, M., Multiple solutions of the forced double pendulum equations, *Analyse nonlinéaire, supplément au vol. 6*, (1989), 259–281.

[FoW2] Fournier, G. and Willem, M., Relative category and the calculus of variations, preprint, 1988.

[Fra1] Franks, J., Generalizations of the Poincaré–Birkhoff theorem, preprint.

[Fuc1] Fučik, S., Remarks on a result by A. Ambrosetti and G. Prodi, *Boll. Un. Mat. Ital.* **11** (1975), 259–267.

[Gho1] Ghoussoub, N., Location, multiplicity, and Morse indices of Min-Max critical points, *Z. Reine und Angew. Math.*, in press.

[GiT1] Gilbarg, D. and Trudinger, N.S., *Elliptic Partial Differential Equations of Second Order,* Springer-Verlag, 1977.

[Gor1] Gordon, W.B., Conservative dynamical systems involving strong forces, *TAMS* **204** (1975), 113–135.

[Gre1] Greenberg, M.J., *Lectures on Algebraic Topology,* Benjamin, 1967.

[Grm1] Gromoll, D. and Meyer, W., On differentiable functions with isolated critical points, *Topology* **8** (1969), 361–369.

[Ham1] Hamilton, R., Harmonic maps of manifolds with boundary, *Lecture Notes* **471**, Springer, 1975.

[Hel1] Helffer, B., Etude du laplacien de Witten associé à une fonction de Morse dégénérée, (1985) preprint.

[Hen1] Henniart, G., Les inégalités de Morse, *Séminaire Bourbaki,* 1983/84, **617**, 43–60.

[Hes1] Hess, P., On a nonlinear elliptic boundary value problem of the Ambrosetti–Prodi type, *Boll. Un. Mat. Ital.* (5) (1980), **17-A**, 187–192.

[Hes2] Hess, P., On multiple positive solutions of nonlinear elliptic eigenvalue problems, *Comm. PDE* **6** (1981), 951–961.

[HeK1] Hess, P. and Kato, T., On some linear and nonlinear eigenvalue problems with indefinite weight functions, *Comm. Partial Diff. Equations* **5** (1980), 999–1030.

[Hin1] Hingston, N., Equivariant Morse theory and closed geodesics, *J. Diff. Geometry* **19** (1984), 85–116.

[Hof1] Hofer, H., Variational and topological methods in partially ordered Hilbert spaces, *Math. Ann.* **261** (1982), 493–514.

[Hof2] Hofer, H., A note on the topological degree at a critical point of mountain pass type, *PAMS* **90** (1984), 309–315.

[Hof3] Hofer, H., Ljusternik–Schnirelman theory for Lagrangian intersections, *Analyse nonlinéaire* **5** (1988), 465–500.

[Hof4] Hofer, H., A geometric description of the neighbourhood of a critical point given by the mountain pass theorem, *J. London Math. Soc.* **31** (1985), 566–570.

[Hus1] Husemoller, D., *Fiber Bundles,* McGraw-Hill, 1966.

[JiW1] Ji, M. and Wang, G.Y., Minimal surfaces in Riemannian manifolds, (1988) preprint. [Jia1] Jiang, M.Y., An existence result for periodic solutions of a class of Hamiltonian systems, *Kexue Tongbao* **33** (1988), 1679–1681.

[Jia2] Jiang, M.Y., A remark on periodic solutions of singular Hamiltonian systems, *Acta Math. Sinica,* in press.

[Jos1] Jost, J., The Dirichlet problem for harmonic maps from a surface with boundary onto a 2-sphere with nonconstant boundary values, *J. Diff. Geometry* **19** (1984), 393–401.

[Jos2] Jost, J., Conformal mappings and the Plateau–Douglas problem in Riemannian manifolds, *J. Reine Angew. Math.* **359** (1985), 37–54.

[JoS1] Jost, J. and Struwe, M., Morse Conley theory for minimal surfaces of varying topological types, *Invent. Math.* **102** (1990), 465–499.

[KaW1] Kazdan, J.L. and Warner, F.W., Remarks on some quasilinear elliptic equations, *Comm. Pure Appl. Math.* **28** (1975), 567–597.

[Kel1] Kelley, J.L., *General Topology,* D. Van Nostrand, 1955.

[Kli1] Klingenberg, W., *Lectures on Closed Geodesics,* Springer, 1978.

[Kra1] Krasnoselskii, M.A., *Topological Methods in the Theory of Nonlinear Integral Equations,* Pergamon, 1964.

[Kui1] Kuiper, N., C^1-equivalence of functions near isolated critical points, Symp. on Infinite-dimensional topology, *Annals of Math. Studies* **69**,

Princeton Univ. Press, 1972.
[Laz1] Lazer, A.C., Introduction to multiplicity theory for boundary value problems with asymmetric nonlinearities, *LN in Math.* **1324** (1988).
[LaM1] Lazer, A.C. and McKenna, P.J., On the number of solutions of a nonlinear Dirichlet problem, *J. Math. Anal. Appl.* **84** (1981), 282–294.
[LaM2] Lazer, A.C. and McKenna, P.J., Critical point theory and boundary value problems with nonlinearities crossing multiple eigenvalues, *Comm. PDE* **10** (1985), 107–150.
[LaS1] Lazer, A.C. and Solimini, S., Nontrivial solutions of operator equations and Morse indices of critical points of Minmax type, *Nonlinear Analysis* TMA, 12.8 (1988), 761–775.
[Lem1] Lemaire, L., Boundary value problems for harmonic and minimal maps of surfaces into manifolds, *Ann. Scuola Norm. Sup. Pisa* **9** (1982), 91–103.
[Li1] Li, S., Multiple critical points of periodic functional and some applications, *ICTP* IC-86-191.
[LiL1] Li, S. and Liu, J.Q., Morse theory and asymptotically linear Hamiltonian systems, *JDE* **78** (1989), 53–73.
[Liu1] Liu, J.Q., A Morse index of a saddle point, preprint.
[Liu2] Liu, J.Q., A generalized saddle point theorem. *J. Diff. Eq.* (1989).
[Liu3] Liu, J.Q., Doctoral thesis, Academy of Science, Beijing, 1983.
[Liu4] Liu, J.Q., A Hamiltonian system of second order, ICTP preprint.
[Llo1] Lloyd, N.G., *Degree Theory,* Cambridge Univ. Press, Cambridge, 1977.
[Lon1] Long, Y., Maslov index, degenerate critical points and asymptotically linear Hamilton systems, preprint.
[Lon2] Long, Y., The structure of the singular symplectic matrix set, preprint.
[Lon3] Long, Y., Multiple periodic solutions of perturbed superquadratic second order Hamiltonian systems, *TMAS* 311 (1989), 749–780.
[LoZ1] Long, Y. and Zehnder, E., Morse theory for forced oscillations of asymptotically linear Hamiltonian systems, preprint.
[LuS1] Lupo, D. and Solimini, S., A note on a resonance problem, *Proc. Royal Soc. of Edinburgh* **102A** (1986), 1–7.
[Mar1] Martin, R.H., *Nonlinear Operators and Differential Equations in Banach Spaces,* John Wiley & Sons, 1976.
[MaP1] Marino, A. and Prodi, G., La teoria di Morse per spazi di Hilbert, *Rend. Sem. Mat. Univ. Padova* **41** (1968), 43–68.
[MaP2] Marino, A. and Prodi, G., Metodi perturbativi nella teoria di Morse, *Boll. Un. Math. Ital. Suppl. Fasc.* **3** (1975), 1–32.
[Maw1] Mawhin, J., Problèmes de Dirichlet, variationnels nonlinéaires, Univ. de Montreal, **104** (1987).
[Maw2] Mawhin, J., Forced second order conservation systems with periodic nonlinearity, *Analyse nonlinéaire,* Suppl. au vol. 6 (1989), 415–434.
[MaW1] Mawhin, J. and Willem, M., Multiple solutions of the periodic BVP for some forced pendulum-type equations, *J. Diff. Equa.* **52**

(1984), 264–287.

[MaW2] Mawhin, J. and Willem, M., Critical point theory and Hamiltonian systems, *Appl. Math. Sci.* **74** Springer-Verlag, 1989.

[Mey1] Meyer, W., Kritische Mannigflatigkeiten in Hilbertmannigflatigkeiten, *Math. Ann.* **170** (1967), 45–66.

[Mil1] Milnor, J., *Morse Theory,* Princeton Univ. Press, Princeton, 1963.

[Mil2] Milnor, J., *Topology from the Differential Viewpoint,* Univ. Press of Virginia, Charlottesville, 1969.

[Mor1] Morse, M., Relations between the critical points of a real function of n independent variables, *TAMS* **27** (1925), 345–396.

[Mor2] Morse, M., The calculus of variations in the large, *Amer. Math. Soc. Coll.* Pub. No. 18, Providence, 1934.

[MoC1] Morse, M. and Cairns, S.S., *Critical Point Theory in Global Analysis and Differential Topology,* Academic Press, New York, 1969.

[MoT1] Morse, M. and Tompkins, C., The existence of minimal surfaces of general critical types, *Ann. of Math.* **40** (1939), 443–472.

[MoT2] Morse, M. and Tompkins, C., Unstable minimal surfaces of higher structure, *Duke Math. J.* **8** (1941), 350–375.

[Ni1] Ni, W.M., Some minimax principles and their applications in nonlinear elliptic equations, *J. d'Analyse Math.* **37** (1980), 248–275.

[Nik1] Nikol'ski, S.M., *Approximation of Functions of Several Variables and Imbedding Theorems,* Springer-Verlag, 1975.

[Nir1] Nirenberg, L., *Topics in nonlinear functional analysis,* Courant Institute Lecture Notes, New York, 1974.

[Nir2] Nirenberg, L., Variational and topological methods in nonlinear problems, *Bull. AMS* **3** (1981), 267–302.

[Nir3] Nirenberg, L., Comments on nonlinear problems, *Le Matematische* **16** (1981).

[Nir4] Nirenberg, L., *Variational Methods in Nonlinear Problems,* LN in Math. 1365, 1989, Springer-Verlag, 100–119.

[Oss1] Osserman, R., *A Survey of Minimal Surfaces,* Van Nostrand, 1969.

[Pal1] Palais, R.S., Morse theory on Hilbert manifolds, *Topology* **2** (1963), 299–340.

[Pal2] Palais, R.S., Homotopy theory of infinite dimensional manifolds, *Topology* **5** (1966), 1–16.

[Pal3] Palais, R.S., Ljusternik–Schnirelman theory on Banach manifolds, *Topology* **5** (1966), 115–132.

[Pal4] Palais, R.S., Critical point theory and the minimax principle, *Global Analysis,* Proc. Symp. Pure Math. **15** (ed. Chern, S.S.), AMS, Providence (1970), 185–202.

[PaS1] Palais, R.S. and Smale, S., A generalized Morse theory, *BAMS* **70** (1964), 165–171.

[Pit1] Pitcher, E., Inequalities of critical point theory, *BAMS* **64** (1958), 1–30.

[PuS1] Pucci, P. and Serrin, J., The structure of the critical set in the mountain pass theorem, *TAMS* **91** (1987), 115–132.

[PuS2] Pucci, P. and Serrin, J., Extensions of the mountain pass theorem, *J. Funct. Anal.* **59** (1984), 185–210.

[Qi1] Qi, G.J., Extension of mountain pass lemma, *Kexue Tongbao* **32** (1987).

[Rab1] Rabinowitz, P.H., Variational methods for nonlinear eigenvalue problems, *Eigenvalues of Nonlinear Problems,* Ed. Cremonese, Roma (1974), 141–195.

[Rab2] Rabinowitz, P.H., A bifurcation theorem for potential operators, *J. Funct. Anal.* **25** (1977), 412–424.

[Rab3] Rabinowitz, P.H., Periodic solutions of Hamiltonian systems, *Comm. Pure Appl. Math.* **31** (1978), 157–184.

[Rab4] Rabinowitz, P.H., Multiple critical points of perturbed symmetric functionals, *TAMS* **272** (1982), 753–770.

[Rab5] Rabinowitz, P.H., Minimax methods in critical point theory with applications to differential equations, *CBMS Reg. Conf. Ser. in Math.* **65** AMS (1986).

[Rab6] On a class of functionals invariant under a Z^n action, CMS Report #88-1, Univ. of Wisconsin, Madison (1987).

[Ree1] Reeken, M., Stability of critical points under small perturbations, Part 2, analytic theory, *Manuscripta Math.* **8** (1973), 69–92.

[Rot1] Rothe, E., Morse theory in Hilbert space, *Rocky Mountain J. Math.* **3** (1973), 251–274.

[Rot2] Rothe, E., Critical point theory in Hilbert space under regular boundary conditions, *J. Math. Anal. Appl.* **36** (1971), 377–431.

[Rot3] Rothe, E., On the connection between critical point theory and Leray Schauder degree, *J. Math. Anal. Appl.* **88** (1982), 265–269.

[Rot4] Rothe, E., Critical point theory in Hilbert space under general boundary conditions, *J. Math. Anal. Appl.* **2** (1965), 357–409.

[Ryb1] Rybakowski, K.P., *The Homotopy Index and Partial Differential Equations,* Springer-Verlag, 1987.

[SaU1] Sacks, J. and Uhlenbeck, K., The existence of minimal immersions of 2 spheres, *Annales of Math.* **113** (1981), 1–24.

[SaU2] Sacks, J. and Uhlenbeck, K., Minimal immersions of closed Riemann surfaces, *TAMS* **271** (1982), 639–652.

[SaZ1] Salamon, D. and Zehnder, E., Floer homology, the Morse index, and periodic orbits of Hamiltonian equations, preprint.

[ScJ1] Schwartz, J.T., *Nonlinear Functional Analysis,* Gordon and Breach, 1969.

[ScR1] Schoen, R., Conformal deformation of a Riemannian metric to constant scalar curvature, *J. Diff. Geom.* **20** (1984), 479–495.

[Sch1] Schoen, R. and Yau, S.T., Existence of incompressible minimal surfaces and the topology of three-dimensional manifolds with nonnegative scalar curvature, *Annales of Math.* **110** (1979), 127–142.

[Shi1] Shiffman, M., The Plateau problem for minimal surfaces of arbitrary topological structure, *Amer. J. Math.* **61** (1939), 853–882.

[Sik1] Sikarov, J.C., Points fixes d'un symplectomorphisme homologue d'identité. *J. Diff. Geom.* **22** (1985), 49–79.

[Sol1] Solimini, S., Existence of a third solution for a class of BVP with jumping nonlinearities, *Nonlinear Analysis TMA*, (1983), 917–927.

[Sol2] Solimini, S., Morse index estimates in Min-Max theorems, *Manuscripta Math.* **63** (1989), 421–454.

[Sol3] Solimini, S., On the solvability of some elliptic PDE with linear part at resonance, *JMAA* **117** (1986), 138–152.

[Spa1] Spanier, E.H., *Algebraic Topology,* McGraw-Hill, 1966.

[Str1] Struwe, M., Infinitely many critical points for functionals which are not even and applications to nonlinear BVP, *Manuscripta Math.* **32** (1982), 753–770.

[Str2] Struwe, M., On the critical point theory for minimal surfaces spanning a wire in R^n, *J. Reine Angew. Math.* **349** (1984), 1–23.

[Str3] Struwe, M., A Morse theory for annulus type minimal surfaces, *J. Reine Angew. Math.* **368** (1986), 1–27.

[Str4] Struwe, M., On the evolution of harmonic mappings, *Commet. Math. Helvetici* **60** (1985), 558–581.

[Str5] Struwe, M., *Plateau Problem and the Calculus of Variations,* Princeton Univ. Press, 1988.

[Szu1] Szulkin, A., A relative category and applications to critical point theory for strongly indefinite functionals, preprint.

[Tau1] Taubes, C.H., A framework for Morse theory for the Yang–Mills functional, preprint.

[Tau2] Taubes, C.H., Minimax theory for the Yang–Mills–Higgs equations, *Comm. Math. Phys.* **97** (1985), 473–540

[Tia1] Tian, G., On the mountain pass theorem, *Kexue Tongbao* **29** (1984), 1150–1154.

[Tro1] Tromba, A., On Plateau problem for minimal surfaces of high genus in R^n, preprint.

[Tro2] Tromba, A., Degree theory on oriented infinite dimensional varieties and the Morse number of minimal surfaces spanning a curve in R^n, 1, *TAMS* **290** (1985), 385–413; 2, *Manuscripta Math.* **48** (1984), 139–161.

[Tro3] Tromba, A., A general approach to Morse theory, *J. Diff. Geom.* **12** (1977), 47–85.

[Uhl1] Uhlenbeck, K., Morse theory on Banach manifolds, *J. Funct. Anal.* **10** (1972), 430–445.

[Uhl2] Uhlenbeck, K., Morse theory by perturbation methods with applications to harmonic maps, *TAMS* **267** (1981), 569–583.

[Vit1] Viterbo, C., Indice de Morse des points critiques obtenus par minimax, *Analyse nonlinéaire* **5** (1988), 221–226.

[WaT1] Wang, T.Q., Ljusternik–Schnirelman category theory on closed subsets of Banach manifolds, preprint.

[WaZ1] Wang, Z.Q., Equivariant Morse theory for isolated critical orbits and its applications to nonlinear problems, *LN in Math.* **1306**, Springer, 1988, 202–221.

[WaZ2] Wang, Z.Q., On a superlinear elliptic equation, *Analyse nonlinéaire* **8** (1991), 43–58.

[WaZ3] Wang, Z.Q., Multiple solutions for indefinite functionals and applications to asymptotically linear problems, *Math. Sinica, New Series* **5** (1989), 101–113.

[WaZ4] Wang, Z.Q., A note on the deformation theorem, *Acta Math. Sinica* **30** (1987), 106–110.

[Warl] Ward, J.R., A boundary value problem with periodic nonlinearity, *Nonlinear Analysis, TMA* **10** (1986), 207–213.

[Wasl] Wasserman, A.G., Equivariant differential topology, *Topology* **8** (1969), 127–150.

[Weil] Weinstein, A., Bifurcation and Hamilton's principle, *MZ* **159** (1978), 235–248.

[Wei2] Weinstein, A., Critical point theory, symplectic geometry and Hamiltonian systems, Proc. 1983 Beijing Symp. on DD4, Gordon Breach, (1986), 261–288.

[Witl] Witten, E., Supersymmetry and Morse theory, *J. Diff. Geom.* **17** (1982), 661-692.

[Wul] Wu, S.P., The nontrivial solution for a class of quasilinear equations, *Applied Math. A.J. of Chinese Univ.* **3** (1988), 339–345.

[YaS1] Yakubovich, V.A., Starzhinskii, V.M., *Linear Differential Equations with Periodic Coefficients,* John Wiley & Sons, 1975.

[Yanl] Yang, X.F., The Morse critical groups of minimax theorem, preprint.

Index of Notation

df	differential of f	19
f_a	level set of f, not above the level a	20
K	critical set	19
K_a	critical set with critical value a	21
(PS)	Palais-Smale condition	20
$exp(\cdot)$	exponential map	72
Δ	Laplacian operator	142
Δ_M	Laplace-Beltrami operator	230
Δ	tension operator	230
∇	gradient operator	141, 229
$mes(\cdot)$	measure	235
$\|\Lambda\|$	measure of Λ	175
$\oplus$	direct sum	180
W^T	transpose of the matrix W	182
$\wedge^1\Lambda$	loop space on Λ	204
$\sharp A$	cardinal number of the set A	216
$Fix(\cdot)$	fixed point set	216
$\wedge$	exterior product	277
i_ω	interior product	277
Id	identity operator	99

INDEX

Arnold conjecture
  on fixed points, 216
  on Lagrangian intersections, 217

Banach manifold, 14
Betti number, 3
bifurcation, 129, 161
blow up analysis, 232
Bott 79, 206

cap product, 9
category, 105
  relative category, 109
conformal group, 360
convex set, 60
  locally, 60
Courant Lebesgue lemma, 268
critical group, 32
critical manifold, 69
critical orbit, 67
critical point, 18
  w.r.t. a locally convex closed set, 62
critical set, 18
critical value, 18
cuplength, 9
cup product, 9

Deformation lemma, 21
Deformation retract, 20
  strong, 21
Deformation theorem
  first, 29
  equivariant first, 67
  second, 23
  equivariant second, 68
degenerate critical point, 43
  non, 33, 41

Euler characteristic, 6

Finsler manifold, 18
Finsler structure, 15
Fredholm operator, 47, 97

G-action, 66
G-cohomology, 75
G-critical group, 76
G-equivariant, 66
G-space, 66
Galerkin approximation, 111
general boundary condition, 55
genus, 96
  cogenus, 96
gradient flow, 19
Gromoll-Meyer pair, 48
Gromoll-Meyer theory, 43

Hamiltonian system, 179
handle body theorem, 38
harmonic map, 229
harmonic oscillation, 285
heat flow, 229
Hilbert Riemannian manifold, 19
Hilbert vector bundle, 70
Hodge theory, 274
homology group, 3
  relative, 3
homotopy group, 12
  relative, 12
Hurewicz isomorphism theorem, 13
hyperbolic operator, 41

invariant function, 111
isolated critical manifold, 69
isolated critical orbit, 74
isolated critical point, 43

Jacobi operator, 251
jumping nonlinearity, 164

Künneth formula, 5,

Landesman-Lazer condition, 153
Leray-Schauder degree, 99
link
  homological, 84
  homotopical, 83
Ljusternik-Schnirelman theorem, 105
locally convex set, 60

Marino-Prodi theorem, 53
  G-equivariant, 80
Maslov index, 183
maximum principle, 143
  strong, 143
minimal surface, 260
minimax principle, 87
Morse decomposition, 250
Morse index, 33
Morse inequality, 36, 79
Morse lemma, 33
Morse-Tompkins-Shiffman theorem, 271
Morse type number, 35
mountain pass point, 90

Nemytcki operator, 141
normal bundle, 70

Palais-Smale condition, 20
  w.r.t. a convex set, 62
  (PS)*, 117
Palais theorem, 14
pendulum, 209
periodic solution, 179
perturbation
  on critical manifold, 131
  Uhlenbeck's method, 136
Plateau problem, 260
Poincaré-Hopf theorem, 99
projective space
  real, 6, 11
  complex, 6, 111
pseudo gradient vector field, 19

regular point, 18
regular set, 18
regular value, 18

saddle point reduction, 188
shifting theorem, 50
Sobolev embedding, 141
Sobolev space, 141, 231
splitting theorem, 44
strong resonance, 156
subordinate classes, 10
subsolution, 145
supersolution, 145
symplectic form, 215
symplectic matrix, 183

tangent bundle, 15
  cotangent bundle, 15

variational inequality, 65, 177
vector bundle, 15

Witten complex, 282

Progress in Nonlinear Differential Equations and Their Applications

Editor
Haim Brezis
Département de Mathématiques
Université P. et M. Curie
4, Place Jussieu
75252 Paris Cedex 05
France
and
Department of Mathematics
Rutgers University
New Brunswick, NJ 08903
U.S.A.

Progress in Nonlinear Differential Equations and Their Applications is a book series that lies at the interface of pure and applied mathematics. Many differential equations are motivated by problems arising in such diversified fields as Mechanics, Physics, Differential Geometry, Engineering, Control Theory, Biology, and Economics. This series is open to both the theoretical and applied aspects, hopefully stimulating a fruitful interaction between the two sides. It will publish monographs, polished notes arising from lectures and seminars, graduate level texts, and proceedings of focused and refereed conferences.

We encourage preparation of manuscripts in some form of TeX for delivery in camera-ready copy, which leads to rapid publication, or in electronic form for interfacing with laser printers or typesetters.

Proposals should be sent directly to the editor or to: Birkhäuser Boston, 675 Massachusetts Avenue, Cambridge, MA 02139

PNLDE1 Partial Differential Equations and the Calculus of Variations, Volume I
Essays in Honor of Ennio De Giorgi
F. Colombini, A. Marino, L. Modica, and S. Spagnolo, editors

PNLDE2 Partial Differential Equations and the Calculus of Variations, Volume II
Essays in Honor of Ennio De Giorgi
F. Colombini, A. Marino, L. Modica, and S. Spagnolo, editors

PNLDE3 Propagation and Interaction of Singularities in Nonlinear Hyperbolic Problems
Michael Beals

PNLDE4 Variational Methods
Henri Berestycki, Jean-Michel Coron, and Ivar Ekeland, editors

PNLDE5 Composite Media and Homogenization Theory
Gianni Dal Maso and Gian Fausto Dell'Antonio, editors

PNLDE6 Infinite Dimensional Morse Theory and Multiple Solution Problems
Kung-ching Chang

PNLDE7 Nonlinear Differential Equations and their Equilibrium States, 3
N.G. Lloyd, W.M. Ni, L.A. Peletier, J. Serrin, editors